Werner Buckel

Superconductivity

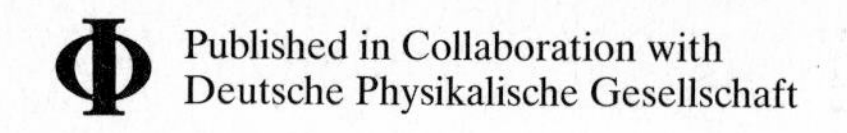

Distribution

VCH Verlagsgesellschaft, P. O. Box 101161, D-6940 Weinheim (Federal Republic of Germany)

Switzerland: VCH Verlags-AG, P. O. Box, CH-4020 Basel (Switzerland)

United Kingdom and Ireland: VCH Publishers (UK) Ltd., 8 Wellington Court, Wellington Street, Cambridge CB1 1HZ (England)

USA und Canada: VCH Publishers, Suite 909, 220 East 23rd Street, New York, NY 10010-4606 (USA)

ISBN 3-527-27893-1 (VCH, Weinheim) ISBN 0-89573-898-8 (VCH, New York)

Werner Buckel

Superconductivity

Fundamentals and Applications

Weinheim · New York · Basel · Cambridge

Prof. Dr., Dr. h.c. Werner Buckel
Physikalisches Institut der Universität Karlsruhe
P. O. Box 6980
D-7500 Karlsruhe 1
Federal Republic of Germany

Published jointly by
VCH Verlagsgesellschaft mbH, Weinheim (Federal Republic of Germany)
VCH Publishers Inc., New York, NY (USA)

Editorial Director: Walter Greulich
Production Manager: Myriam Nothacker

Library of Congress Card No. applied for.

British Library Cataloguing-in-Publication Data:
Buckel, Werner
Superconductivity
1. Superconductivity
I. Title
537.623

ISBN 3-527-27893-1

Deutsche Bibliothek Cataloguing-in-Publication Data:
Buckel, Werner:
Superconductivity : fundamentals and applications / Werner Buckel. [Aus dem Dt. von F. Hampson]. – Weinheim ; Basel (Switzerland) ; Cambridge ; New York, NY : VCH, 1991
Einheitssacht.: Supraleitung <engl.>
ISBN 3-527-27893-1 (Weinheim ...) brosch.
ISBN 0-89573-898-8 (New York) brosch.

Printed on acid-free paper.

Typesetting: Filmsatz Unger & Sommer, D-6940 Weinheim
Printing and bookbinding: Progressdruck GmbH, D-6720 Speyer
Printed in the Federal Republic of Germany.

Heike Kamerlingh Onnes

I would like to thank Professor Dr. C. J. Gorter, Kamerlingh Onnes Laboratorium, Leiden, for allowing me to use the above photo in this work

To my venerable teacher
Dr. Rudolf Hilsch
with my deepest thanks

Foreword to the English Edition

Several excellent introductory texts on superconductivity are already available in English language, for instance, those written by Professor M. Tinkham (1975) or Professor P. DeGennes (1966). Special topics, such as applications or superconducting electronics are treated in monographs. However, a short, yet comprehensive, introduction to the experimental aspects of superconductivity on a student textbook level does not exist as far as I know. Therefore, I believe this book can bridge a gap in the literature.

Especially since the sensational discovery of the new superconducting oxides with transition temperatures as high as 125 K, the interest in superconductivity and its possible applications has increased remarkably. I hope this book can help people who are interested in becoming acquainted with the phenomenon of superconductivity.

The aim of this book is to describe the experimental aspects of superconductivity and to explain them on the basis of our present understanding of condensed matter. The basic ideas are made clear without involving too much formalism.

I would like to thank Mr. Frank Hampson who translated the book and VCH Publishers for their decision to publish this English edition.

Karlsruhe, September 1990 Werner Buckel

Foreword to the 4th German Edition

There were mainly references to further developments in electrical technology and measuring techniques in the "Outlook for the 3rd Edition (1984)". Superconductivity appeared to be well understood and the theory of Bardeen, Cooper and Schrieffer with its extension by Eliashberg provided a completely satisfactory description of the phenomenon.

In 1984, as the 3rd Edition appeared, more than 10 years had passed by without its being possible to raise the transition temperature T_c above the then record value of 23.2 K for Nb_3Ge in spite of great efforts by materials physicists. We had gradually come to accept the idea that superconductivity would remain restricted to very low temperatures ($T <$ ca. 30 K).

So when new superconductors with transition temperatures of ca. 100 K were discovered in 1987 it was a scientific sensation. The pioneer work for this fascinating development, which is still taking place, was carried out by J.G. Bednorz and K.A. Müller in the IBM laboratory in Rüschlikon (Switzerland). They received the 1987 Nobel prize for physics for their work. Since the discovery of the new superconductors two years ago several thousand publications have appeared on these new materials.

It would, therefore, appear appropriate to devote a section (Section 1.3) to these recent developments. The basic properties of superconductors remain unchanged. It has, hence, proved possible to retain the total concept of the book. The presentation in other chapters has been made somewhat more concise in order that the volume of material should not grow too much.

I would again like to thank my colleagues in Karlsruhe very much for valuable discussions. I particularly thank Dr. Politis for keeping me up to date with the newest developments. As representatives of their research groups I thank Messrs. P. Chaudhari, IBM Watson Research Center, Yorktown Heights, B. Batlogg, AT & T Bell Laboratories Murray Hill, T. Geballe, Stanford University, V.V. Moshalkov, Lomonossov University, Moscow and Z.X. Zhao, Academy of Sciences, Peking very much for continually informing me of their own work. I must also thank the VCH-Verlag for the pleasant cooperation.

But I would particularly like to thank Messrs Bednorz and Müller for their constant readiness to inform me about their work at the earliest opportunity. I was, thus, able to participate very intensely in this exciting development.

Karlsruhe, January 1989 Werner Buckel

Foreword to the 1st German Edition

It was impossible to explain superconductivity satisfactorily for almost 5 decades. We now have a microscopic theory which comprehends a wide range of phenomena and even describes some of them quantitatively. So that it is possible, at least in principle, to understand superconductivity.

The technological exploitation of superconductivity has begun with the construction of large superconducting magnets. Further applications in electrical technology, e.g. for power transmission, are under intense study. In some fields of electrical measurement technology superconductivity has really brought about a breakthrough by increasing the sensitivity by several orders of magnitude of, for instance, the measurement of magnetic fields.

So in the future interest in this phenomenon is not going to be restricted to the physicist. Rather, more and more engineers will be confronted with it. Its application will ensure that superconductivity will also increasingly come into the field of vision of the technically interested public.

This present introduction is intended for all these interested "nonspecialists". An attempt has been made to present our basic understanding of superconductivity clearly and with deliberate suppression of mathematical formulations. The manifold phenomena involved are discussed against the background of this understanding. The applications are also treated in detail.

Such an introductory treatment can naturally only present a limited number of considerations and facts. Any selection of these must, of necessity, be very subjective. An attempt has been made to present, with the omission of much detail, a comprehensive picture of superconductivity and, in particular, its quantum nature. It did not seem appropriate here to follow the historical development. The phenomena are ordered and treated according to their internal relationships. There is no doubt that many distinguished pioneer investigations have not been appropriately honored. Neither does the literature list provide a representative cross section of the many thousands of publications on superconductivity that have been appeared. The intention has been the more limited one of providing the reader with access to the original literature. There are anyway a whole range of excellent monographs that can be consulted concerning particular questions.

This book will have fulfilled its purpose if it can introduce superconductivity to a broader range of interested people. Perhaps it can, as a brief summary, also provide a little aid to those who address questions of superconductivity directly.

Many people have aided me in the writing of this book by their constant readiness to discuss with me in detail the problems that arose. I thank them all most sincerely. In particular my dear colleague Falk who was untiringly ready to answer and discuss my questions. My heartfelt thanks also to my colleagues, both in Karlsruhe and Jülich, amongst these, in particular, Dr. Baumann, Dr. Gey, Dr. Hasse, Dr. Kinder and Dr. Wittig. I would also like to thank Dr. Appleton (EEDIRDC), Dr. Schmeissner (CERN), Dr. Kirchner (Munich), Prof. Rinderer (Lausanne), Dr. Eßmann (Stuttgart) and Dr. Voigt (Erlangen) together with the Siemens, Vakuumschmelze and General Electric companies for allowing me to reproduce illustrations. I am very grateful to the Physik Verlag for the pleasant cooperation.

My particular thanks go to my dear wife for bearing it all with great patience when I spent evenings and Sundays working on this book.

Jülich, August 1971 Werner Buckel

Contents

Introduction

It was Heike Kamerlingh Onnes[1] in 1908 who succeeded in liquifying helium, the only remaining noble gas to be liquified. He described this important experiment in a lively contribution delivered to the 1st International Refrigeration Congress in Paris in October 1908 [1]. The liquification of helium made accessible a new temperature range in the vicinity of absolute zero (ca. 1 K to 10 K[2]). The first successful experiment had required the total resources of the institute; but Onnes was soon able to perform experiments at these temperatures. First of all he began to investigate the electrical resistance of metals.

The understanding of the mechanism of electrical conductivity was very incomplete at that time. It was known that it must be the electrons which transported the charge. The temperature-dependence of the electrical resistance of many metals had also been determined and it had been discovered that in the region of room temperature the resistance falls linearly with temperature. However, it was found that the reduction became ever smaller as the temperature was reduced. There were, thus, in principle, three possibilities available for discussion.

1. The resistance could continuously fall to zero as the temperature falls (Fig. 1, curve 1);
2. it could fall to a fixed limiting value (Fig. 1, curve 2) or
3. it could pass through a minimum and approach infinity at very low temperatures (Fig. 1, curve 3).

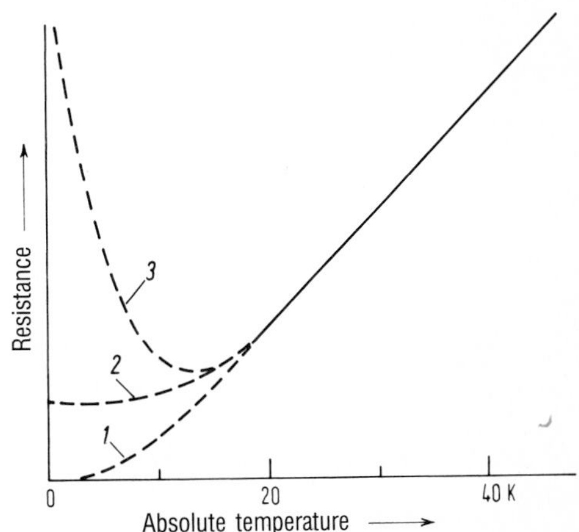

Fig. 1. Possible temperature dependences of the electrical resistance at low temperatures.

The notion that electrons ought to be bound to their atoms at sufficiently low temperatures spoke for the third possibility. This would end their freedom of mobility. The first possibility, whereby the resistance approaches zero at low temperature, was favored by the great reduction with falling temperature.

---

1 Heike Kamerlingh Onnes, 21.09.1853–21.02.1926, Head of the Cryogenic Laboratory of the University of Leiden, which he founded and brought to world-wide fame, 1913, Nobelprize for physics.

2 K (Kelvin) is the unit of absolute temperature. The conversion of degrees Celsius (°C) and absolute temperature is laid down by: a K = (a − 273.16 K)°C

Because it was possible even then to obtain them in a very pure form Onnes first investigated samples of gold and platinum. He found that on approaching absolute zero the electrical resistance tended towards a fixed value, the residual resistance, so that the behavior corresponded to the 2nd possibility. The size of the residual resistance depended on the degree of purity of the sample. These findings are illustrated in Fig. 2. On the basis of these results Onnes tended to the opinion that an ideally pure platinum or gold would have a vanishingly small resistance at the temperature of liquid helium. He described these experiments and reflections at the 3rd International Congress of Refrigeration in Chicago in 1913. He said there[1]: "Allowing a correction for the additive resistance I came to the conclusion that probably the resistance of absolutely pure platinum would have vanished at the boiling point of helium". This hypothesis was also supported by quantum physics, which was then in a state of hectic development. Einstein[2] had proposed a model of solid bodies where the vibrational energy of the atoms would fall exponentially at very low temperatures. Since Onnes was of the opinion – which we now know was completely correct – that the resistance of very pure samples would only result from these movements of the atoms, the above quoted hypothesis followed.

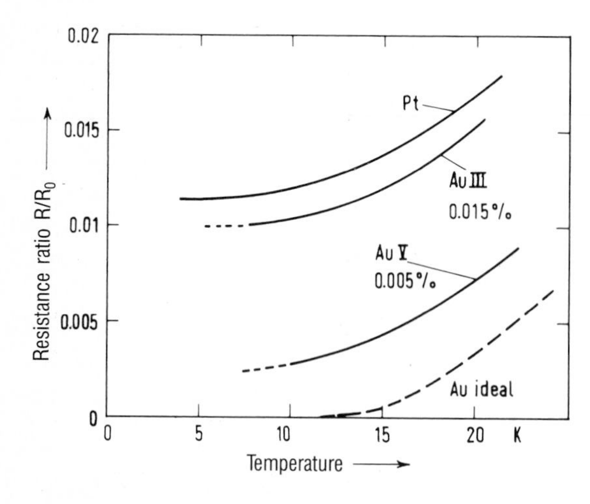

Fig. 2. Electrical resistance of various metal samples at low temperature – effect of impurities on the residual resistance.

Onnes decided to test these hypotheses by an investigation of mercury, the only metal in those days that he could hope to bring to a high degree of purity by multiple distillation. He estimated that with his apparatus he would just be able to measure the resistance of mercury at the boiling point of helium but that this would rapidly fall towards zero at even lower temperatures.

The first experiments appeared to confirm this opinion. The resistance of mercury was truely unmeasurably small below 4.2 K. In his lecture in 1913 Onnes described this phase of his thinking as follows: "With this beautiful prospect before me there was no more

1 H.K. Onnes: "Report on researches made in the Leiden Cryogenic Laboratory between the Second and Third International Congress of Refrigeration" [2].

2 A. Einstein, 14.03.1879–18.04.1955, 1921, Nobel prize for the explanation of the photoelectric effect by hypothesizing light quanta.

question of reckoning with difficulties. They were overcome and the result of the experiment was as convincing as could be."

But he soon recognized, in further experiments with improved apparatus, that the effect observed corresponded in no way with the expected reduction in resistance. For the resistance change took place over a temperature range of only a few hundredths of a degree, that is it was more in the nature of a resistance jump than a continuous decrease. Figure 3 illustrates the curve published by Onnes [3]. He himself said of it: "At this point (somewhat below 4.2 K) within some hundredths of a degree came a sudden fall not foreseen by the vibrator theory of resistance, that had framed, bringing the resistance at once less than a millionth of its original value at the melting point... Mercury has passed into a new state, which on account of its extraordinary electrical properties may be called the superconductive state." [2].

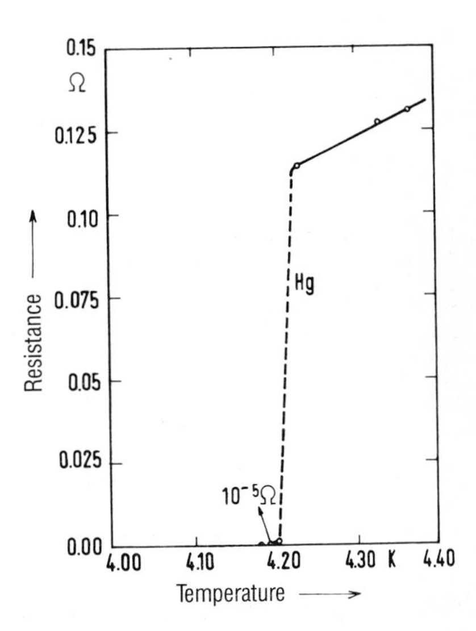

Fig. 3. Superconductivity of mercury (after [3]).

A name had, therefore, been found for this new phenomenon. The discovery occurred unexpectedly in experiments designed to test a well-founded hypothesis. It was soon found that the purity of the sample was not of importance for the disappearance of the resistance. The sufficiently carefully and critically carried out experiment had led to the discovery of a new state of matter [1)]. Our penetration into virgin scientific territory not infrequently takes place in this manner.

The importance, which the rest of the scientific world attached to this discovery, can be seen with the award of the Nobel prize for physics to Kamerlingh Onnes in 1913. Then, however, it was impossible to imagine what a wealth of basic questions and interesting possibilities would arise as a result of this observation and that it would be about half a century before superconductivity would be understood, at least in principle.

---

1 The experiments were carried out by von Holst, a young coworker of Onnes. His name must be mentioned here.

Even today superconductivity is a pure low-temperature phenomenon. What this means can be best understood if we look in general at how temperature is related to physical phenomena.

Since the development of the kinetic theory of heat – one of the most magnificent products of the human intellect – in the first part of the 19th century, temperature has been associated with the random motions of the elementary building blocks of a body. The higher the temperature of the body, the greater the random motion of its building blocks – expressed physically, the greater the energy involved in this random motion[1].

The elementary building blocks of matter are subject to many ordering forces in addition to the random statistical motion. The physicist speaks of interactions between the building blocks. A whole range of phenomena result from the competition between ordering interactions and random thermal motion. When, as the temperature is increased, the energy of thermal motion becomes large enough in comparison to a particular ordering interaction, then the ordered state of the material that existed at the lower temperature is destroyed. All phase transitions and the build-up of complex systems, such as the atomic shells, are subject to these laws. We can assign every physical interaction to a temperature range according to its strength. Figure 4 reproduces this classification for some of the best-known interactions.

We begin at the upper end of the scale with the strongest interactions that we know at the moment between particles, the nuclear forces. The name of this force already indicates that this force holds together the building blocks of the atomic nucleus, the nucleons (protons and neutrons) and, thus, makes possible the formation of atomic nuclei. The following figures will give at least some idea of the strength of this nuclear force. An energy of 7.5×10^5 kWh (kilowatt hours) is required in order to break down the atomic nuclei of just 4 g helium, each of which is made up of 2 protons and 2 neutrons, into their components. That corresponds to the average daily electric power consumption of ca. 100 000 households.

Many millions of degrees are required in order to make the random thermal energy comparable with this interaction energy brought about by the nuclear force. Such temperatures prevail in the interiors of the fixed stars. The temperature of the core of our own sun, for example, is ca. 20 million degrees. Nuclear transfomations are possible at these temperatures; however, it is the fusion of light nuclei to heavier ones that is preferred, i.e. the production of atomic nuclei.

Today we know that the fantastic amounts of energy which the sun has been radiating into space for several thousand million years[2] are the result of the production of helium nuclei from hydrogen. If the radiation rate remains constant this process can satisfy the energy requirements of the sun for many more thousands of millions of years. The solar prominences, where masses of gas are catapulted many thousands of kilometers above

---

1 The physical quantity "temperature" is not completely described by these energetic considerations, rather the concept of entropy is also essential. However, the energy considerations are sufficient if it is kept in mind that thermal motion is always completely random.

2 If it were desired to produce the amount of energy radiated by the sun by burning carbon as in the case of conventional power stations for the generation of electricity then it would be necessary each day to burn an amount approximating to the weight of the earth. This trivial example gives some idea of the unimaginable extent of energy production within the sun.

the surface of the sun, provide us with some inkling of the violence of this energy production. These energy outbursts are, however, very small in comparison with the total output of the sun.

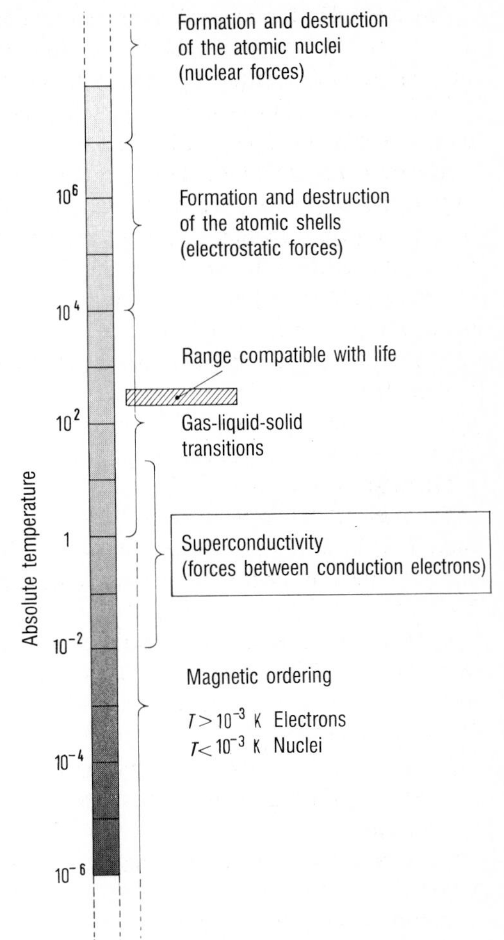

Fig. 4. Temperature ranges of various interactions. We have access to the entire temperature range from ca. 10^7 K down to ca. 10^{-5} K for physical experiment. The highest temperatures are obtained in short-term gas discharges, with the hot plasmas made up of ions and electrons compressed by means of magnetic fields. The lowest temperatures of ca. 10^{-8} K are obtained in the system of the nuclear moments of copper. Professor Pobell and his colleagues at the University of Bayreuth have been able to cool the copper lattice itself to about 10^{-5} K with these nuclear moments.

If we were to succeed in employing the fusion of hydrogen to helium for energy production on earth, then all the energy problems of mankind would be solved for the foreseeable future. We are unfortunately very far from being able to overcome the problems involved in this task. But it is possible to say even now that a solution only appears possible with the aid of superconducting magnets. Even today very large superconducting magnets have been constructed for research purposes (see Section 9.1.4.2).

The next temperature range represented in Fig. 4 is that from about 10^4 to 10^7 degrees. In this range electronic shells of the atoms are gradually built up as the temperature is reduced, whereas at even higher temperatures only naked nuclei and free electrons exist – we speak of a very hot plasma. The ordering interactions responsible for the formation of the atomic shells, what are known as the Coulomb interactions, result from the electrostatic attraction of charges. The energies associated with rearrangements of the outer shells of electrons are known to us in the heat of reaction of chemical reactions. They are about 10 million-fold less than the energies involved in nuclear reactions.

The conversions between the gaseous, liquid and solid states of matter occur between about 1 K and 10^4 K. Here the ordering forces that are responsible are of very differing natures, but they all bring about basically the same phenomena. As the temperature is reduced a liquid separates out from the vapor and finally the liquid freezes to a solid, usually highly ordered, crystalline body. The condensation and freezing of water is a familiar example.

This interatomic interaction is particularly small in the case of helium – here it is, as with all noble gases, what are known as van der Waals forces [1]. The boiling point of helium is only 4.2 degrees above absolute zero, that is almost one hundred times lower than that of water. Here it will only be mentioned in passing that the stepwise penetration to ever lower temperatures was achieved by means of the liquification of various gases (air, hydrogen, helium). Today it is possible to maintain continuously temperatures of a few thousandths of a degree using a mixture of ^{3}He and ^{4}He.

As would be expected ever new weak interactions make their appearance as the temperature is reduced. The interactions characteristic of the range under ca. 1 K are those of the elementary magnetic dipoles of the atoms or molecules with each other or with their environments. They lead to magnetic processes of ordering such as are known to us in the case of a few substances of the ferromagnetic group at higher temperatures (e.g. for iron at 1043 K = 770 °C).

These ordering interactions are first effective for the magnetic moments of the electronic shells. It is only at even lower temperatures that they order the hundred to thousand-fold smaller nuclear moments. Just these systems of elementary magnets can be used as a working substance to penetrate these temperature regions in the same way as gas liquification was utilized as a cooling process in the temperature range of the gas – liquid – solid transitions. The lowest temperature yet reached was attained with a system of nuclear moments (copper nuclei) and lies at approximately 10^{-8} K. That is only one hundred millionth of a degree above absolute zero. In view of such temperatures the question of the point of such experiments perhaps becomes more evident. The answer can only be that such experiments enable us to learn ever finer details of the structure of matter [2].

Until two years ago superconductivity was limited to the range between about 10^{-3} and 20 K on this vast temperature scale. A burst of development, that has led today to temperatures over 100 K and has still not ended, will be described in Section 1.3. The drastic increase of almost an order of magnitude in the transition temperatures, throws open yet again the question of the interaction responsible. This will be discussed in Chapter 2.

In this overview we want to establish that superconductivity too, independant of all its remarkable properties which will be treated in the chapters that follow, is ultimately

---

1 Van der Waals, Netherlands physicist, 23.11.1837–07.03.1923, University of Leiden, Nobel prize 1910.

2 Nuclear physical investigations on magnetically oriented atomic nuclei revealed in 1958 that the, until then, universally accepted symmetry principle, whereby inanimate nature was not endowed with any handedness (left or right handedness), was not completely generally applicable (violation of parity). This is a very impressive example of the very close relationships that can exist in physics between apparently very different fields.

dependent on a transition from a disordered to an ordered state. It has recently been extended to the temperature of liquid nitrogen, which we view as being readily accessible. Thus, there is no doubt that the large-scale technological exploitation of superconductivity, which has already begun with the construction of large magnets, will now be extended more quickly to other fields.

However, this overview is intended to demonstrate something rather different too. It is intended to give at least an indictation of how it is possible for the human intellect to describe, with the aid of suitable concepts, natural phenomena whose immediate perception is completely impossible. Figure 4 includes the narrow temperature range from ca. $-70\,°C$ to $+\ 100\,°C$ which encompasses organic life. Our "natural" receptors for "hot" and "cold" are limited to an even narrower range. Neither do they constitute objective measuring instruments, for the feeling of heat or cold is dependent on the previous history of the observer. On a hot sticky summer's day a room at ca. 20 °C will be experienced as being pleasantly cool. After a walk in the snow objectively the same room will be experienced as being "very hot". Science has to abstract itself from all these subjective sensations, be they – a fact that must not be denied – nevertheless, so important for human life.

We are now in a position to be able to generate and work at any temperature between 10^{-6} K and 10^{6} K. This mastery of nature – and we have by no means come to the end – is the result of a fascinating intellectual exploration. As in earlier centuries adventurers explored the earth, now in the scientific and technological age the human intellect explores fields that are even further from our biologically appointed limits. Every scientific book ought to impart a whiff of this adventure.

1. Some basic facts

The very first observations of superconductivity in mercury threw up two basic questions:

1. How large is the reduction in resistance as superconductivity begins? Or to put it another way: How correct is it to speak of a disappearance of the electrical resistance?

2. Is superconductivity a specific property, perhaps of mercury alone, or can other metals exhibit it?

In the years following 1911 Kamerlingh Onnes carried out many elegant investigations concerning these two questions. Even though it would be very fascinating to follow the development of an unexpected discovery, it should better not be carried through. Rather we will, in this section, deal with the two questions from today's viewpoint and answer them as far as it is possible.

1.1 The disappearance of the electrical resistance

A conventional method of resistance measurement was employed for the first investigations of superconductivity. The electrical potential was measured across a sample with a current flowing through it. This made it possible to establish that the resistance fell to less than one thousandth as superconductivity commenced[1]. So it was only justifiable to speak of the resistance disappearing in that the resistance fell below the sensitivity limit of the apparatus and could no longer be detected.

It must be very clearly realized that it is fundamentally impossible to demonstrate in an experiment the assertion that the resistance has fallen to exactly zero. An experiment can only ever deliver an upper limit for the resistance of a superconductor.

Now it is naturally important for the understanding of such a new phenomenon to test with the most sensitive method possible whether a residual resistance can be detected in the superconducting state. The aim, therefore, is to measure extremely low resistance. As early as 1914 Onnes began to use what is absolutely the best method. For Onnes observed the decay of the current in a closed superconducting circuit. The energy stored in such a current would, if a resistance were present, be gradually converted into Joule heat. So all that is necessary is to observe such a current. If it decays with time there is certainly a resistance still present. If it is not possible to detect such a decay, then the observation time and geometry of the superconducting circuit allow the calculation of an upper limit for the resistance[2].

This method can be made many orders of magnitude more sensitive than the conventional current-potential drop measurement. Its principle is illustrated in Fig. 5. We have

---

1 This method can be extended by a few powers of ten by the use of larger currents and particularly sensitive potential difference measurement (see Section 9.5).

2 This method basically corresponds to that used for the measurement of a very large resistance. Here the discharge is observed of a capacitor connected across the unknown resistance. That is, the drop in an electrical field is determined as a function of time. The decay of a magnetic field with time is utilized in the case of small resistances.

a ring of superconducting material, e.g. lead – lead is also a superconductor – above the transition temperature T_c[1], that is in the normal conducting state. A bar magnet is employed to pass a magnetic field through the hole in the ring. The ring is now cooled to a temperature at which it is superconducting ($T < T_c$). Practically nothing happens to the magnetic field passing through the opening. If the magnet is now removed a current will begin to flow through the superconducting ring because any change in the magnetic flux Φ through the ring induces an electrical potential along the ring. This induction potential causes the current to flow.

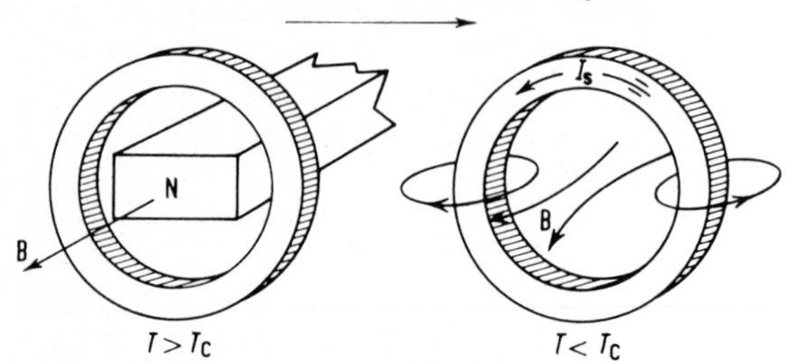

Fig. 5. The generation of a persistent current in a superconducting ring.

If the resistance now would be exactly zero, then this current should continue to flow unchanged as a "persistent current" for as long as the lead ring remains superconducting. The current will decay exponentially if a resistance R is present. So that:

$$I(t) = I_0 \mathrm{e}^{-\frac{R}{L}t} . \tag{1-1}$$

Here I_0 is any current flowing at the time from which time is reckoned. $I(t)$ is the current at time t. R is the resistance and L is the coefficient of self-induction[2] that only depends on the geometry of the ring.

In order to make an estimate let us assume that we have a wire ring with a diameter of 5 cm made from wire which is 1 mm thick. The coefficient of self-induction L of such

---

1 This symbol is an abbreviation of T_{critical}.

2 The coefficient of self-induction L can be defined as the proportionality factor between the induction voltage and the change in the current in the conductor with time:

$$U_{\text{ind}} = -L \frac{\mathrm{d}I}{\mathrm{d}t} .$$

The dimensions of L are voltage × time/current, the unit of L is volt sec/ampere (Vs/A) also known as Henry (Joseph Henry, American scientist, 17.12.1797–13.5.1878).

For a circular ring with a radius of r made up of wire with a diameter $2d$ also of circular cross section ($r \gg d$):

$$L = \mu_0 r (\ln r/d + 0.23)$$

$$\mu_0 = 4\pi \times 10^{-7}\,\text{Vs/Am}.$$

a ring is ca. 1.3×10^{-7} H. If the current in such a ring decays by less than 1% within one hour then it can be concluded that its resistance is less than $4 \times 10^{-13}\ \Omega$[1]. But that is equivalent to a resistance change of more than 8 powers of ten with the onset of superconductivity.

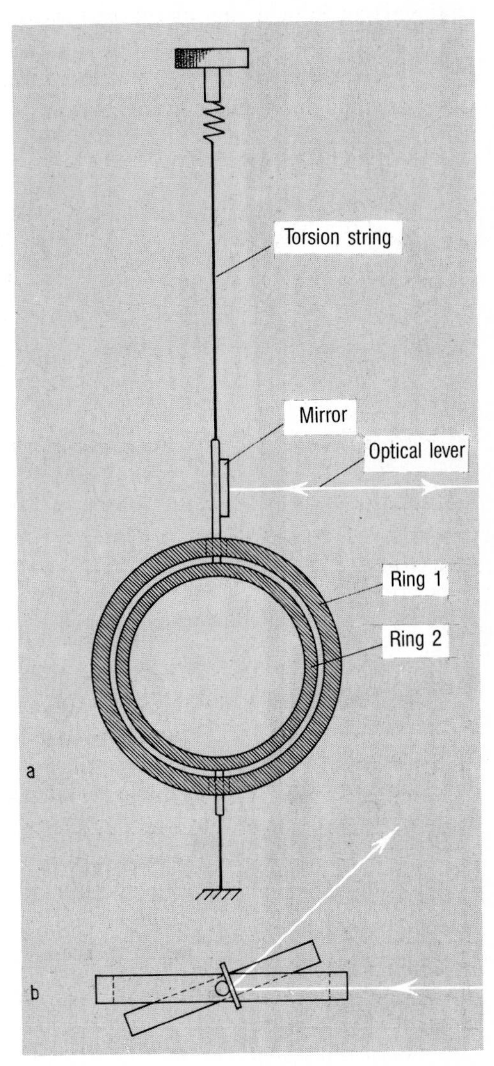

Fig. 6. Device for the observation of a persistent current (according to [5]). Ring 1 is attached to the cryostat. The energy stored in a ring with a persistent current is given by $1/2\, LI^2$. The change with time of this energy is just equal to the Joule heat RI^2 generated by a resistance so that

$$-\frac{\mathrm{d}}{\mathrm{d}t}\frac{1}{2}LI^2 = RI^2$$

which yields the differential equation

$$-\frac{\mathrm{d}I}{\mathrm{d}t} = \frac{R}{L}I$$

whose solution is (1–1).

In all these experiments it is necessary to determine the strength of the persistent current. This was done in the first experiments [4] simply with the aid of a compass needle, by observing its deflection as a result of the magnetic field of the persistent current. A

1
$$R \leqslant \frac{-\ln 0.99 \times 1.3 \times 10^{-7}}{3.6 \times 10^{3}}\,\frac{\mathrm{Vs}}{\mathrm{As}} \cong 3.6 \times 10^{-13}\,\Omega\,.$$

more sensitive apparatus was employed by Onnes and later by Tuyn [5]. It is illustrated schematically in Fig. 6. A persistent current is set up in each of the two superconducting rings by means of an induction process. These currents attempt to maintain the rings parallel to each other. One of the two rings (here the inner) is suspended on a torsion fiber and twisted somewhat out of parallel. This causes the torsion fiber to be twisted as a result of the repulsive force of the persistent current. An equilibrium position is, therefore, occupied where the torsional moments produced by the persistent current and the torsion fiber are equally large. This equilibrium position is very sensitively observed via an optical lever. If the persistent current in the ring decays the optical lever will indictate a change in the equilibrium position. A change in the persistent current has never been observed with experiments of this type.

A beautiful demonstration of this persistent current is illustrated in Fig. 7. A small permanent magnet is lowered towards a superconducting lead dish and, according to Lenz's rule, induces an induction current that repulses the magnet. The induction current supports the magnet at an equilibrium height. This arrangement is known as a "floating magnet". The magnet is supported for as long as the persistent current flows through the lead dish, i.e. for as long as the lead is kept superconducting [1].

Fig. 7. The "floating" magnet to demonstrate the persistent current that is generated by induction when the magnet is lowered. Left: starting position, right: equilibrium position. My thanks are due to Frau Stremme for producing the pictures.

The most sensitive methods of setting an upper limit to the resistance of the superconducting state employ conductor geometries with particularly small coefficients of self induction L and increased periods of observation. Such highly developed apparatus can lower the limit further [2]. We now know that the drop in resistance on entering the superconducting state is at least 14 powers of ten [6]. A metal in the superconducting state can at most have a resistance which is ca. 17 powers of 10 lower than the specific

1 This demonstration can be carried out at liquid nitrogen temperatures using a sheet of $YBa_2Cu_3O_7$ (see Section 1.3).

2 Modern superconducting magnetic field meters (see Section 9.5) allow a further increase in the sensitivity.

resistance[1] of copper our best metallic conductor. Since it is unlikely that anyone can grasp what "17 powers of 10" implies, another comparison will be included. The difference in resistance between a metal in the superconducting state and in the normal conducting state is at least as large as the difference between copper and normally employed insulators.

These results would seem to justify the assumption, at least for the present, that the electrical resistance really does disappear in the superconducting state[2].

How novel, indeed unbelievable, this conclusion is and how greatly this resistance-free flow of current through a metal contradicted, in its time, concepts that were well backed by a great deal of experience, only becomes evident on taking a closer look at the transport of charge through a metal. This will put us in a better position to appreciate the problem we set ourselves in trying to understand superconductivity.

We know that charge is transported through metals by electrons[3]. The concept that a certain number of electrons per atoms in a metal – e.g. one electron, the valence electron in alkali metals – are present free, as if it were, so to say, a gas, was developed early on (Drude 1900, Lorentz 1905)[4]. These "free electrons" also mediate the bonding of the atoms in the metallic crystal. The free electrons are accelerated under the influence of an electric field. After a certain time, the mean collision time τ, they collide with atoms and release the energy acquired in the electric field and are re-accelerated. The existence of free charge carriers, which only interact with the metal lattice on collision, makes the good electrical conductivity of metals understandable.

The increase in resistance (reduction in conductivity) with increasing temperature is also explained with no difficulty. With increasing temperature the random thermal motion of the metal atoms increases; they vibrate with statistical amplitude around their resting positions. This increases the probability of a collision between the electrons and the atoms, i.e. the time between two collisions becomes shorter[5]. Since the conductivity is directly proportional to this time, during which the free electron is accelerated by the field, it decreases as the temperature increases – the resistance increases.

Thus, this model of free electrons, which only transfer energy to the lattice by collisions with the atomic residues (atoms without the electrons moving freely in the metal), provides a plausible explanation for electrical resistance. On the basis of this model it is unthinkable that the collisions with the residual atoms should suddenly become forbidden over a very small temperature range at finite temperature. What mechanism in the superconducting state can lead to there no longer being any energy exchange allowed be-

---

1 The specific resistance is defined as:

$$R = \rho \frac{l}{A}$$

R = resistance, l = length of conductor and A = cross section of conductor.

2 An explicit reminder, that this statement only applies to direct current, is appropriate here.

3 Apart from certain alloys charge transport in metals does not take place by electrolytic migration of metal atoms.

4 P. Drude, 12.07.1863–5.07.1906.
H.A. Lorentz, Netherlands physicist, 18.07.1853–04.02.1929, Nobel prize 1902.

5 This will be readily understood by anyone who has had to cross a square or street teeming with a crowd of excited people.

tween the electrons and the lattice? An explanation would, at first sight, seem completely impossible.

Classical theory provides another great difficulty for the hypothesis of a free electron gas in metals. According to the very general laws of classical statistical mechanics all the degrees of freedom[1] of a system should, on average, contribute $k_B T/2$ (k_B = Boltzmann's constant[2], that is 1.38×10^{-23} Ws/degree) to the internal energy[3] of the system. This also means that the free electrons should contribute per free electron the $3\,k_B T/2$ characteristic for a monatomic gas. However, determinations of the specific heat of metals revealed that the contribution of the electrons to the total energy of the metal is about one thousand times smaller than expected according to the classical laws.

This clearly reveals that the classical treatment of the metallic electrons as a free electron gas cannot provide a complete understanding. Then, however, the discovery of Planck's constant (Planck 1900[4]) opened up a new understanding of physical processes particularly at the atomic level. In the years that followed the quantum theory revealed the comprehensive importance of this new approach, which developed from the discovery made by Max Planck.

It also proved possible for Sommerfeld[5] (1928) to resolve, on the basis of the quantum theory, the discrepancy between the contribution of the free electrons to the internal energy of a metal demanded by classical theory and that observed experimentally.

The basic idea of the quantum theory is that discrete states can be assigned to every physical system. An exchange of physical quantites, such as energy, can only take place by the system passing from one state to another.

This restriction to discrete states is evident in the case of atomic structures. In 1913 Bohr[6] was able to propose the first stable atomic model, which allowed comprehension of a wide range of previously inexplicable observations. Bohr postulated the existence of discrete stable states of the atom. If an atom interacted in any way with its environment, such as by taking up or losing energy (e.g. by the absorption or emission of light), then this is only allowed in discrete steps, when the atom must pass from one discrete state to another. If the energy or other interaction quantity is not available in sufficient amount the state remains stable.

In the last analysis it is this relative stability of quantum states which provides the key for an understanding of superconductivity. We have seen that we require some mechanism which forbids interactions between the electrons of the superconductor, which are carrying the current, and the lattice. If it is assumed that the "superconducting" electrons are in a quantum state a certain degree of stability is to be expected. It was generally accepted by 1930 that superconductivity must be a typical quantum process. But the road

---

1 Every coordinate of a system which appears squared in the total energy is regarded as a thermodynamic degree of freedom. Examples are the velocity, $E_{kin} = (1/2)mv^2$, the displacement x from the resting position in the linear force law $E_{pot} = (1/2)Dx^2$ (D = force constant).

2 L. Boltzmann, Austrian physicist, 20.02.1844-06.09.1906.

3 The internal energy of a physical system is defined as the energy that is measured at the coordinates of the center of gravity, that is in a coordinate system where the total impulse P equals zero.

4 Max Planck, 23.04.1858-04.10.1947, originator of the quantum theory as a result of his discovery of the constant named after him, Nobel prize 1918.

5 Arnold Sommerfeld, 05.12.1868-26.04.1951.

6 Niels Bohr, Danish physicist, 07.10.1885-18.11.1962, Nobel prize 1922.

to a proper understanding of the phenomenon was a long one. There is no doubt that one difficulty lies in the fact that one had become accustomed to quantized phenomena in atomic systems but not in the macroscopic bodies. In order to express this peculiarity of superconductivity it was often referred to as a "macroscopic quantum phenomenon". We will come to understand this term even better later (Chapter 2).

Modern physics has revealed yet another aspect, which must be mentioned here, since it is indispensible for a real understanding of the phenomenon of superconductivity. It has taught us that particle representations and wave representations are complementary descriptions of the same physical object. It can be taken as a simple rule that it is appropriate to describe propagation processes in terms of waves and exchange processes involving interaction with other systems in terms of particles.

Just two examples should serve to elucidate this important point somewhat. Light, for instance, is familiar to us as waves from many diffraction and interference phenomena. On interaction with matter, as in the photoelectric effect (the ejection of an electron from a surface), it exhibits distinct particle character. For we find that independent of the intensity of the light the energy transferred to the electron depends only on the frequency. This, however, is just what would be expected if light is regarded as a stream of particles, where all the particles have an energy dependent on the frequency.

Conversely, the particle description of the electron is the more familiar. We can deflect electrons in electric and magnetic fields and we can evaporate them thermally from metals (hot cathodes). These are all processes which we understand by regarding electrons as particles. L. de Broglie[1] proposed the hypothesis that every particle is also associated with a wave with a wavelength equal to Planck's constant h divided by the momentum p of the particle that is $\lambda = h/p$. The square of the wave amplitude at a point (x,y,z) would then be a measure of the probability of finding the particle at this point.

So the particle is smeared through space. If it is desired to express in the wave representation that a point is a particularly favored place for the position of the particle, then a wave must be set up which has an amplitude at this point which is much higher than it is at all other points – such a wave is known as a wave packet. The velocity with which the wave packet travels in space is then equal to the velocity of the particle.

This hypothesis was magnificently confirmed in the following ways. We can produce diffraction and interference phenomena with electrons. Electron diffraction has come to be an important aid for structural investigation[2]. In the electron microscope we produce images with electron beams and because of the very small wavelength of the electron in comparison with visible light the resolution obtained is very high.

In the case of a wave associated with a moving particle – they are often referred to as matter waves – there exists, as for every wave process, a characteristic differential equation, the fundamental Schrödinger equation[3].

We now have to use this deeper insight into the nature of the electron in the description of metallic electrons. The electrons in a metal also possess wave character. With the aid

1 L. de Broglie, French physicist, 15.08.1892–13.03.1987, Nobel prize 1929.

2 Other particles also possess this wave character. For instance, neutron diffraction is a very important method of investigating special, particularly magnetic, structures.

3 E. Schrödinger, 12.08.1887–04.01.1961, Nobel prize 1933.

of certain simplifying assumptions[1] the Schrödinger equation gives us the discrete quantum states of these electron waves in the form of the relationship between the allowed energies E and what is known as the wave number vector $\vec{k}$. The modulus of $\vec{k}$ is given by $2\pi/\lambda$ and the direction of $\vec{k}$ is the direction of propagation of the wave. This relationship is very simple for a free electron.

It is:

$$E = \frac{\hbar^2 \vec{k}^2}{2m} \tag{1-2}$$

$\left(m \text{ is the mass of the electron; } \hbar = \frac{h}{2\pi}\right)$.

The relationship is illustrated in Fig. 8.

Now the electrons within a metal are not completely free. First they are restricted to the volume of the metal; they are enclosed in the metal as if in a box. This restriction leads to the allowed values of $\vec{k}$ becoming discrete, simply because the allowed electron waves must fulfil certain conditions (boundary conditions) at the walls of the box. It can, for instance, be required that the amplitudes of the electron waves disappear at the boundary.

Secondly, because of the electrostatic forces the electrons inside the metal experience the positively charged atom residues, which are normally disposed on account of the crystalline structure of the metal. The electrons are referred to as being exposed to a periodic potential. By this it is meant that on account of the positive charges of the atom residues the potential energy of the free electrons in their neighborhood is less than it is between the atoms. This periodic potential leads to not all energies being allowed in the $E-\vec{k}$ relationship. Rather there are bands of allowed energies, which are separated by bands of forbidden energies. An example of an $E-\vec{k}$ relationship modified by a periodic potential is illustrated schematically in Fig. 9[2]. We speak of energy bands.

Our electrons now have to be fed into these states. A further important principle has to be observed here, namely that of Pauli[3] (1924). This Pauli exclusion principle states that electrons (more generally all particles with half-integral spins, what are known as fermions[4]) can only occupy each discrete state in quantum physics once. Since the spin of the electron, which has not been considered until now, is a quantum number with only two possible values it follows that it is only possible to place two electrons in each of our discrete $\vec{k}$ levels. So in order to introduce all the electrons into the metal it is necessary to fill up the states as far as those of relatively high energy. The energy up to which they are filled is known as the Fermi energy.

---

1 A rigorous solution of the problem is not possible in the case of a macroscopic piece of metal containing very many particles.

2 Since the periodicity of the metal lattice differs in various directions in space, this relationship varies in general with the direction in space.

3 W. Pauli, 25.04.1900–15.12.1958, Nobel prize 1945.

4 E. Fermi, Italian physicist, 20.09.1901–28.11.1954, Nobel prize 1938.

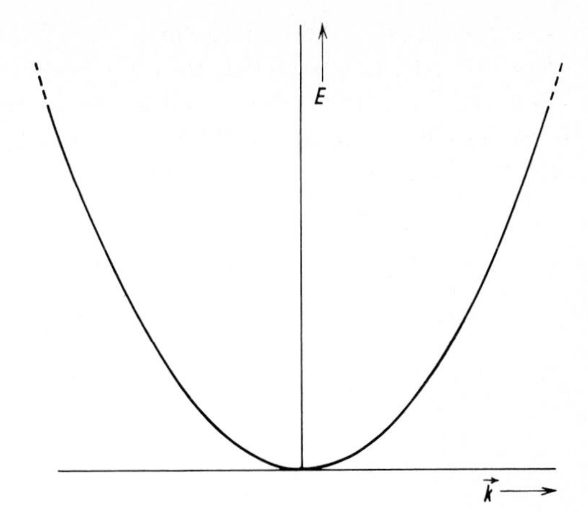

Fig. 8. Relationship between energy and momentum for a free electron.

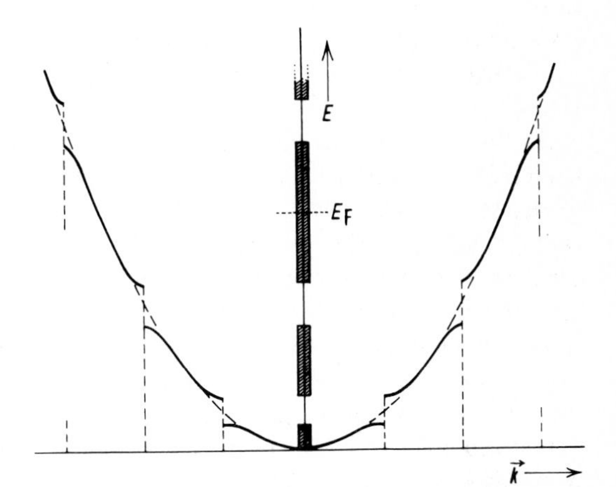

Fig. 9. Relationship between energy and momentum for an electron in a periodic potential.

A metal can now be characterized by the fact that this Fermi energy lies within a permitted energy band, i.e. that this band is only partially filled [1]. The Fermi energy for this case is indicated in Fig. 9. The occupation of the states is determined by the Fermi function, the distribution function for fermions. This Fermi function, which takes account of the Pauli exclusion principle, is:

$$F = \frac{1}{e^{\frac{E-E_F}{k_B T}} + 1}, \qquad (1\text{-}3)$$

k_B is the Boltzmann constant.

1 An insulator is characterized by the fact that it only has completely filled bands. It is easy to understand that the electrons in a filled band cannot be accelerated by an electric field because there are no free states into which they can be excited.

This Fermi function is plotted in Fig. 10 for the case $T = 0$ (broken line) and for the case $T \neq 0$ (solid line). The Fermi function is somewhat smeared at finite temperatures. The smearing amounts to about the mean thermal energy, that is equal to ca. 1/40 eV[1] at room temperature. At finite temperature the Fermi energy is that energy where the distribution takes the value 1/2. It is the order of a few eV. This means, however, – and this is very important – that the smearing at the Fermi energy is very small at ordinary temperatures. Such a system of electrons is known as a "degenerate electron gas".

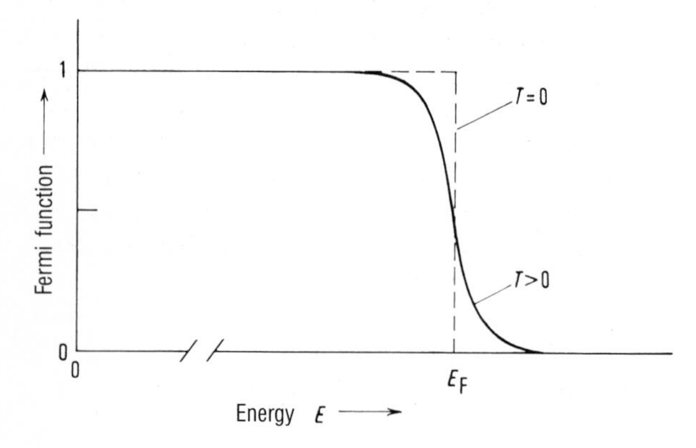

Fig. 10. Fermi function. E_F amounts to a few eV, the temperature smearing, on the other hand, a few times 10^{-3} eV. The abscissa is broken to indicate this.

These considerations enable us to understand the very small contribution of the electrons to the internal energy. For it is only a very few electrons, namely those lying in the smeared region at the Fermi energy, which are able to take part in thermal exchange processes. All the others cannot be excited by means of thermal processes because they cannot find a vacant state which they could occupy after being excited.

Anyone who wishes to understand modern solid state physics must become accustomed to thinking in terms of quantum states and their occupation. Acclimatization to this somewhat abstract concept is also indispensible for an understanding of superconductivity. For this reason – and also to practise with the many new concepts – we will now briefly consider the occurrence of electrical resistance. The electrons will now be treated as waves, permeating the crystal in all directions. A current occurs when somewhat more waves are propagating in the direction of the current than in the opposite direction. The interaction between the atom residues now consists of a scattering of the electron waves. This scattering corresponds to the collisions of the particle representation. It can – and this is new for the wave representation – not take place at the strictly periodic lattice. The states of the electrons determined with the aid of the Schrödinger equation are stable quantum states. Only a disturbance of the periodic potential, whether it be as a result of thermal vibrations of the atoms, of defects in the crystal lattice or impurity atoms, can cause a scattering of the electron waves, that is a redistribution of the quantum states. The scattering by thermal vibrations yields the temperature-dependent part of the resistance and that by the lattice defects and impurity atoms yields the residual resistance.

---

1 eV is an energy unit mainly used for atomic and nuclear processes. 1 eV = 1.6×10^{-19} Ws.

After our first brief and necessarily simplified excursion into the modern theoretical treatment of the conduction process we can now turn to our actual problem, the existence of a resistance-free transport of charge in the superconducting state. The wave-mechanical way of consideration itself does not make the persistent current less incomprehensible. We have merely altered the manner in which we speak. We now have to ask: What are the mechanisms which, at a finite temperature and over a very narrow temperature range, can forbid all energy exchange with the lattice via scattering. We do not seem to have got anywhere at all. But we will see that these considerations are of very decisive importance for an understanding of superconductivity. What is required, in addition to this, is to take into account a particular interaction of electrons with each other. In the treatment up to now we have considered the quantum states for a single electron and acted as if these states were not altered when we populated them with electrons. However, if there are interactions between electrons this is no longer rigorously true. We must rather ask, what states the system of electrons with interactions has, in other words what collective states exist. This is the key to the understanding and also to the difficulties of the superconducting state. It is a typical quantum and collective phenomenon.

1.2 Superconducting elements, alloys and compounds

Onnes was able to demonstrate soon after the discovery of the superconductivity of mercury that other metals, e.g. tin and lead, could also be made superconducting. Over the next decades new superconductors were discovered amongst the most varied groups of substances. Here a first overview will be given of this multiplicity of individual examples. The new oxides, which have been the subject of so much excitement very recently, will be treated in the next Section 1.3. Some individual interesting groups of materials, e.g. superconductors of the β-tungsten structure and the organic superconductors, will be described in more detail in Section 8.7.

Even a glance at the elements which are known today to possess superconducting phases (Table 1) reveals that superconductivity is not a rare property of metals. Some metals are only superconducting in high pressure phases; they are listed at the end of the table. Crystal structures and melting points are included in the table to reveal how different superconducting elements can be in their other properties. As we will see the Debye temperature is of particular importance for superconductivity.

The transition temperatures for the elements lie between a few hundredths and ca. 10 K. There is no connection discernible between the temperature of transition and other characteristics such as, for instance, crystal structure or melting point, with which it would be possible to separate the superconducting from the nonsuperconducting metals. It is merely noticeable that only cesium of the metals in the first main group of the periodic table (P.T), the typical univalent metals, has been found to exhibit superconductivity until now, even though most of these metals have been investigated down to about 0.1 K. From investigations of very dilute alloys of the noble metals with transition temperatures in the mK range it can be concluded that pure Au would be superconducting at ca. 0.2 mK [8]. These experiments yielded T_c values of ca. 10^{-6} mK for Cu and Ag.

The question as to whether particular metals, for instance some in the 1st main group of the P.T., do not, in principle, become superconducting in the purest form and at whatever low temperature cannot yet be answered with certainty. Cs, for example, was only found to be superconducting in a new high pressure phase. The theoretical considerations, which we will become more closely acquainted with, do not provide for any compulsion for the assumption that all metals ought, in principle, to become superconducting. On the other hand, it must be admitted that it would be very difficult to detect superconductors with low transition temperatures (of the order of $T_c < 10^{-2}$ K). The tiniest contamination with paramagnetic atoms (e.g. Mn, Co etc. at concentrations of less than 1 ppm) can, as can the tiniest magnetic fields (e.g. a fraction of the earth's magnetic field) completely suppress superconductivity in such cases. It is, therefore, possible to

Table 1. Superconducting elements [7]. The transition temperatures given in brackets refer to other crystalline modifications. Some of the figures, particularly those for the Debye temperature, can only be regarded as approximations.

	Element	T_c in K	Crystal structure	Melting point in °C	Debye temp. in K
1	Al	1.19	f.c.c.	660	420
2	Be	0.026	hex.	1283	1160
3	Cd	0.55	hex.	321	300
4	Ga	1.09	orth.	29.8	317
		(6.5; 7.5)			
5	Hf*	0.13	hex.	2220	
6	Hg	4.15	rhom.	−38.9	90
		(3.95)	tetr.		
7	In	3.40	tetr.	156	109
8	Ir	0.14	f.c.c.	2450	420
9	La	4.8	hex.	900	140
		(5.9)	f.c.c.		
10	Mo	0.92	b.c.c.	2620	460
11	Nb	9.2	b.c.c.	2500	240
12	Np**	0.075	ortho.		
13	Os	0.65	hex.	2700	500
14	Pa	1.3			
15	Pb	7.2	f.c.c.	327	96
16	Re	1.7	hex.	3180	430
17	Rh***	3.2×10^{-4}	f.c.c.	1966	269
18	Ru	0.5	hex.	2500	600
19	Sn	3.72	tetr.	231.9	195
		(5.3)	tetr.		
20	Ta	4.39	b.c.c.	3000	260
21	Tc	7.8	hex.		351
22	Th	1.37	f.c.c.	1695	170
23	Ti	0.39	hex.	1670	426
24	Tl	2.39	hex.	303	88
25	U (α)	0.2	orth.	1132	200
26	V	5.3	b.c.c.	1730	340
27	W	0.012	b.c.c.	3380	390
28	Zn	0.9	hex.	419	310
29	Zr	0.55	hex.	1855	290

* C. Probst: Dissertation, TU Munich (1974)
** J. Smith, and Hunter Hill: Bull. Am. Phys. Soc. Ser II., Band 21 383 (1976)
*** Ch. Buchal, F. Pobell, R. M. Mueller, M. Kubota, and J. R. Owers-Bradley: Phys. Rev. Lett. *50*, 64 (1982)

(Continuation of Table 1.)

Elements that only become superconducting under pressure or in high-pressure phases.				
29	As	0.5	p ≅ 120 kbar	(a)
30	Ba	5.1	p > 140 kbar	(b)
		(1.8)	p > 55 kbar	
31	BiII	3.9	p = 26 kbar	(c)
	BiIII	7.2	p > 27 kbar	
	BiV	8.5	p > 78 kbar	
32	Ce	1.7	p > 50 kbar	(d)
33	Cs	1.5	p ≅ 100 kbar	(e)
34	Ge	5.4	p > ca. 110 kbar	(f)
35	Lu	0.1–0.7	p: 80–ca. 130 kbar	(g)
36	P	4.6–6.1	p > ca. 100 kbar	(h)
37	Sb	3.6	p > 85 kbar	(i)
38	Se	6.9	p > ca. 130 kbar	(k)
39	Si	6.7	p > ca. 120 kbar	(f)
40	Te	4.5	p > 43 kbar	(l)
41	Y	1.5–2.7	p: 120–160 kbar	(e)

a) I. V. Berman, and N. B. Brandt: JETP *10*, 55 (1969)
b) J. Wittig, and B. T. Matthias: Phys. Rev. Letters *22*, 634 (1969)
c) A. Eichler, and J. Wittig: Z. angew. Physik *25*, 319 (1968)
d) J. Wittig: Phys. Rev. Letters *21*, 1250 (1968)
e) J. Wittig: Phys. Rev. Letters *24*, 812 (1970)
f) J. Wittig: Z. Physik *195*, 215 (1966)
g) Chr. Probst, W. Wiedemann, and J. Wittig: 9th Annual Meeting of the EPS High Pressure Research Group, June 1971. Umea, Sweden
h) J. Wittig, and B. T. Matthias: Science *160*, 994 (1968)
N. B. Brandt, and I. V. Berman: LT 11, St. Andrews 1968, Vol. *2*, 973
i) J. Wittig: J. Chem. Phys. Solids *30*, 1407 (1969)
k) J. Wittig: Phys. Rev. Letters *15*, 159 (1965)
l) B. T. Matthias, and J. L. Olsen: Phys. Letters *13*, 202 (1964)

take the view that we have not detected superconductivity in many metals simply because we have not investigated these metals in a sufficiently pure state at a sufficiently low temperature. It is impossible, in principle, to contradict the statement "All metals would become superconducting in a sufficiently pure state and at a sufficiently low temperature". It is only possible to confirm it by demonstrating that all metals[1] can be superconducting. The question remains open at the moment.

It can also be seen from the Table that superconductivity is dependent on the arrangement of the atoms. One and the same metal has different transition temperatures for different crystal structures. It can happen, as in the case of bismuth (Bi), that one modification is not superconducting right down to very low temperatures (down to $T \approx 10^{-2}$ K), while several other modifications exhibit superconductivity.

It has also been found that a crystalline structure is not a necessary condition for superconductivity. It has been demonstrated that "amorphous" samples, such as can be prepared from some metals by condensation of the metal vapor on a very cold surface,

1 The ferromagnetic metals (e.g. Fe, Ni, etc.) must be omitted from consideration; they cannot become superconducting on account of their strong magnetism.

can exhibit superconductivity – sometimes at very high transition temperatures. We will return to these "amorphous" superconductors, which we are now coming to understand better, in Chapter 8 (Section 8.5).

The distribution of the superconducting elements within the periodic table is illustrated in Fig. 11. Two distinct groups are immediately recognizable.

1. The nontransition metals, including the high pressure phases of the elements Si, Ge, P, As, Sb, Bi, Se and Te.

2. The transition metals, where an inner shell (3d, 4d or 5d levels) is being filled up as the atomic number increases[1].

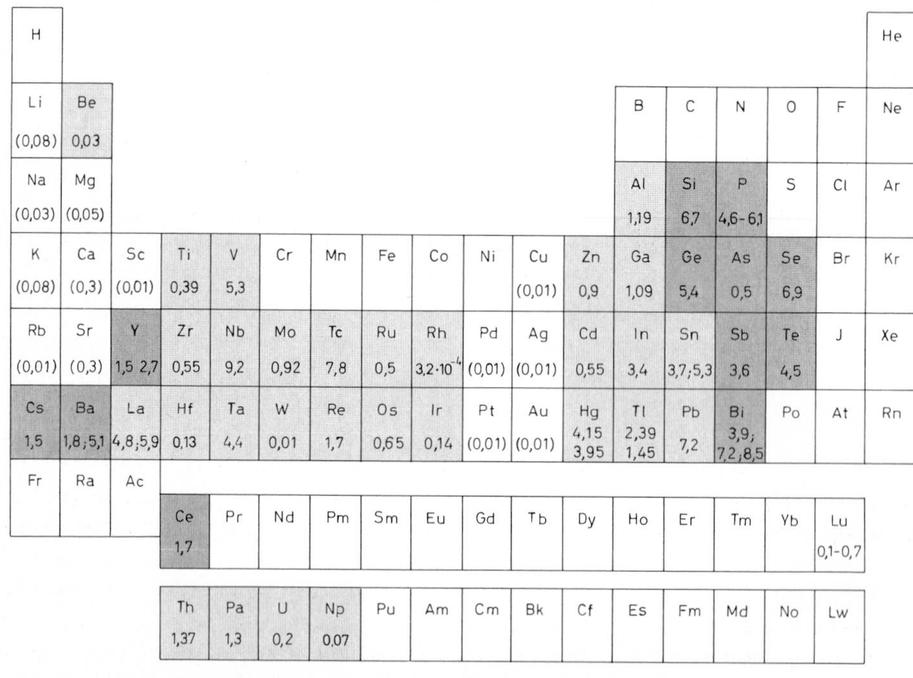

Fig. 11. Distribution of superconductors in the periodic system. The elements that only become superconducting in their high pressure phases have a dark shade. The numbers in brackets give the temperatures to which elements have been cooled without becoming superconducting.

The ferromagnetic metals cannot become superconducting on account of their strong magnetism. We also think we now undertand why there are only a few superconductors with very low transition temperatures in the 8th group of the periodic table [9].

The picture becomes even more confusing when very many more than 1000 superconducting alloys and compounds are also taken into consideration [7]. We find superconducting compounds, the superconductivity of neither of whose two components has yet been detected. CuS with $T_c = 1.6$ K is an example. Metal hydrogen systems have recently been found with suprisingly high transition temperatures [10]. Here the hydrogen is present interstitially in the lattice (see also 8.7.2). Another remarkable superconductor,

1 It is the 4f and 5f levels that are filled up in the case of the lanthanides and actinides.

Table 2. Superconducting compounds [7].

Substance	T_c in K	Crystal structure
CuS	1.6	hex., B18
Bi_2Cs	4.75	f.c.c., C15
BiNa	2.25	tetr., $L1_0$
$Pd_{70}Ag_{30}$ $^{+}$H	16	f.c.c.
$(SN)_x$	0.26	Bundles of $(SN)_x$ chains
V_3Ge	6.0	β-tungsten A15
V_3Ga*	14.2–14.6	β-tungsten A15
V_3Si	17.1	β-tungsten A15
Nb_3Au	11.0–11.5	β-tungsten A15
Nb_3Sn	18.0	β-tungsten A15
Nb_3Ge	23.2	β-tungsten A15

* Values of T_c up to around 20 K could be obtained after particularly careful annealing – G. Webb RCA, Princton, USA 1971.

the polymer $(SN)_x$ was discovered in 1975 [11]. The polymerization process produces monocrystalline fibers of $(SN)_x$ chains. This is the first polymer exhibiting metallic conduction down to very low temperatures. It becomes superconducting at 0.26 K.

According to a hypothesis proposed by W.A. Little [12] organic materials ought to include superconductors with very high transition temperatures. This hypothesis has not yet been confirmed. But organic superconductors have been found; they will be described in 8.7.3.

The superconductors with the β-tungsten structure of type Nb_3Sn are still an interesting group. Before the discovery of the new superconducting oxides a representative of this group, namely Nb_3Ge, had the highest known transition temperature of 23.2 K [12a] for more than a decade (Table 2). This group is dealt with in somewhat more detail in Section 8.7.1.

During the more than 75 years that there has been increasing interest in the investigation of superconductivity, there has been no lack of attempts to find rules for its occurrence. It was pointed out very early on that the atomic volume, i.e. the volume an atom occupies within the metal, could be of importance. If this atomic volume of the element is plotted against the atomic number then it can be seen that the superconductors are found preferentially in areas of smaller atomic volume [1]. These considerations could be of a certain importance for the understanding of the superconducting, high-pressure phases. Hydrostatic pressure reduces the atomic volume. Some elements, such as Ba and even Cs, which possess a particularly large atomic volume in their normal state, become superconducting under sufficiently high pressure. It must be kept in mind, however, that these elements can undergo phase transitions at high pressure, whereby the ordering of the atoms and other parameters important for superconductivity can change.

Matthias put forward [13] a very fruitful, empirical rule for the magnitude of the transition temperature. This Matthias rule states that the mean number of valence electrons in a substance is a decisive quantity for superconductivity. All electrons in incomplete

1 See "Roberts Report", B.W. Roberts, General Electric Report No. 63-RL-3252 and N.B.S Technical Note 482.

valence shells are counted as valence electrons, i.e. the number of valence electrons is identical with the group in the P.T. in which the element is to be found. The mean number of valence electrons is taken as being the arithmetic mean of all valence electrons. In the case of the transition metals there are, according to the Matthias rule, distinct maxima in T_c for special numbers of valence electrons n_v between 3 and 8. For $n = 5$ this is manifest in the relatively high transition temperatures of V, Nb and Ta. A particularly high transition temperature is observed for technetium in the 7th group of the periodic table. The elements in the 4th and 6th groups of the P.T. have, in contrast, very low transition temperatures. An increase in transition temperature with increasing numbers of valence electrons is plainly evident in the case of the group of nontransition metals.

The Matthias rule is especially fruitful in the field of alloys. In Fig. 12 the numbers of valence electrons of some alloys [1] are plotted against their transition temperatures. Two distinct maxima of T_c are visible at mean valence electron numbers of ca. 4.7 and 6.5. The compounds with the β-tungsten structure with particularly high T_c have a mean n_v of about 4.7.

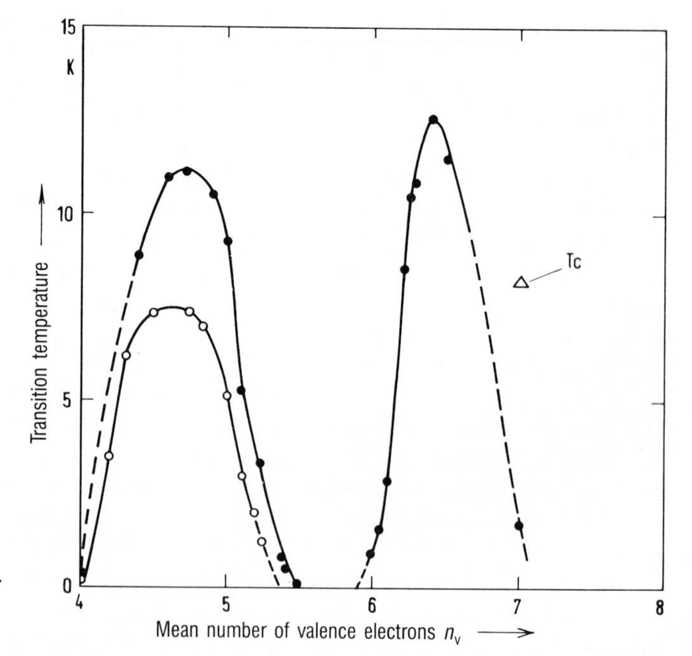

Fig. 12. Transition temperatures of some alloys of the transition metals as a function of the mean number of valence electrons (after [14]).
• • Zr-Nb-Mo-Re,
o o Ti-V-Cr.

As previously mentioned we initially only wish to learn from this wealth of information that superconductivity is a very general property of metals. Its theoretical explanation must only involve very general properties of metals. We now believe, on the basis of the BCS theory (see Chapter 2), that we understand the classical superconductors, in princple at least. We are in the position of being able to calculate the transition tempera-

1 Since other parameters alongside the numbers of valence electrons influence superconductivity it is only possible to compare alloys, whose crystal structures are the same or similar and whose components do not lie too far apart in the P.T.

ture of a simple superconductor, such as aluminium, from the general parameters of metals.

The electron-phonon interaction (see Chapter 2) on which the microscopic theory of superconductivity of Bardeen, Cooper and Schrieffer is based, once seemed adequate for a deeper understanding of superconductivity. With the discovery of the superconducting oxides the question of the decisive interaction allowing the understanding of transition temperatures higher than 100 K is open once more.

1.3 The new superconducting oxides

In the September 1986 issue of Zeitschrift für Physik B J.G. Bednorz and K.A. Müller published a paper entitled "Possible high T_c superconductivity in the Ba-La-Cu-O system" [15]. The authors started from the hypothesis that substances with a pronounced Jahn-Teller effect [1] could also be superconductors with particularly high transition temperatures. They began with a study of compounds based on nickel oxide, since Ni^{3+} in an octahedron of oxygen atoms exhibits a large Jahn-Teller effect. However, they found no superconductors in this group of substances. They then passed on systematically to copper oxide. Cu^{2+} also has a large Jahn-Teller effect in an octahedron of oxygen atoms.

After a few months Bednorz and Müller had samples which exhibited a steep drop in the electrical resistance even above 30 K. They were also able to demonstrate, even in this first publication, that the drop in resistance was displaced to lower temperatures when a large current was passed. This is a characteristic of superconductivity. Had superconductors been found with $T_c > 30$ K? This would have been a breakthrough after more than 10 years of stagnation.

This work attracted remarkably little attention. Doubts were raised as to whether the phenomenon really was superconductivity. The samples were mixtures of several phases, including insulating substances. Hence, they had extremely large specific resistances. It was quite possible that some phase change had caused the drop in resistance [2]. So that more convincing evidence was required that these samples were superconducting.

This was provided by Bednorz, Müller and Tagashige who demonstrated the Meissner-Ochsenfeld effect [16]. The authors submitted this publication to the editors of Europhysics Letters on 22nd October 1986. This opened the gate to superconductors with higher transition temperatures. Figure 13 reproduces the decisive results of this publication. Above 40 K both samples exhibited the small amount of paramagnetism well-known for metals, which is almost temperature-independent. At about 30 K, that is in

---

1 The Jahn-Teller effect is the name given to the displacement of an ion from a highly symmetrical state with respect to its surroundings. This abolishes the degeneracy of the states of the ion and reduces its total energy. A strong Jahn-Teller effect is an expression of a strong electron-phonon interaction. So that the hypothesis of Müller and Bednorz was certainly compatible with the BCS theory.

2 In the mid-1940s it was observed that there was a sharp drop in the resistance of metallic solutions of sodium in ammonia below 70 K; this was attributed at first to superconductivity. The phenomenon was, in fact, the result of the precipitation of sodium fibers from the solution (R.A. Ogg Jr.: Phys. Rev. *69*, 243 and 668 (1946); *70*, 93 (1946)).

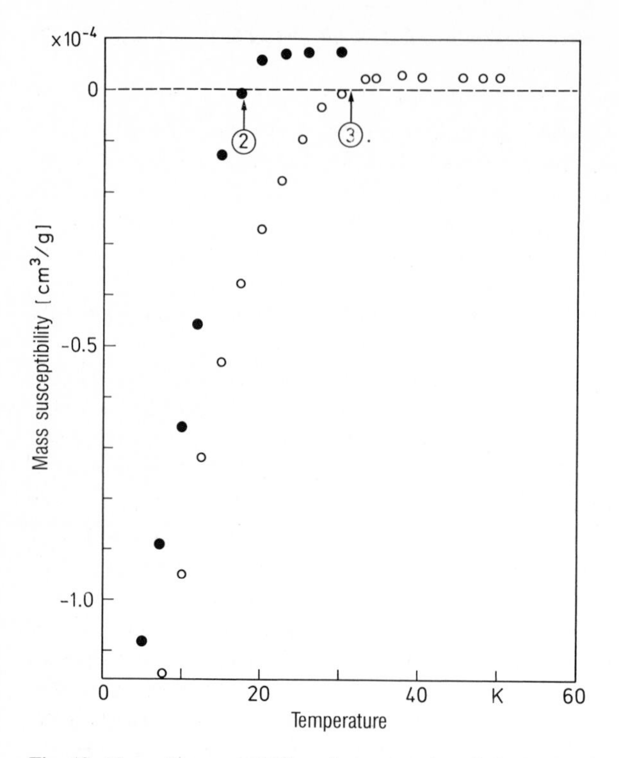

Fig. 13. Magnetic susceptibility of two samples of the La-Ba-Cu-O system as a function of the temperature (after [16]).

the same temperature range where the drop in resistance occurred, increasing diamagnetism was observed on cooling in a magnetic field – the magnetic field was expelled from the samples – a phenomenon corresponding to the Meissner-Ochsenfeld effect.

The reason why this result came as such a surprise to the specialists was that Bernd Matthias and his coworkers had made a systematic investigation of metallic oxides in the middle of the 1960s [17]. They investigated substances based on the oxides of transition metals, such as W, Ti, Mo and Bi. They discovered some extraordinarily interesting superconductors, e.g. in the Ba-Pb-Bi-O system, but no particularly high transition temperatures [17].

The "gold rush" began at the end of 1986 and beginning of 1987, as a Japanese group, led by S. Tanaka, was able to confirm the results of Bednorz and Müller in their entirety. Now in numerous laboratories all over the world scientists began to study these new oxides. This extraordinary burst of scientific effort was soon crowned with success. It was found that superconductors with transition temperatures above 40 K could be prepared from the system La-Sr-Cu-O [18].

A few weeks later transition temperatures higher than 80 K were observed in the system Y-Ba-Cu-O [19]. At this phase announcement of results was made more frequently at press conferences than in scientific journals. The media devoted themselves particularly zealously to these developments. With superconductivity at temperatures above the boiling point of liquid nitrogen (T = 77 K) it was possible to enthuse about the large-scale, industrial application of this phenomenon.

The superconductors of the systems La-Sr-Cu-O and Y-Ba-Cu-O were also discovered simultaneously and independently in Beijing by Z.X. Zhao and his coworkers [20]. These rapid developments were also kept up with in Germany. Transition temperatures above 100 K were realized in the system Y-Ba-Cu-O [21].

Within the space of one year (1987) the properties of the new materials were studied very precisely and the results published in a large number of papers. What is known as the "1,2,3 compound" $YBa_2Cu_3O_7$ is a model substance. It is certainly the best investigated of such substances to date. Nevertheless, there is still a lack of convincing results for some important parameters e.g. the energy gap (see Section 3.3).

The development of these new superconductors is still rapidly underway. In the spring of 1988 there were reports of new superconductors in the systems Bi-Sr-Ca-Cu-O [22] with T_c values of up to 110 K and in Tl-Ba-Ca-Cu-O with T_c values of over 120 K [23]. These results have been confirmed and are, thus, scientifically established. Reports have been made by many groups of the existence of superconducting phases at much higher temperatures. These observations – known as USO (*U*nconfirmed *S*uperconductivity *O*bservations) – have not as yet been confirmed by further experiments. But all in all they suggest that nature might yet have more secrets in this direction.

The properties of these novel materials, such as the critical magnetic field and critical current density, will, insofar as data are available, be dealt with in the appropriate sections. Their structures and some specific problems will be discussed in Section 8.7.4 [1)].

Since this discussion of superconducting materials has left us with many unanswered questions it is now time to stop listing facts and to form an initial picture of what we today understand, in principle, concerning the superconducting state. These basic hypotheses certainly represent the basis of the phenomenon correctly. They are able to describe a large number of results from more basic principles and sometimes quantitatively at that. As already stated there are still problems concerning the relationship with the characteristic properties of metals.

1 In an introduction such as this it is only possible to refer to a few examples from the several thousand publications which have now appeared on the subject of the new materials. No claim can be made as to a representative selection.

2. The superconducting state

2.1 The electron-phonon interaction and Cooper pairs [1)]

Even the consideration of superconductivity within the scale of ordering processes (Introduction) provides a presumption that a specific interaction leads to the occurrence of a new state of the metal. In order to understand superconductivity it was necessary to discover what this interaction could be. It was only then that it was possible to develop an atomistic [2)] theory of superconductivity.

The difficulties involved in such a theory were extraordinarily large. On the basis of the striking change in electrical conductivity and, as we will see (Chapter 4), in the magnetic properties as superconductivity starts, it is possible to conjecture that an ordering process in the system of conduction electrons is basically reponsible. As we have already seen these conduction electrons (see Section 1.1) have, on account of the Pauli exclusion principle, considerable energies amounting to several eV. An eV corresponds to the mean thermal energy $k_B T$ of about 11 000 degrees. However, the transition to the superconducting state takes place at only a few degrees. So it is necessary to discover an interaction that can lead to order in the electron system, independent of the high energies of the electrons.

Now a whole range of interactions could occur between the conduction electrons of a metal. It had been thought that the Coulomb repulsion of the electrons could lead to a spatial orientation of the electrons into lattice-like regions (Heisenberg 1947) [24]. It is also possible to imagine a magnetic interaction (Welker 1929) [25]. The electrons as they move through the metallic lattice with considerable velocity [3)] produce a magnetic field as does a current and can interact with each other via this magnetic field. Further interactions can have their origin in the structure of the electronic states (permitted energy bands (see Section 1.1)) [26].

None of these attempts led to anything like a satisfactory atomistic theory of superconductivity. Only in 1950/51 did Fröhlich [27] and Bardeen [28] simultaneously and independently suggest an interaction of the electrons via the vibrations of the lattice, which was later to lead to a basic understanding of superconductivity within the framework of our other knowledge concerning the metals. Starting from this interaction Bardeen, Cooper and Schrieffer were able in 1957 [29] to propose an atomistic theory of superconductivity – commonly known as the BCS theory – which was in a position to explain a wealth of results quantitatively and which, above all, had an extremely stimulating effect. As a result of the impetus provided by this theory a large number of new experiments were carried out in the years following 1957; these have not only increased our understanding of superconductivity very considerably they have also – and this is no exaggeration – fundamentally altered it. It could well be that efforts to come to a theoretical un-

---

1 The discovery of the new superconducting oxides with transition temperatures above 100 K makes it seem likely that, in addition, other interactions, responsible for the ordering effect, are at work in these.

2 Such theories are often called "microscopic" theories as well.

3 The electrons lying in states at the Fermi limit E_F have energies of a few eV. If the velocities of these electrons are calculated, then they are found to be of the order of one hundredth the velocity of light.

derstanding of the new superconductors will lead to quite new insights. However, this is an open question at the moment.

It turned out that the way from the description of the new interaction (1950) to the development of an acceptable theory (1957) was a very difficult one. It must be counted a particularly fortunate accident that almost simultaneously to the theoretical formulation of this new interaction and its possible implications for superconductivity there was also a startlingly unequivocal demonstration of the basic correctness of its premises. For an investigation had been made of various isotopes of a superconductor, where it was found that the transition temperature T_c to superconductivity was dependent on the atomic mass. And not only that – the experimentally determined dependence corresponded exactly to that which was to be expected from the theoretical considerations developed by Fröhlich and Bardeen (see Section 3.1).

This demonstrated that, in spite of all the formal difficulties associated with the theory, the basis was evidently correct. This superb confirmation of the new basic idea was very probably of decisive importance for the development that followed.

Now how are we to understand this interaction between the electrons, which is mediated by the lattice vibrations? We will now discuss some models for this interaction. It must, however, be pointed out at the start that these models only have a very limited predictive power if we wish to draw further conclusions from them.

Let us start with a static model. The lattice of atomic residues, in which the conducting electrons move like a Fermi gas, has elastic properties. The atomic residues are not fixed rigidly in their resting positions, but can be displaced from their resting positions. As we have discussed they vibrate randomly around these resting positions at finite temperatures. Now if we introduce two negative charges into this lattice of atomic residues and neglect – a drastic simplification and somewhat unrealistic – all the other electrons, then the negative charges of our two electrons will influence the lattice by attracting the surrounding positive electrical charge. Thus: The lattice is polarized by the negative charge. This situation is illustrated schematically in Fig. 14. The polarization implies a concentration of positive charge in the neighborhood of the polarizing negative charge in comparison to the even distribution of the positive charges. The second electron can sense with its polarization the polarization of the first electron. It experiences an attraction to the site of the polarization and, hence, to the first electron. We have described an attractive interaction between two electrons mediated by the polarization of the lattice.

It is possible to describe a mechanical analogue of this static attractive interaction. We represent the elastically deformable lattice of atomic residues by an elastic membrane, such as a thin sheet of rubber in tension, or the surface of a liquid [1]. We now lay two spheres on the membrane – in the liquid model they must not be wetted by it. When they are distant from each other each will deform the membrane on account of its weight (Fig. 15a). This corresponds to the polarization of the lattice. Now it is plain to see, even without calculation, that the energy of the whole system (membrane plus two spheres) would be reduced if both spheres lay in a single trough. They would both sink lower

1 On account of the surface tension, energy is necessary to deform the surface of a liquid from its equilibrium state.

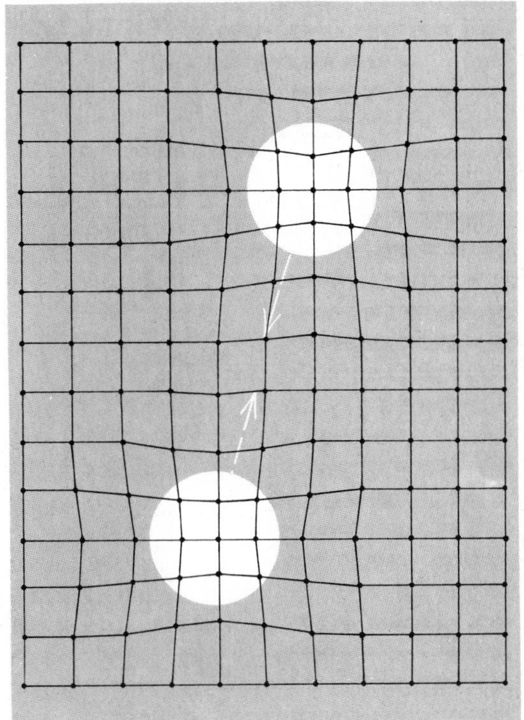

Fig. 14. The polarization of the lattice of atomic residues by the electrons. In a *static* model this polarization cannot overcome the repulsion of the electrons as a result of their equal charges. It can only reduce this repulsion greatly.

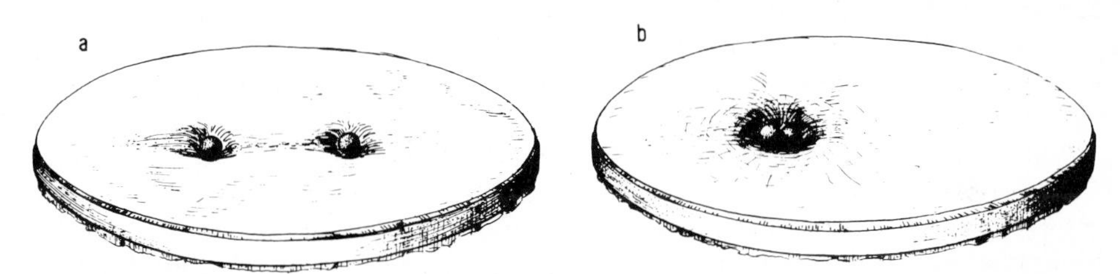

Fig. 15. The attraction of spheres on an elastic membrane. Configuration a is unstable and goes over to b.

(Fig. 15b), which corresponds to a reduction of their potential energy in the gravitational field. This means that the total energy of the system is reduced [1)]. Thus, there is an interaction between the spheres which is mediated by the elastic membrane and leads to a bound state, i.e. to a state where the spheres are as close as possible to each other in space.

The model makes it clear to us that attractive forces can be set up by elastic deformations. But that is all. Our electrons have considerable velocities within the metals. They do not polarize the lattice statically. It could rather be assumed that, on their journey through the lattice, a polarization is set up along the path they take; this would depend

1 The energy difference must naturally be transported away. The new equilibrium is approached via an oscillation of the membrane, in which the difference in the mechanical energy between the starting and final states is transformed into heat by frictional effects.

decisively on how rapidly the lattice can respond to the polarizing effect of the electron. It would depend on the time in which the lattice of atomic residues can undertake a displacement. This, however, depends, in the case of an elastic system, on the natural frequencies. We have achieved a fundamental advance by this general introduction of a dynamic element. We now understand, at least qualitatively, how, under otherwise equal conditions, the strength of the polarization and, thus, of the interaction can depend on the natural frequency of the lattice and, hence, on the mass of the atomic residues. Heavy isotopes vibrate somewhat more slowly, that is have a lower lattice frequency. They are only able to react more slowly to the polarizing effect of the electrons, i.e. the polarization is less. We would, therefore, expect that the interaction would be weaker and that the temperature of transition to the superconducting state would be lower. The temperature of transition decreases with increasing isotopic mass. This is in accordance with the experimental findings (see Section 3.1). But is must be explicitly pointed out that the considerations discussed here are of a heuristic nature and do not allow of quantitative conclusions. A quantum mechanical approach is required to provide us with conclusions concerning which frequencies of the lattice are decisive for the interaction.

It is true that we have introduced dynamic elements into our interaction via polarization of the lattice, but we have adopted the proposition that polarization of an electron can lead to a reduction in the energy level of a second from static considerations. In order to carry our dynamic model somewhat further we can imagine that the second electron follows the polarization track of the first and, hence, lowers its energy because the lattice is in an already polarized state.

We are now faced fundamentally with two possibilities. The two electrons can fly with the same momentum. We would then have a structure that we could aptly consider to be a particle, namely an electron pair. Such a pair would, however, have a total momentum, namely twice that of a single electron. The other possibility is that the electrons have opposing momenta. But one can fly in the path of the other. However, it is more difficult to imagine a new particle, an electron pair, in this case. However, if we only think abstractly, we then find it evident that in the first case there is the requirement that the two electrons have the same momentum and, hence, are correlated by $\vec{p}_1 = \vec{p}_2$. The requirement $\vec{p}_1 = -\vec{p}_2$ is just as unequivocal a correlation. We are, therefore, completely justified in calling *these* two rigorously correlated electrons a pair too. This electron pair has a total momentum of zero. Such an electron pair is known as a "Cooper pair", because it was Cooper [30] who was first able to demonstrate that such a correlation leads to a reduction of the total energy. If we take into account the intrinsic rotational momentum of the electron, which is important for the statistical behavior of the new particle, then a Copper pair is made up of two electrons with equal and opposite momenta and equal and opposite intrinsic rotational momenta [1),]

$$\text{Cooper pair: } \left\{ \vec{p}_\uparrow, \ -\vec{p}_\downarrow \right\} .$$

The correlation to Cooper pairs is energetically favored by the polarization of the positive lattice.

1 The intrinsic rotational momentum of a particle, such as the electron, is known as its spin. The electrons of a Cooper pair, therefore, have opposite spins.

Since the possibility of pair correlation is the decisive basis for the atomistic theory of superconductivity and, hence, for the understanding of the superconducting state, this phenomenon will now be considered from another, more general point of view. It is possible to understand the formation of electron pairs in the lattice on the basis of a very general formalism, namely exchange interaction.

It is a triviality that systems which exchange quantities of any description are interacting with each other. This is a quite general statement. The point is that in the case of quantum mechanical exchange interactions, the exchange can lead to an attraction between two physical systems. For example, two particles can experience attraction by exchanging a third particle, leading to the binding of the two particles together.

It is easy to understand repulsion as a result of exchange at the classical level. Two persons throwing a ball backwards and forwards experience such a repulsion. That is immediately evident and can easily be proved by placing the persons on easily movable trucks or trolleys, which can move along the line joining the two persons. These trolleys will move away from each other as the ball is thrown backwards and forwards, whereby the repulsion is solely the result of the exchange of the ball and the exchange of momentum associated with it.

We will not try to demonstrate the attractive exchange interaction on the basis of such a simple model. Rather we will discuss two examples from modern physics. Both examples are fundamentally different but they have one thing in common: The interaction is produced by the exchange of particles between two systems.

We know that two hydrogen atoms form a hydrogen molecule and that this molecule is very strongly bonded. An energy of 26×10^4 W s/mol (62.5 kcal/mol) is required to break this bond, i.e. to dissociate 2 grams of H_2. How can we understand this strong binding of two, of themselves neutral, hydrogen atoms to form an H_2 molecule? To make the principle clear let us consider a somewhat simpler system namely H_2^+, a singly positively charged hydrogen molecule. This molecule is made up of two hydrogen nuclei, two protons and an electron. The two possible states of this system, when the two protons are a great distance apart, are illustrated in Fig. 16a and b. The electron is at one of the two protons. Now if we bring the two protons together, then quantum mechanics predicts that there will be a certain probability of the electron "hopping" from one proton to the other; in our terminology, of being "exchanged". The probability of being exchanged increases rapidly as the separation is decreased. The electron belongs equally to either proton as indicated in Fig. 16c. The decisive prediction of quantum mechanics concerning this problem is that this exchange reduces the total energy of the system. This means that smaller separations R are energetically favored. The protons are bonded by the common electron. The equilibrium distance is a result of the requirement that the attractive force of the exchange is equal to the repulsive force of the two positive protons.

The reduction in energy as a result of electron exchange can be particularly simply understood with the aid of the uncertainty principle, a very fundamental principle of modern physics. This principle lays down that the two parameters of a particle, its momentum and its position, cannot both be determined equally accurately. We cannot have a more accurate determination of the two quantities than is laid down by the relationship:

$$\Delta p_x \, \Delta x = \hbar \, . \qquad (2\text{-}1)$$

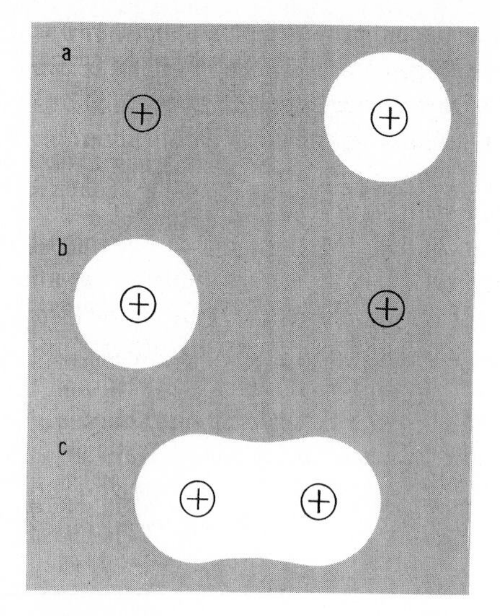

Fig. 16. The bonding energy of the H_2^+ molecule. The models are not to scale.

This means for our system that we can reduce the smearing of the momentum Δp_x, if we allow the electron to be at both protons, because we then increase the uncertainty of its position Δx. But this makes the energy uncertainty smaller, i.e. the energy of the electron is reduced [31].

If this reduction in energy is greater than the increase because of the coulombic repulsion of the two positive protons, there is a net attraction. We can see, therefore, that the bonding of the H_2 molecule is a typical quantum mechanical effect. The considerations discussed above form the basis for the understanding of chemical bonding.

The nuclear forces, that is those forces which hold together protons and neutrons in atomic nuclei, can also be explained in terms of particle exchange. Yukawa [1)] proposed, in the 1930s, that nuclear forces should be regarded as exchange forces. For this purpose he had to propose a particle having a mass equal to several hundred electron masses. This particle, known as the π-meson, was later discovered to be a real particle. So the attraction between the nucleons can be understood quantum mechanically as an exchange interaction with the π-meson being the particle exchanged.

The attractive interactions between the conduction electrons in a metal can be understood on the basis of comparable considerations. Quite new particles can be exchanged in the metal, known as phonons. The phonons are nothing more than the elementary vibrational states of the lattice. Any given complex vibrational process undergone by the lattice can be analyzed into harmonic waves. This analysis corresponds to the Fourier analysis [2)]. Now for a macroscopic body these harmonic waves have defined energies. They also possess particular wavelengths and, hence, since $|\vec{p}| = h/\lambda$ defined momenta. It is permissible to regard them as particles and to name them phonons or acoustic quanta.

1 Hideki Yukawa, Japanese physicist, 23.1.1907–08.9.1981, Nobel prize 1949.

2 In general Fourier analysis is understood as being the analysis of a vibration into the separate harmonic vibrations of the body. These stationary characteristic vibrations are produced in our analysis into harmonic waves as "standing waves" by the superposition of 2 waves with momenta of the same magnitude and opposing direction.

Thus, an electron in a lattice can interact with another electron by exchanging acoustic quanta, that is phonons. This is referred to as an electron-electron interaction mediated by phonons. The exchanged phonons are said to be virtual because they are only in existence during the exchange from one electron to the other, but do not have the possibility of passing away from the electron into the lattice as real phonons[1]. This reveals a fundamental difference to the exchange interaction in the H_2 molecule. The electrons that are exchanged there are real particles.

The interaction is illustrated schematically in Fig. 17. Under certain conditions, which obtain in superconductors, it can be so great that it overcomes the repulsion of the electrons as a result of electrostatic forces[2]. We can then achieve the pair correlation described.

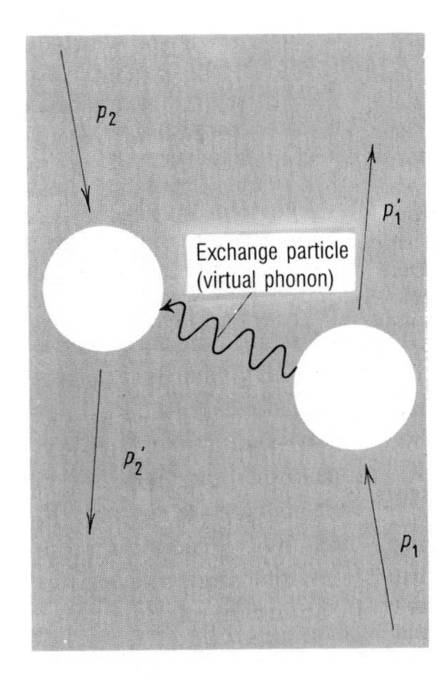

Fig. 17. Electron-electron interactions via phonons.

The mean separation at which this pair correlation becomes effective is between 100 and 1000 nm[3] for a pure superconductor. This distance is referred to as the coherence length ξ_{Co} of the Cooper pair.

---

1 If real phonons are to be generated by the electron this would be a process that produces electrical resistance, since it would be possible for this system to transfer energy from the electrons to the lattice (see Section 1.1).

2 It must be remembered that the electrostatic repulsion is very greatly screened by the positive charges of the residual atoms.

3 The interaction produced by the exchange of phonons is so weak that the electrons of a Cooper pair cannot be localized to an uncertainty of less than 10^{-4} cm. On account of the uncertainty principle, less uncertainty would result in the kinetic energy of the electrons being greater than the binding energy of the pair. Naturally these graphic descriptions rapidly run into difficulties. One could, for example, ask whether the high Fermi velocity does not disturb the whole pair relationship. The answer to this question is "no", although we will not try to justify this. The simple particle picture is inadequate here.

This coherence length ξ_{Co} can also be regarded as the average size of a Cooper pair and we can say with great simplification: In a pure superconductor a Cooper pair has an average size of 10^2 to 10^3 nm. This size is large compared with the mean distance between two conduction electrons which amounts to a few 10^{-1} nm. The Cooper pairs overlap very greatly. In the region of one pair there are 10^6 to 10^7 other electrons which are themselves correlated into pairs. One would intuitively expect that the ensemble of a group of particles, that are so penetrated into each other, would possess peculiar properties. We will talk about that in the next section.

2.2 The macroscopic occupation of the ground state and the energy gap

We have discussed the possibility of pair formation amongst the conduction electrons in very great detail. This was in order to become used to the idea of the presence of such pairs in superconductors. However, the extraordinary properties of a superconductor can only be understood if we first consider the ensemble of the Cooper pairs.

For these pairs are by no means independent. On the contrary they are particularly rigidly correlated. We must, for instance, accept from many experiments the condition that all Cooper pairs are in the same quantum state. It is just this *macroscopic occupation of a single* quantum mechanical state that is the reason for the remarkable properties of a superconductor.

What is the significance then of this requirement that all Cooper pairs take up the same quantum state? It means that all pairs have the same values for all physical quantities. Consider, for example, the momentum of the pairs. The total momentum of a Cooper pair $\{+\vec{p}, -\vec{p}\}$ is zero when there is no external influence. This applies to every pair; our condition is automatically fulfilled and seems not to have contributed anything new. The situation changes immediately when we submit the ensemble of the Cooper pairs to an electric field, for example, by putting a potential across a superconductor. The Cooper pairs will be accelerated in the electric field, they acquire momentum. But this momentum has to be exactly the same for every pair. So our requirement does not allow a Cooper pair to exchange momentum simply by interacting with the lattice. This Cooper pair would then enter a different state and that is just what we have forbidden it to do. In order to remove a Cooper pair from the ensemble of all Cooper pairs we have to break it apart, i.e. destroy it. To do this we require a certain energy, namely the binding energy of the Cooper-pair correlation. If this energy cannot be provided our Cooper pair will not interact with the lattice[1)].

But this means nothing more or less than the existence of a resistance-free transport of charge through the lattice. We arrived at this possibility as a result of the requirement that all Cooper pairs must be in the same state and can only leave this state if the pair correlation is broken. It becomes very evident here that the resistance-free transport, a characteristic property of superconductivity, is determined by the stability of a quantum mechanical state.

---

1 In Section 8.6 we will meet superconductors for which this simple argument is not valid.

This stability is naturally not unlimited. If we carry on increasing the collective momentum of the Cooper pairs, we will reach a critical value where the kinetic energy of a pair taken up in the field is equal to the binding energy of a Cooper pair. On increasing the momentum and, hence, the kinetic energy further the pair can break up. So the interaction with the rest of the metal begins again above a certain critical momentum. The existence of a critical momentum for the system of Cooper pairs is the same thing as the existence of a critical current density for the superconductor. We will learn about the critical parameters and the stability of the superconducting state during the treatment of the thermodynamic properties of a superconductor. Our picture of the superconducting state has brought us, without difficulty, to the existence of such critical parameters.

Now we learned in Section 1.1 that, as particles with odd half-integral spins, electrons are subject to Fermi statistics. All experience has shown us that such particles can only occupy a state defined by all the quantum numbers once. Here we are speaking of a situation where a state is occupied by very many particles and these are composed of electrons. We must ask ourselves how this is possible. The answer is simply: "We are no longer dealing with electrons but with new particles, namely Cooper pairs". It is the total spin that is responsible for the statistical behavior of a Cooper pair. This is zero and, thus, no longer half-integral. Particles with integral spins obey Bose-Einstein statistics [1]. The Pauli exclusion principle does not apply to them. Such particles – they are known as bosons – can occupy a state as often as desired. Indeed the statistics predict just the opposite, a sort of "anti Pauli exclusion principle" where the tendency to occupy a state will be the greater, the greater the occupation of the state is already. That means in other words: If very many bosons are already in a particular state, then further particles are particularly likely to enter this state. This behavior can lead to a "condensation" taking place to a particular state. The state is first occupied by a few particles, thus, increasing the tendency for others to enter the same state, which again increases the tendency to occupy and so on. The correlation to electron pairs is, thus, the basic requirement for the ability of our electrons all to enter *a single* state macroscopically as electron pairs. The Fermi character of the single electron is naturally important for all excitations of the system that lead to the breaking apart of the Cooper pairs.

As quantum mechanical particles, when Cooper pairs acquire a momentum $\vec{P}$, they must acquire an associated wavelength $\lambda = h/|\vec{P}|$. The requirement discussed above that all Cooper pairs have the same momentum is the same thing as saying that all Cooper pairs have the same wavelength. We must naturally also require the same energy, i.e. the same frequency for the particle waves for all Cooper pairs. That is immediately evident if we maintain that all Cooper pairs occupy *one* quantum state.

The strict correlation of the Cooper pairs also determines another quantity, namely the phase. A classical wave possesses a well-defined phase. The importance of phase in all interference phenomena, i.e. in all superpositions of waves, is well-known.

Quantum mechanics predicts that the phase of a wave, which corresponds to a state, can be defined the more accurately the greater the occupation of this state. Since we have so many Cooper pairs in a single state we can describe this state by means of a classical

1 N.S. Bose, 1.10.1894–4.2.1974. The exchange of ideas with Albert Einstein led to the development of Bose-Einstein statistics.

wave with a well-defined phase[1]. We can assign the ensemble of Cooper pairs *one* wave function Ψ. The square of the amplitude of this wave function gives us the Cooper-pair density $n_s/2$. Since a Cooper pair is made up of two electrons, n_s is the number of "superconducting electrons". The fact that we can describe the ensemble of Cooper pairs with *one* wave function is often referred to as phase coherence or strict phase correlation. This property of the ensemble of Cooper pairs is particularly clearly evident in the quantization of the magnetic flux (see Section 3.2) and in the Josephson effects (see Section 3.4).

We can summarize the picture of the superconducting state we have built up so far as follows: In the superconducting state a fraction of the conduction electrons are correlated as Cooper pairs. These pairs possess *one* quantum state. Their ensemble can be described by means of *one* wave function, whose square is proportional to the density n_S of the "superconducting electrons". We can add here that below a certain temperature T_c this density n_S rises continuously as the temperature falls from a value zero to a particular value at $T = 0$. The "superconducting electrons" are also known as the "supercomponent" of the electron system. These statements concerning the state of superconductivity also remain valid for the new superconductors. However, in these substances it is not electrons but "defect electrons" that are responsible[2].

We will follow this qualitative description of the state of superconductivity on the basis of the BCS theory by explaining some relationships which are of particular importance for the question of the energy exchange with the electron system in the superconducting state. The BCS theory demands that this energy exchange can only take place in two ways in the superconducting state. On the one hand, Cooper pairs can break up and, on the other hand, unpaired electrons can take up or pass on energy[3]. It is important for both processes to know in what states unpaired electrons are available in the superconducting state. Because of the Pauli exclusion principle the free electrons can only make transitions in a free state. Both the breaking up of Cooper pairs and the uptake or release of energy from the system of unpaired electrons is dependent on the multiplicity of states available for electrons.

The question is what is the situation concerning this multiplicity of states for unpaired "normally conducting" electrons in the superconducting state? One would presume that the interactions between the electrons that leads to the formation of pairs would also alter the quantum mechanical states of the unpaired electrons. This is, in fact, true and becomes very evident if we consider the density of states of electrons in the neighborhood of the Fermi energy. The density of states is the number $\Delta Z(E)$ of the states per energy

---

1 This can be expressed as follows by saying: The ensemble of Cooper pairs makes up a field. This statement will become clear if we consider, for example, an electromagnetic wave. The radiation of electromagnetic waves from a radio transmitter is generally described as an electrical field varying in space and time. According to the electromagnetic quantum picture we have to say that very many quanta are radiated in the same state. We can then describe these quanta by means of a classical wave. This corresponds exactly to the generally accepted picture of a field.

2 A defect electron is a hole in an otherwise filled energy band. Such a hole can be regarded as a positive particle (cf. [32] for example).

3 The kinetic energy, which represents the motion of the ensemble of Cooper pairs, is not considered here, since it can only be transmitted to the rest of the metal by the break up of the Cooper pairs.

interval ΔE and per mol of the metal[1] for the limit $\Delta E \rightarrow 0$. This density of states depends on the lattice structure of the substance in a complex manner.

Here we will consider the influence of the electron-phonon interaction on the density of states in the superconducting states on the basis of the simple model of free electrons in a box, that is without the specific complications of the periodic potential of the atomic residues that leads to a band structure (see Section 1.1).

All three directions in space are equivalent to a free electron. The kinetic energy of an electron is given by:

$$E_{kin} = \frac{1}{2m}(p_x^2 + p_y^2 + p_z^2) \,. \tag{2-2}$$

The surfaces of constant energy are spheres in momentum space. We now have to ask what is the number of states lying in a spherical shell of thickness dE and having an energy E. This number is proportional to the volume of a corresponding spherical shell in momentum space (Fig. 18).

$$\mathrm{d}Z(|\vec{p}|) \propto \vec{p}^2 \, |\mathrm{d}\vec{p}| \,. \tag{2-3}$$

If we go from the modulus of the impulse $|\vec{p}|$ to the energy as variable then because $E \propto \vec{p}^2$ we get

$$\mathrm{d}Z(E) \propto \sqrt{E} \, \mathrm{d}E \tag{2-4}$$

and the density of the states is

$$N(E) = \mathrm{d}Z(E)/\mathrm{d}E \propto \sqrt{E} \,. \tag{2-5}$$

The volume of the metal under consideration is included in the constant of proportionality. This density of states $N(E)$ is represented in Fig. 19. The occupation of the states is obtained if we multiply the density of states by the Fermi function $F(E)$ (see Equation (1-3)[2]) (full line in Fig. 19).

We only require an area of ca. 10^{-2} eV around the Fermi energy for the purposes of our considerations, because the electron-phonon interactions are of this order of magnitude. In this narrow region we can, as a good approximation, regard the density of states $N(E)$ as being constant and equal to $N(E_F)$.

We can now represent the influence of the interaction leading to pair forming in this picture on the density of states for the unpaired electrons. The interaction mediated by

---

1 It is also possible to refer the density of states to one atom or to 1 cm^3 of the metal.

2 The Fermi energy E_F is obtained from the requirement:

$$n = \int_0^\infty N(E)\, F(E)\, \mathrm{d}E$$

i.e. the hatched area must equal n the total number of electrons.

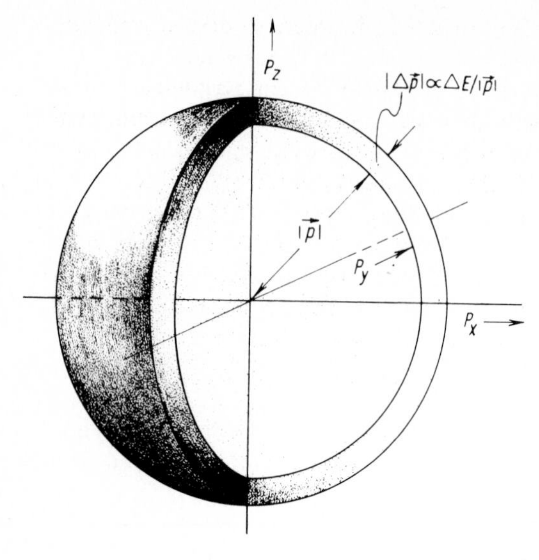

Fig. 18. The density of states of free electrons. The density of states is constant for free particles in momentum space. Therefore, the number of states in a particular volume of this momentum space is proportional to this volume.

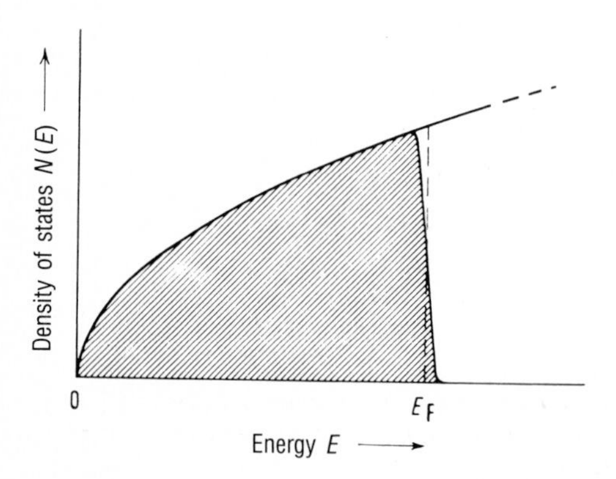

Fig. 19. The density of states of free electrons. Occupation of the states hatched. The blurring at the Fermi energy is exaggerated for clarity. The occupation illustrated would represent a temperature of ca. 1000 K. $E_F = 5$ eV.

phonons leads to a very narrow energy range around the Fermi energy (a few 10^{-3} eV) being forbidden for the electrons, an "energy gap" is formed. The states that lie in this gap in the normally conducting state are displaced to its boundaries by the effect of the interaction. The shape of the maxima appearing in the density of states depends on the approximation to the theory chosen [1]. The shape given by the BCS theory is reproduced in Fig. 20. The energy gap is a function of temperature. It aproaches zero as $T \to T_c$. The Curves *1* and *2* show the BCS density of states for $T = 0$ and for $0 < T < T_c$. The occupation of these electronic states is indicated by the hatching. For $T \rangle 0$ we find elec-

1 The curve is given in the BCS theory by:

$$N_s(\epsilon) = N_n(0)\ |\epsilon|/\sqrt{\epsilon^2 - \Delta^2} \quad \text{for} \quad |\epsilon| \geqslant \Delta$$

where $N_s(\epsilon)$ is the density of states in the superconducting state. The energy ϵ is measured from the Fermi energy ($\epsilon = E - E_F$), and Δ is the half-width of the energy gap. When $|\epsilon| < \Delta$ then $N_s(\epsilon) = 0$.

trons with thermal energies that are greater than 2 Δ, i.e. there is a certain amount of occupation of the states above $E_F + \Delta$.

What can we conclude from this representation? Firstly, we can now understand that the energy difference between the normally conducting and the superconducting state is surprisingly small, expressed another way, the heat of transition is very small[1]. Only a very small number of the conducting electrons are changed in energy as a result of the interaction, namely that portion of them that find themselves in states within the energy gap when conductivity is normal. Since the energy gap is only 10^{-3} eV wide and the Fermi energy is several eV, these only amount to about 10^{-3} of all the conducting electrons. This allows a question to be answered that, for a long period, put difficulties in the way of coming to an understanding of the problem (see Section 2.1).

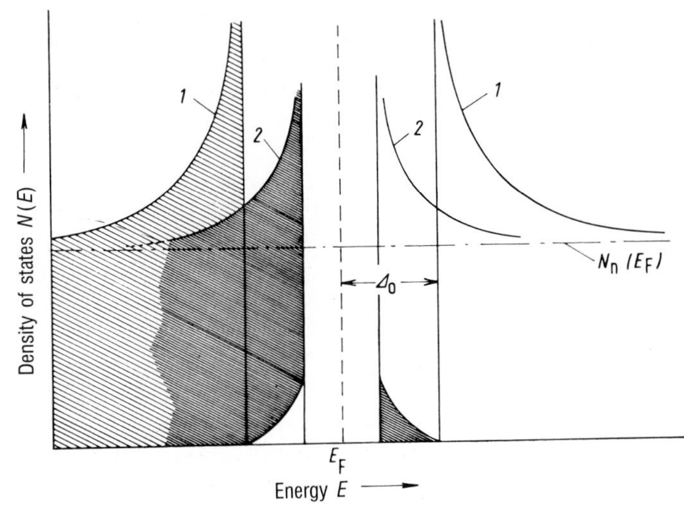

Fig. 20. Density of states for unpaired electrons in superconducting state according to the BCS theory. Curve 1: $T = 0$. Curve 2: $0 < T < T_c$. The occupied states are depicted by hatching. Δ_0 = half energy gap at $T = 0$ K, e.g. for Sn: $\Delta_0 = 5.6 \times 10^{-4}$ eV.

This picture also provides us with predictions concerning the possible energy exchange in this system. Consider Curve *1*. All electrons sit in states below E_F. An energy contribution of 2Δ is necessary in order to "excite" this system. So the binding energy of the Cooper pairs is reflected in the density of states of the normal components. In this type of superconductor[2] we can only break up a Cooper pair with a minimum energy of 2Δ.

But it must be clearly emphasized that it is the density of states for single electrons that is represented in Fig. 20. It is not possible to introduce the Cooper pairs mean-

1 For Pb (T_c = 7.2 K) the energy of transition on entering the superconducting state is 23.5×10^{-3} Ws/mol, while the evaporation of liquid helium at 4.2 K requires ca.90 Ws/mol.

2 In the case of the superconductors with no energy gap we will encounter a superconducting state without any energy gap in the excitation spectrum. This does not mean that the energy of binding of the Cooper pair is zero.

ingfully into this representation. But we can, nevertheless, understand a range of other observations on the basis of this picture. Only one example will be mentioned here. If we irradiate a superconductor with electromagentic radiation, then a strong absorption will only take place when

$$h\nu \geqq 2\Delta \, . \tag{2-6}$$

The existence of an energy gap was first demonstrated experimentally as a result of absorption measurements. We will deal with the energy gap and its measurement in more detail in Section 3.3.

In the first few years after the discovery of the BCS theory the existence of an energy gap in the density of states of the individual electrons was considered to be a fundamental characteristic of the superconducting state. Today we also know that superconductivity can occur without an energy gap (gapless superconductivity). The fundamental requirement for superconductivity is the existence of Cooper pairs. Under certain conditions, e.g. in superconductors containing paramegnetic impurities, such Cooper pairs can exist without an energy gap being present in the density of states of the individual electrons (see Section 8.3).

Finally, a remark must be made concerning the statistical character of Cooper pairs. Are the Cooper pairs bosons as we assumed in order to explain the fact that they occupy only one single state "macroscopically"? Or are Cooper pairs fundamentally fermions because exchange rules for fermions are employed to describe their behavior? One meets various opinions with regard to this point. But, in truth, it does not seem difficult to clarify the position.

It is only possible to describe a particle as a boson or as a fermion so long as processes are being described in which the particle maintains its identity. There are no problems in the case of particles such as light quanta (bosons) or electrons (fermions). But Cooper pairs are more complicated from this point of view. They only maintain their identity in the "ground state". As soon as we attempt to isolate them from the ensemble of Cooper pairs they break up into two electrons, that is two fermions. For this reason it is certainly confusing to describe the transition to the superconducting state as a Bose-Einstein condensation. This would be to give the impression that Cooper pairs could also exist as a gas of more or less independent particles. This is not the case[1)].

As long as we are only considering processes where the Cooper pairs stay in the ground state, e.g. persistent current experiments (Section 1.1) or the Josephson effects (Section 3.4) then we can designate Cooper pairs as particles with the statistical characteristics of bosons. So that it is immediately clear that Cooper pairs, like bosons, can occupy one and the same quantum state macroscopically. But in all processes where Cooper pairs are excited so as to fall apart into electrons, such as in their interaction with radiation, whose quanta are able to break up Cooper pairs, we must remember that the electrons that are produced are fermions. That is the reason for the difficulty in assigning Cooper pairs to *one* type of statistics. We must always ask ourselves, what is the process we wish to consider.

---

1 In the case of the bipolaron model of the new superconductors a transition is again discussed which is connected with a Bose-Einstein condensation.

3. Experiments for the direct confirmation of the basic understanding of the superconducting state

In the previous section we have built up a picture of the superconducting state on the basis of the BCS theory, the very successful atomistic theory of superconductivity. Naturally it is fundamentally the wealth of all the observations that are quantitatively or even only qualitatively explained by the theory that constitutes the complete justification for the basic conception. Amongst these observations there are some that particularly directly illustrate especial properties of the superconducting state. These will be treated in this next section, on the one hand, to increase your trust in the somewhat complicated concepts of the superconducting state and, on the other, to broaden these concepts on the basis of concrete experiments.

Some of the experimental facts, such as the behavior of the specific heat or the isotope effect, were introduced before the development of the BCS theory. Other experiments, such as the measurement of the tunnel effect and the Josephson effect, were first stimulated by the atomic theory. Flux quantization constitutes a special case in that it was put forward by F. London [1] long before the BCS theory was proposed [33], but it was only observed experimentally after the development of this theory and in its quantitative results provides particularly convincing evidence for the BCS theory.

3.1 The isotope effect

The question as to whether nuclear mass has an effect on superconductivity, whether – put another way – superconductivity depends on the lattice of residual atoms or whether it is limited to the electron system, was investigated by Onnes as early as 1922 [34]. There were then only available to him the types of lead that are found in nature ($\langle m \rangle$ = 206 uranium derived lead and $\langle m \rangle$ = 207.2 natural lead). He was unable to detect any difference in transition temperature with the accuracy available to him. Neither was it possible to detect an influence of atomic mass on T_c in later experiments with lead (E. Justi 1941) [35].

It was the advent of modern nuclear physics that first allowed the preparation of isotopes of large mass difference in adequate quantities in nuclear reactors. Thus, the dependence of the transition temperature of mercury on the nuclear mass was established almost simultaneously in 1950 by Maxwell [36], on the one hand, and Reynolds, Serin, Wright and Nesbitt [37], on the other. Some of the results are summarized in Table 3.

We have already mentioned that these results were so decisive for the development of superconductivity because they appeared at just the right time to confirm the idea of electron-phonon interaction so excellently. Even the first more or less qualitative consid-

1 A very successful phenomenological theory of superconductivity was developed by the brothers Fritz and Heinz London in the 1930s. F. London: "Une conception nouvelle de la supraconductibilité", Paris, Hermann u. Cie., 1937.

erations of Fröhlich and Bardeen predicted that the transition temperature T_c ought to be inversely proportional to the root of the atomic mass:

$$T_c \propto m^{-1/2}. \tag{3-1}$$

Table 3. Isotope effect for mercury*.

Mean atomic weight	199.7	200.7	202.0	203.4
Transition temperature T_c in K	4.161	4.150	4.143	4.126

* Results of Reynolds et al., Phys. Rev. *78,* 487 (1950).

This dependence also remains in the BCS theory which appeared 7 years later. The BCS theory predicts a dependence of the transition temperature of the following form:

$$T_c = 1.13 \frac{\hbar \omega_D}{k_B} \exp \left(\frac{1}{N(E_F)\, V^*} \right) \tag{3-2}$$

where $\hbar$ = Planck's constant/2π; ω_D = Debye frequency; k_B = Boltzmann constant; $N(E_F)$ = density of states of the electrons at the Fermi energy; V^* = constant characterizing the electron-phonon interaction.

When the BCS theory was first formulated it was assumed that V^*, the parameter characterizing the interaction, was constant. So that the dependence of the transition temperature only comes about from the Debye frequency ω_D. If this Debye frequency is regarded as the frequencies of the atom residues and it is assumed that they are undergoing harmonic motion, then [1]:

$$\omega_D \propto \frac{1}{\sqrt{m}} .$$

So that

$$T_c \propto m^{-1/2} \qquad m = \text{atomic mass} . \tag{3-1}$$

This relationship is obeyed very well by a wide range of superconductors. The results for tin are illustrated in Fig. 21. Tin is a particularly good example because it allows a relatively large difference in nuclear mass, namely from $(m) = 113$ to $(m) = 123$. The results obtained by several laboratories are included in Fig. 21 [38]. The broken line corresponds to the exponent $-1/2$ in Equation (3-1). The agreement between experimental results and theoretical expectations is very good.

1 The frequency of the harmonic motion is given by $\omega = (D/M)^{1/2}$, where D is the force constant in the linear force law $K = Dx$ (K = force, x = displacement). This force constant depends on the bonding of the atoms in the metal and ought not to be altered by variation of the isotopic mass.

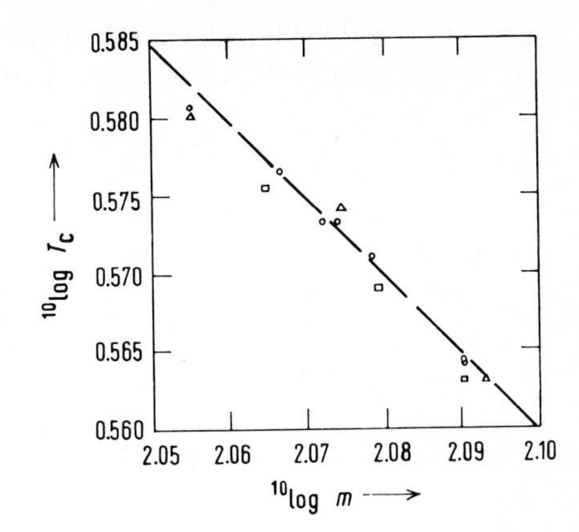

Fig. 21. Isotope effect for tin. ○ Maxwell, □ Lock, Pippard and Shoenberg; △ Serin, Renolds and Lohman (after [38]).

This is almost surprising, looked at from the present-day point of view, because the theory leading to Equation (3–1) employs very drastic simplifications. The agreement reveals that these simplifications are evidently justified for a large number of superconductors.

The results obtained on measuring this isotope effect are summarized for several superconductors in Table 4. While the nontransition metals possess the expected index $\beta = 1/2$ to a good degree of accuracy, the transition metals diverge very considerably from this value. These deviations can be taken as being established in spite of the considerable difficulties involved[1]. For uranium a value of $\beta = -2.2$ has even been reported, that is an isotope effect of the opposite sign [39]. We now have to ask ourselves how these deviations can be explained. It might be imagined in the case of a substance such as ruthenium (Ru), where no change is observed in T_c on changing the isotopic mass, that it is not the electron-phonon interaction that is responsible for the superconductivity but some other interaction concerning say the band structure. Conjectures of this kind have been put forward several times.

Table 4. Isotope effect for various superconductors.

Element	Hg	Sn	Pb	Cd	Tl	Mo	Os	Ru
Isotope exponent β*	0.50	0.47	0.48	0.5	0.5	0.33	0.2	0.0

* β was obtained from the experimental results by fitting to the relationship $T_c \propto m^{-\beta}$. The values given are taken from the book "Superconductivity" by R.D. Parks, Marcel Dekker Inc., New York 1969, page 126.

1 Because of the small difference in T_c involved it is not very easy to carry out the experiment with the necessary accuracy. The T_c measurements have to be made in different samples. This means that it is necessary to make all other influences on T_c such as internal strains, impurities and lattice defects which could all affect T_c sufficiently equally for all the samples, in order to be able to observe the effect of the isotopic mass alone.

On the other hand, when you consider the great success of the BCS theory an attempt must be made to find an explanation within its framework. This is quite possible if the characteristic parameter V^* for the interaction in Equation (3–2) is analyzed somewhat more closely. This interaction is basically made up of the *attractive* electron-phonon interaction and the *repulsive* Coulomb interaction between the electrons. If both these interactions are introduced into the theory explicitly, as became possible as the theory was developed, then you obtain for T_c the improved but more complicated formula [40]:

$$T_c \propto \omega_D \exp\left(-\frac{\lambda^* + 1}{\lambda^* - \mu^*(1 + \lambda^* \langle\omega\rangle/\omega_D)}\right). \tag{3-3}$$

Here the electron-phonon interaction is characterized by λ^* and the Coulomb interactions by μ^*. $\langle\omega\rangle$ is a particular mean value of all the frequencies of the lattice [1].

We will not look at this equation more closely. The decisive thing is that, as is to be expected in such an improved analysis, the lattice frequencies also enter the exponent explicitly. So the influence of the factor ω_D on T_c can be affected more or less depending on the size of λ^* and μ^* (see Eq. (3–3)). Deviations of the size of β in $T_c \propto m^{-\beta}$ from 1/2, or even the complete absence of a m-dependence of T_c on the value of m cannot, therefore, be employed as evidence against the importance of electron-phonon interactions in these superconductors. On the other hand, it is also not possible at present to say with certainty whether the assumptions made concerning λ^* and μ^* in order to explain the anomalous isotope effects are really justified. We have again come up against the same limit as in Section 1.2. We do not have enough quantitative understanding of the connection between superconductivity and the other properties of metals [2].

The tunnel experiments on superconductors have brought about a significant advance (Section 8.7.2). The electron-phonon interaction appears in the current potential characteristics of tunnel diodes at sufficient intensities. A careful analysis of such characteristics allows the determination of λ^* and μ^*.

The isotope effect, however, reveals to us quite directly the influence of the lattice vibrations and confirms, with its quantitative agreement with the theory, at least in the case of many superconductors the decisive importance of the electron-phonon interaction. Here the advances in nuclear physics, which have made possible the production of new isotopes in the nuclear reactor, have had a fundamental effect on the development of superconductivity. This is but one of the numerous examples in connection with superconductivity of how quite different physical fields have a mutual effect on each other's advance.

Efforts are naturally being made to understand the new superconductors within the framework of the BCS theory. The relatively high frequencies of the light oxygen atoms offered themselves as starting points here. They would make the factor ω_D in (3–3) large

---

1 The following results provide a certain impression of the order of magnitude of these parameters. λ^* varies from 0 to 2 whereby the approximation becomes uncertain for large values of λ^*. μ^* lies in the range between about 0.1 and 0.2 and $\langle\omega\rangle/\omega_D$ is about 0.6.

2 B.T. Matthias [41] has interpreted these deviations from 0.5 in a different manner.

and, hence, cause T_c to be raised. From this point of view it would be of extraordinary interest to measure the isotope effect which occurred on replacing ^{16}O with ^{18}O.

Very careful determinations with ca. 75% of the ^{16}O replaced by the heavier isotope did not reveal that the isotopic mass had any effect [42]. The simplified BCS theory for $YBa_2Cu_3O_7$ predicted a reduction of T_c by 3.5 K. The transition temperature remained unchanged within the limits of a measurement of ca. 0.1 K.

From what has been said above this is not evidence against the importance of the electron-phonon interaction in these substances. But the simple explanation involving a large ω_D factor is disproven. In the case of $La_{1.85}Sr_{0.15}CuO_4$ an isotope effect of oxygen was detected with an exponent $\beta^* = 0.16$ [43].

3.2 Flux quantization

We will start from the experiment presented in Fig. 5 (see Section 1.1). We have produced a persistent current in a superconducting ring by means of an induction process. According to all experience this persistent current will remain constant so long as the ring remains superconducting. The system "ring with current" is *stable*, it is in a fixed state with respect to time. We can bring the ring into other states by cooling below the transition temperature in other magnetic fields. We will then set up other persistent currents on switching off the magnetic field. The lowest of all these states energetically is the one with no persistent current.

From our experience of macroscopic phenomena we would now expect that induction processes with suitable magnetic fields would allow the setting up of any desired persistent current. On the other hand, we are familiar, from the quantum mechanical treatment of physical systems, with the idea that states that do not change with time, i.e. stationary states, are governed by quantum conditions. In the Bohr model of the atom, for instance, the stationary electron states are characterized by the quantum conditions for angular momentum, whereby the angular momentum of the electron takes up whole multiples of $\hbar$ [1].

So that we must presume that our superconducting ring with its persistent current can only take up *discrete* states which are determined by some sort of quantum conditions. The quantum jumps can be very small. This means that in the case of a macroscopic system there can be a practically continuous variation in the quantity observed, here the current or its magnetic field.

This assumption was made by F. London [33]. He came to the conclusion that the magnetic flux [2] through a superconducting ring could only take up multiple values of the "flux quantum" Φ_0^L. Where Φ_0^L is defined by:

1 The angular momentum $\vec{L}$ of a point-like particle of mass m in an orbit of radius r is given by:

$$\vec{L} = mr^2\,\vec{\omega} \qquad (\vec{\omega} = \text{angular velocity of orbit})$$

2 Precisely speaking it is not the magnetic flux through the ring that is quantized, but the fluxoid. As we will see the supercurrent also enters into the fluxoid.

$$\Phi_0^L = \frac{h}{e} \cong 4 \times 10^{-11} \mathrm{T} \times \mathrm{cm}^2 \tag{3-4}$$

$$1\ \mathrm{T} = 1\ \mathrm{Tesla} = 1\ \mathrm{Vs/m}^2 = 10^4 \mathrm{G} = 10^4 \mathrm{Gauss}^{1)}.$$

As we will see later F. London arrived at this quantity of the flux quantum because he assumed that single electrons carried the supercurrent. Today, we know from the BCS theory that in the superconducting state it is electron pairs, Cooper pairs, which are the new particles, that determine the behavior of the state of superconductivity. So we must expect that the elementary flux quantum will be a factor of 2 smaller than proposed by F. London since the current is carried by particles with a charge $2e$. The flux quantum must, therefore, be:

$$\Phi_0 = \frac{h}{2e} \cong 2 \times 10^{-11} \mathrm{T} \times \mathrm{cm}^2 . \tag{3-5}$$

This prediction of the BCS theory has now been confirmed to the hilt.

Measurements of flux quantization were published by 2 groups in 1961 – Doll and Näbauer [44] in Munich and Deaver and Fairbank [45] in Stanford – that revealed that the size of the flux quantum was $\Phi_0 = h/2e$. These superlative experiments not only provided an impressive proof of the existence of Cooper pairs in the superconducting state they also exerted a great influence on the development of superconductivity. We will meet the flux quantum time and again.

Because of their especial importance and not least because of the excellence of the experimental technique we will describe these experiments in more detail.

The aim was to test whether it is actually true that the magnetic flux through a superconducting ring can only take up discrete values $n\Phi_0$ ($n = 1, 2, 3, \ldots$). To do this persistent currents had to be set up in superconducting rings by means of various magnetic fields and the magnetic flux created by these measured so accurately that the resolution revealed the individual quantum jumps. These experiments are extremely difficult on account of the smallness of the flux quantum. In order to achieve a reasonably large relative change in the flux it is necessary to produce states with relatively few flux quanta, that is states with 0, 1, 2,... flux quanta. It is necessary to employ very small superconducting rings to do this, otherwise the fields required to produce the persistent currents would be too small. We call these fields "freezing fields", because the flux that they create through the hole in the ring is "frozen in" as superconductivity begins. If the hole is only 1 mm^2 then one flux quantum is created by a field of 10^{-9} T.

Both groups, therefore, employed very small samples in the form of fine tubes with a diameter of only ca. 10 μm (10^{-3} cm). For this diameter the creation of a flux quantum $\Phi_0 = h/2e \approx 2 \times 10^{-11}$ T cm^2 requires a field of $\Phi_0/\pi r^2 = 2.5 \times 10^{-5}$ T. Such fields can be handled experimentally if external fields, such as that of the earth, are carefully screened out.

Doll and Näbauer employed lead cylinders that had been condensed onto a quartz fiber (Fig. 22). The persistent current was initiated in this lead cylinder in the usual manner

1 Carl Friedrich Gauss, german mathematician, physicist and astronomer, 30.04.1777 - 23.02.1855.

by cooling in a freezing field B_e parallel to the cylinder axis and switching off this field at $T < T_c$ after superconductivity had been established. The persistent current now made the lead cylinder into a magnet. In principle, it is possible to determine the quantity of the flux that has been frozen in from the torque that is produced when a measuring field B_M is applied at right angles to the axis of the cylinder. This is done by suspending the sample on a quartz fiber. The displacement can be detected via a mirror with an optical lever. However, the torques were so small that a static determination appeared impossible even with the finest suspension fiber. Doll and Näbauer overcame this difficulty with an extremely elegant measurement technique, that can be described as an autoresonance method.

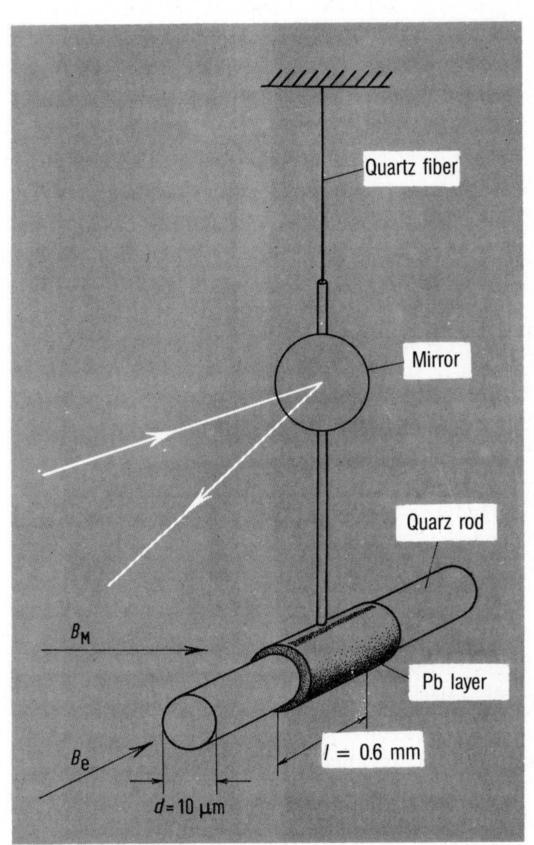

Fig. 22. Schematic representation of the sample of Doll and Näbauer (after [44]). The quartz rod, with the small lead cylinder formed from a condensed layer, oscillates in liquid helium.

It exploits the small torque that is exerted by the measurement field on the lead cylinder to set up a torsional vibration in the system. When resonance occurs the amplitudes are large enough to be recorded with ease. The resonance amplitude is proportional to the size of the exciting torque that is to be determined. For excitation the measurement field B_M must be reversed with the frequency of the vibration. In order to be certain that the stimulation always takes place at exactly the resonance frequency the field reversal is activated by the system itself via the optical lever and a photocell.

Figure 23 reproduces the results obtained by Doll and Näbauer. The ordinate is the resonance amplitude divided by the measurement field, that is a quantity which is proportional to the torque being sought. The abscissa is the freezing field employed on each occasion. If the flux in the superconducting cylinder were continuously variable then the resonance amplitude observed would vary proportionately to the freezing field employed (broken line in Fig. 23). The experiment emphatically reveals a different behavior. No flux at all is frozen in up to a freezing field of 1×10^{-5} T. The superconducting lead cylinder takes up its lowest state with $n = 0$. It is only when the freezing field is greater than 1×10^{-5} T that a state is taken up which does have a frozen-in flux. It is the same for all freezing fields from 1×10^{-5} and ca. 3×10^{-5} T. The resonance amplitude is constant in this range. The flux calculated from this amplitude and the apparatus constant amounts to about one flux quantum $\Phi_0 = h/2e$. Further quantum jumps are observed at larger freezing fields.

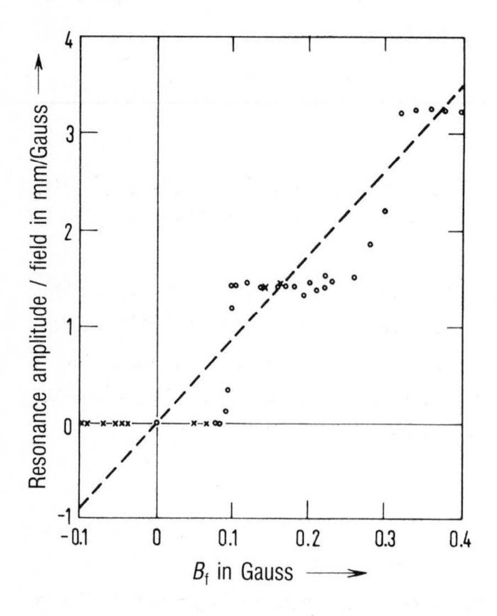

Fig. 23. Results of Doll and Näbauer on flux quantization in a Pb cylinder (after [44]) (1 G = 10^{-4} T).

This experiment reveals unequivocally that the magnetic flux through a superconducting ring can only adopt discrete values. The quantitative result is an excellent confirmation of the existence of Cooper pairs.

The results of Deaver and Fairbank – Fig. 24 gives an example – also demonstrated the quantization of the magnetic flux in a hollow superconducting cylinder and yielded $\Phi_0 = h/2e$ for the size of the elementary flux quantum. Deaver and Fairbank employed a completely different method for measuring the frozen-in current. They vibrated the superconducting cylinder[1] backwards and forwards with an amplitude of 1 mm along its length at a frequency of 100 Hz. This caused the induction of a voltage in two small measurement coils surrounding the ends of the rod; the currents due to this voltage were

---

1 Tubes were employed having a length of ca. 0.9 mm, an inner diameter of 13 μm and a wall thickness of 1.5 μm.

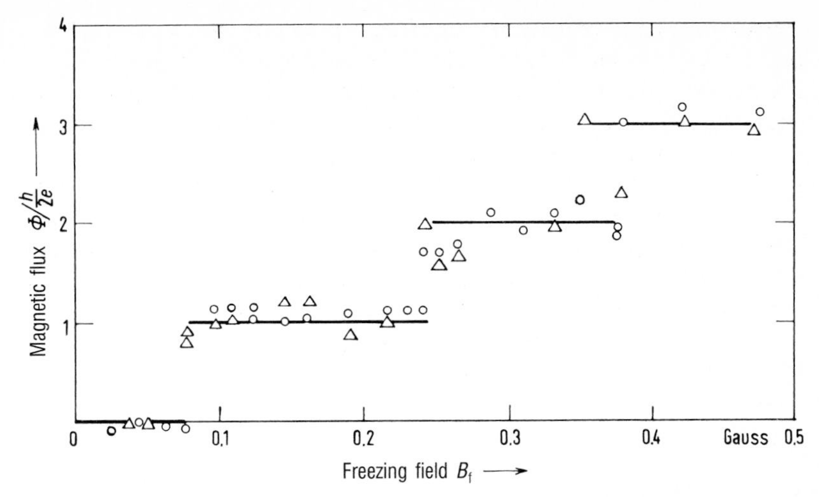

Fig. 24. Results of Deaver and Fairbank on flux quantization in a Sn cylinder (after [45]) (1 G = 10^{-4} T).

large enough to be measured after amplification. Figure 24 shows a plot of the multiples of elementary fluxes Φ_0 in the tube as a function of the freezing field. The states with 0, 1 and 2 flux quanta are readily recognized.

As already mentioned F. London [33] had already predicted the quantization of the magnetic flux in a superconducting ring for theoretical reasons as early as 1950. Quantum mechanics yields this quantization if we start from the credible assumption that a superconducting ring with a persistent current is a system that is in a quantum state. The quantum conditions for circulation of a single particle, the Cooper pair, carrying the persistent current must have the following form:

$$\oint \vec{p}_C \, d\vec{l} = nh \qquad n = 0, 1, 2, \ldots \tag{3-6}$$

The momentum of the Cooper pair, p_C, integrated over one circuit must be exactly a multiple of h. This condition is identical with the quantum conditions for the Bohr model of the atom [1].

We now have to combine these quantum conditions for the individual Cooper pairs with the magnetic flux through the superconducting ring. This is done by means of a general quantum mechanical rule for the description of charged particles in a magnetic field. This rule states that two momentum-like quantities have to be distinguished for such motions, firstly what is known as the canonical momentum $\vec{p}$ which determines the wavelength of the matter wave $\lambda = h/|\vec{p}|$ and which, therefore, forms part of the quantum conditions and secondly the quantity $m\vec{v}$, which may be termed the kinetic momentum

---

1 This condition can also be viewed in the wave picture. This says that the waves in one circuit must exactly end in phase, that $2\pi r = n\lambda$. When $\lambda = h\ |\vec{p}|$ then this gives the special case of the circuit $2\pi r|\vec{p}| = nh$, which corresponds to the somewhat more general conditions $\oint \vec{p} \, d\vec{l} = nh$.

or just the $m\vec{v}$ momentum. The $m\vec{v}$ momentum determines the kinetic energy $E_{\text{kin}} = (m\vec{v})^2/2m$ of a particle[1].

The rules of quantum mechanics can be obeyed by a particle in a magnetic field if account is taken of the fact that $\vec{p}$ and $m\vec{v}$ are related in the following manner:

$$\vec{p} = m_q \vec{v} + q\vec{A} \,. \tag{3-7}$$

m_q and q are the mass and charge of the particle. $\vec{A}$ is the vector potential. $\vec{A}$ is related to the magnetic field $\vec{B}$ as follows[2]:

$$\operatorname{curl} \vec{A} = \vec{B} \,. \tag{3-8}$$

With this condition for the treatment of charged particles in the magnetic field and the quantum conditions of Equation (3-6) it follows that

$$n\,h = \oint m_q \, \vec{v}_s \, d\vec{l} + q \oint \vec{A} \, d\vec{l} \,. \tag{3-9}$$

As with the quantum conditions in Equation (3-6) the integration takes place along a closed path around the hole in the ring.

We can now relate the velocity $\vec{v}_s$ of the particle to the current density in the ring. For:

$$\vec{j}_s = n_q \, q \, \vec{v}_s \,. \tag{3-10}$$

where n_q is the density of particles of charge q, $\vec{v}_s$ their velocity. The subscript s in $\vec{v}_s$ indicates that superconducting particles are involved. We then obtain from Equation (3-9):

$$n\,h = q \left[\oint \frac{m_q}{n_q q^2} \, \vec{j}_s \, d\vec{l} + \oint \vec{A} \, d\vec{l} \right] . \tag{3-11}$$

and with $\operatorname{curl} \vec{A} = \vec{B}$[3]:

$$\oint \vec{A} \, d\vec{l} = \iint_S \operatorname{curl} \vec{A} \, d\vec{S} = \iint_S \vec{B} \, d\vec{S} = \Phi_S \,. \tag{3-12}$$

---

1 The necessity of such a distinction becomes at least plausible if it is remembered that a magnetic field exerts a force on a charged particle which is always perpendicular to $\vec{v}$ and, thus, changes just the direction but not the modulus of $\vec{v}$. The magnetic field, therefore, changes the momentum but not the energy of the particle. So the magnetic field has to be included in the relationship between energy and momentum. It is not possible to derive the form of Equation (3-7) using this argument.

2 curl $\vec{a}$ is a vector, whose components $(\operatorname{curl} A)_x, \ldots$ are obtained from A_i as follows:

$$(\operatorname{curl} \vec{A})_x = \frac{\partial A_z}{\partial y} - \frac{\partial A_y}{\partial z} \,; \quad (\operatorname{curl} \vec{A})_y = \frac{\partial A_x}{\partial z} - \frac{\partial A_z}{\partial x} \,; \quad (\operatorname{curl} \vec{A})_z = \frac{\partial A_y}{\partial x} - \frac{\partial A_x}{\partial y} \,.$$

3 The linear integral of a vector field A along a closed curve that surrounds the area A can be replaced by an area integral from rot $\vec{A}$ over the area A (Stokes's law).

We, therefore, have quantum conditions of the form:

$$\frac{nh}{q} = \oint \frac{m_q}{n_q q^2} \vec{j}_s \, d\vec{l} + \Phi_S \, . \tag{3-13}$$

The quantity on the right hand side is known as the "fluxoid"[1].

As we will see in Chapter 5, the currents in superconductors generally only flow in a very thin layer (thickness a few 10^{-6} cm) below the surface of the superconductor. So if we choose an integration path inside the superconductor the line integral over will equal zero and we have the quantum conditions for the magnetic flux. The linear integral over the current density is normally only a small correction. We are entirely justified – except for individual cases with particular geometries[2] – in speaking of the quantization of the magnetic flux in a superconducting ring.

The experiment yielded the value $2e$ for q. This means that when we consider our derivative the current is carried by particles with a charge $2e$, that is by the Cooper pairs[3].

The quantization of the flux also demonstrates, the strict phase correlation of the Cooper pairs to each other. We assume and experiment has confirmed this assumption that all Cooper pairs, that carry the supercurrent, have the same quantum number n. On passing over to another quantum number n' all Cooper pairs must enter the new quantum state. If the quantum number n could be increased by 1 for just one single Cooper pair, while leaving all the rest in the same state, then the associated change in the flux would be smaller by many powers of ten[4].

We have, until now, treated the flux quantization for a superconducting ring that is for a superconducter with a hole in it. The particular importance of the fluxoid is that it is not limited to such superconductors with a hole but always appears when a magnetic flux goes through a superconductor. In superconductors of type 2 (Section 5.2) we will meet substances where an externally applied magnetic field penetrates the superconductor. This can only occur in the form of fluxoids or bundles of fluxoids.

Finally fluxoids are required in the description of the Josephson effect in Section 3.4. At this point we wished to establish experimentally the existence of Cooper pairs and of the strict phase correlation between them.

Experiments on flux quantization with the new superconductors [46, 47] have also demonstrated the importance of pairs, i.e. current carriers with two elementary charges. This excludes theoretical models where more than 2 charges are postulated as the charge carrier.

---

1 The charge q of the Cooper pairs is negative. However, the sign of the charge is unimportant for the flux quantization; it is only of importance that both terms on the right-hand side of (3-13) be added.

2 For instance, the current term acquires importance when the hole in the ring is very small and the integration path is along the internal surface of the ring.

3 The quantity $m_q/n_q\, q^2$ is independent of whether single electrons or Cooper pairs carry the persistent current. In the case of electron pairs $m_q = 2m_e$, $q = 2e$ and $n_q = n_e/2$.

4 The change in the flux would not then be $\Delta\Phi = h/2e$ but $\Delta\Phi' = (1/z_C) \times (h/2e)$, where z_C is the number of Cooper pairs that carry the persistent current.

3.3 The energy gap

Figure 20 (see Section 2.2) represents the densities of states of unpaired electrons in the superconducting state for two temperatures. The existence of a region of forbidden energy (energy gap) has provided us with a simple understanding of the reason why Cooper pairs cannot interact with the lattice below a certain critical kinetic energy. Various methods can be employed to measure the size of the energy gap and these will be treated briefly in what follows. We will just treat the tunnel experiment in greater detail, since it leads us directly to the very important Josephson effects.

3.3.1 Absorption of electromagnetic radiation

We have already referred to the possibility of determining the energy gap by measuring the absorption of electromagnetic radiation. The first experimental evidence of the energy gap in the term scheme of single electrons at $T < T_c$ was provided in 1957 by Glover and Tinkham [48] from observations of the infrared transparency of thin superconducting films.

As early as the beginning of the 1930s it had been pointed out [1] that it should be possible to break up the state of order of a superconductor at a temperature below the transition temperature by means of electromagnetic radiation of suitable frequency. This ought to be revealed in an anomaly in the absorption. Successful experiments were not possible in the 1930s because the necessary wavelength range was scarcely accessible to experiment. Assume a binding energy of E_B of ca. 10^{-3} eV [2] then the corresponding frequency for radiation quanta of this energy is $\nu = E_B/h \approx 2.4 \times 10^{11}\ s^{-1}$. That is a wavelength of ca. 1 mm for which there were neither methods of generation nor methods of detection in the 1930s. It was a good 20 years later that this method was successfully employed to measure the energy gap, whose size could by then be predicted from the BCS theory [3] [48]. We now have sufficiently good experimental aids for the wavelength range from ca. 500 μm to 3 cm to use this type of determination to measure the energy gap.

We will not go more deeply into this method whose quantitative interpretation is difficult. Figure 25 gives just one example for such a determination [49]. The radiation is led into a small cavity made of the material under investigation. Inside the cavity the radiation undergoes very many reflections before it exits again. The greater the absorption at the walls of the cavity the less will be the intensity of the radiation when it is detected. The determination is carried out by measuring the intensity for each wavelength at a fixed temperature (here ca. 1.4 K) in both the normal and the superconducting states I_s and I_n. For this purpose the superconductivity can be disturbed by means of a magnetic field of suitable strength. The differences between these intensities is the difference in reflection in the two states. In Figure 25 this difference with respect to the intensity in the

1 See W. Meissner: Handbuch der Experimentalphysik by Wien and Harms, Volume XI, Part 2, page 260, Akademische Verlagsgesellschaft Leipzig 1935.

2 The energy gap $2\Delta_0$ for tin is 1.1×10^{-3} eV (Fig. 20).

3 The existence of an energy gap in the term scheme of a superconductor below T_c was postulated in 1946 by Daunt and Mendelssohn [50].

normally conducting state is plotted as a function of the frequency. At low frequencies an appreciable difference can be observed between the reflectances in the s- and the n-conducting states. The reflectance is greatest in the s-conducting state. A greater difference between the two states occurs at a particular frequency and the difference becomes zero at higher frequencies. The interpretation is that the large reduction occurs when the quantum energy of the radiation is sufficient to break up the Cooper pairs. This implies

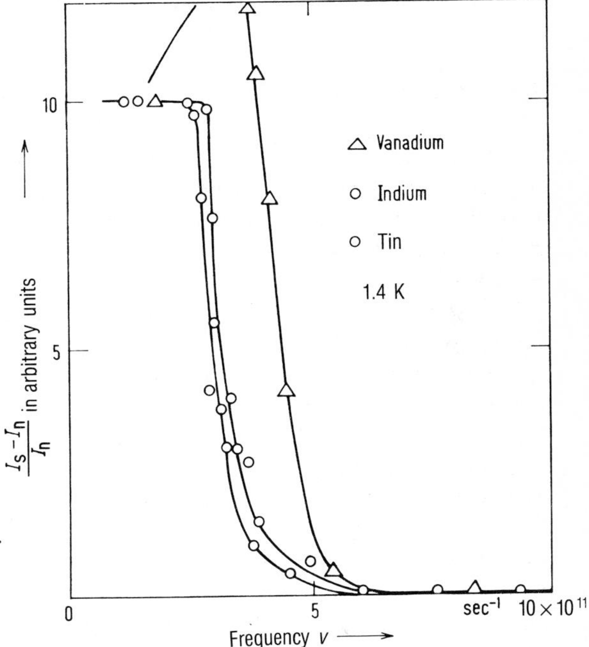

Fig. 25. Absorption of electromagnetic waves by superconductors. Transition temperature T_c: V 5.3 K; In 3.42 K; Sn 3.72 K; measurement temperature: 1.4 K. In the case of vanadium a freqency dependence was detected for $\nu < 2\Delta/h$ whose origin it is not possible to discuss here (after [49]).

Table 5. The energy gap $2\Delta_0$ for some superconductors, measured in the units $k_B T_c$.

Element	Measuring method Tunneling effect	 Ultrasound	 Light absorption
Sn	3.5 ± 0.1	–	3.5
In	3.5 ± 0.1	3.5 ± 0.2	3.9 ± 0.3
Tl	3.6 ± 0.1	–	–
Ta	3.5 ± 0.1	3.5 ± 0.1	3.0
Nb	3.6	4.0 ± 0.1	2.8 ± 0.3
Hg	4.6 ± 0.1	–	4.6 ± 0.2
Pb	4.3 ± 0.05	–	4.4 ± 0.1

The results were taken from:

a) R.D. Parks: "Superconductivity", p. 141 and p. 216.

b) D.H. Douglas Jr. and L.M. Falicov: "Progress of Low Temperature Physics", Volume *4*, 97 (1964). North-Holland, Amsterdam.

an additional absorption [1]. At energies $h\nu \gg 2\Delta$ the energy gap has practically no effect on the absorption because the radiation quanta excite the electrons greatly over the gap. Table 5 lists some values of the energy gap at $T = 0$ from investigations of this sort.

3.3.2 Absorption of ultrasound

Ultrasound also interacts with the system of conduction electrons in a metal. A wave of ultrasound can be regarded as a stream of coherent phonons. Until recently it was only possible to generate ultrasound up to a maximum frequency of ca. $3 \times 10^{10}\ s^{-1}$. Most determinations were made from a few times 10^6 to $10^7\ s^{-1}$. These frequencies are appreciably lower than the width of the energy gap [2]. Therefore, amongst other mechanisms the absorption is basically determined by the number of unpaired electrons. This number falls rapidly with temperature below T_c. Figure 26 reproduces the result of such a determination [51]. Since the numbers of free electrons at any temperature depend on the width of the energy gap, the results of such absorption determinations can be used to ascertain the width of the energy gap by comparison with the expected theoretical dependence. The curve to be expected from the BCS theory for $2\Delta(0) = 3.5k_BT_c$ is included in Fig. 26.

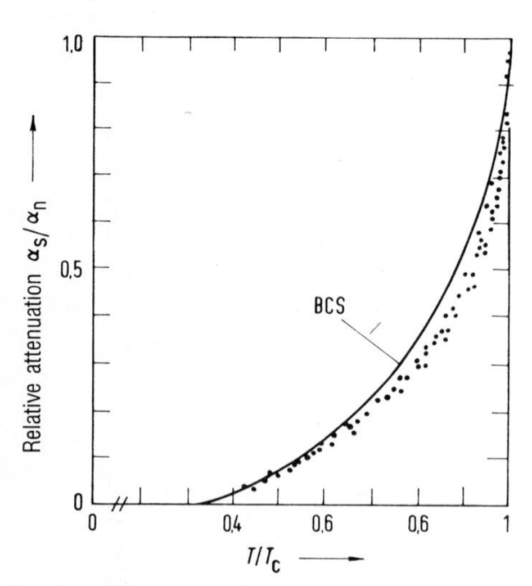

Fig. 26. Ultrasound absorption in superconducting tin and indium. The curve that has been drawn in gives the results produced by the BCS theory for an energy gap $2\Delta\ (0) = 3.5k_BT_c$ (after [51]).

We cannot go further into the analysis of such measurements here since they are not very simple. Suffice it to say that the sound waves possess a great advantage over the electromagnetic waves, they can penetrate deeper into the metal, while only a very thin surface layer, namely the skin depth, is accessible to high-frequency, electromagnetic waves.

1 Such "absorption edges" have also been observed in semiconductors. However, the range of forbidden energies is very much greater there, e.g. ca. 0.8 eV for germanium. So that the reduction in absorption is at ca. 1000-fold lower wavelengths, i.e. at wavelengths of ca. 1 µm that is in the near infrared.

2 It is only in the immediate neighborhood of T_c where the energy gap $2\Delta\ (T)$ approaches zero that the sound energy at this frequency will be comparable with $2\Delta\ (T)$.

3.3.3 Tunnel experiments

The possibility of using tunneling experiments to determine the energy gap was suggested in 1961 by I. Giaever[1] [52, 53]. We will treat this method in detail because it provided a wide range of new results in addition to a determination of the width of the energy gap.

This method is based on the observation of a tunneling current through a thin insulating layer between a reference sample and the superconductor that is under investigation. The experimental design is illustrated schematically in Fig. 27a. Two metallic conductors, e.g. two layers of aluminium, are separated by a very thin layer of insulator, e.g. Al_2O_3. Al_2O_3 is a very good insulator, i.e. no free electrons can exist in it[2]. We would, therefore, expect that no current would flow through the insulator. However, experiment shows that a small but finite electrical current is possible through such an insulator layer if only it is thin enough. The electrons can pass through the barrier that a thin insulating layer represents without their energy being sufficient to pass over the barrier. They pass through the barrier as through a tunnel. Such currents are, therefore, known as "tunneling currents".

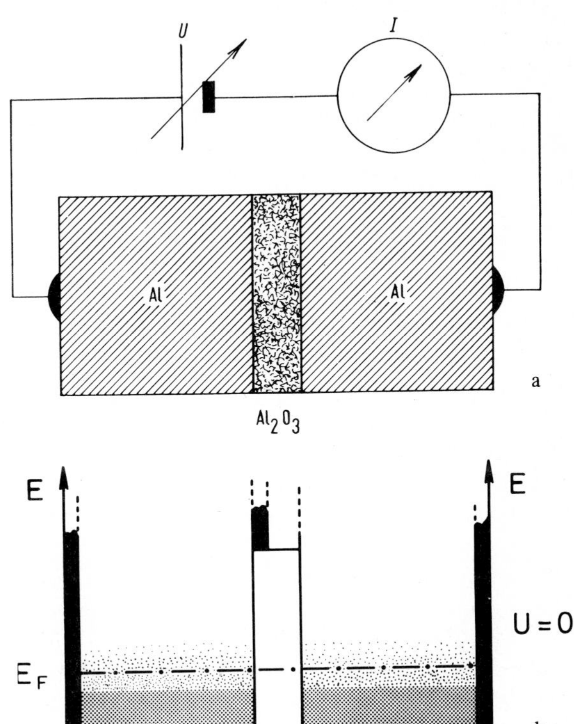

Fig. 27. a) Arrangement for measurement of a tunneling current (schematic); b) diagram of allowed energy levels (black) and their occupation (dotted).

1 In 1973 I. Giaever received the Nobel prize for physics together with L. Esaki and B.D. Josephson.

2 According to Section 1.1 we must say more accurately that an allowed energy band in Al_2O_3 is completely filled, and that the next one, separated from it by a large region of forbidden energies, is completely empty. Free electrons can only be present in a partially filled band. In Al_2O_3 they would have to possess an energy so large that it practically never occurs thermally. That is under laboratory conditions there are no electrons in this band at thermal equilibrium.

On account of its importance, e.g. in the field of semiconductor physics too, we will look at this tunneling current a little more closely. For this purpose Fig. 27b illustrates the permitted and forbidden energies at about the level of the Fermi energy in the three parts of the tunnel experiment. The solid vertical columns represent the permitted energy bands. The dots represent the occupation. The temperature smearing of the occupation is indicated. The next permitted and vacant energy levels of the insulator lie much higher.

The tunneling effect can be understood without complicated calculation if we remember the wave nature of our particles. If a wave impinges on the interface of a material, into which it cannot penetrate, then the wave must be *totally reflected*[1]. Whereby it is intuitively clear that the wave must penetrate to a certain extent into the forbidden region. It explores, as it were, the possibility of its existence in this substance. Here its amplitude decays exponentially and the more rapidly the greater the difference between the energy of the wave and a permitted energy, in other words the higher the barrier is[2]. This naturally very qualitative point of view makes it immediately clear that in the case of a thin enough barrier there is a finite probability of a wave passing through the barrier, that is when the thickness of the barrier is comparable with the decay distance of the amplitude in the forbidden range. For then a finite amplitude penetrates through and can exit into the permitted region. The amplitude there is naturally very small, i.e. the probability of the passage of a particle reduces rapidly with the wave amplitude as the thickness of the barrier increases. The tunneling probability is, therefore, a function of the energetic height[3] and thickness of the barrier.

Now our experiment involves electrons, that is fermions. The Pauli exclusion principle applies to them (see Section 1.1). In order to cross the barrier the electron must find an empty state on the other side. If all the states there are already occupied then no transport will occur even if the barrier is thin enough. So, the number of particles crossing the barrier depends on three quantities:

1. The numbers of electrons encountering the barrier,
2. the probability that they will tunnel through the barrier,
3. the numbers of empty states available on the other side.

These three quantities must be included in the quantitative description of the tunneling current of electrons.

Figure 27b represents the tunneling experiment *without* applied voltage. If electrons can be exchanged between two systems the equilibrium state is laid down as being the situation when the Fermi energies have the same level, that is in a horizontal straight line in our representation. The net exchange of particles is zero in this state. An equal number of electrons tunnel from left to right as from right to left[4].

Now we apply a voltage $U \neq 0$ to our experiment. The potential drop is practically completely at the insulating layer. This means that the Fermi energies to the left and right

---

1 When a light wave passes from an optically dense to an optically less dense medium (e.g. from glass into air) we have such total reflection for a certain range of angles of incidence.

2 The wave mechanical calculation does not provide a stationary solution in the region of forbidden energy, such as would correspond to a wave, but an exponentionally decaying amplitude.

3 The energetic height of the barrier must be measured from the energy of the particle, in the case of Fig. 27b, therefore, effectively from E_F.

4 The equilibrium conditions are clear for our symmetrical arrangement. However, they apply to any substance where there is electron exchange.

of the insulating layer differ by an energy of eU. The tunneling currents in the two directions no longer cancel each other out. A net current of I flows.

In order to clarify the size of the current and its dependence on the voltage U a representation has been chosen in Fig. 28 that also includes the densities of states. In the immediate region of the Fermi energy we can regard the density of states in the free electron model as being approximately constant. The tunneling experiments for the voltages $U = 0$, $U = U_1$ and $U = U_2 > U_1$ are illustrated in Figs. 28 a, b and c. The occupied states

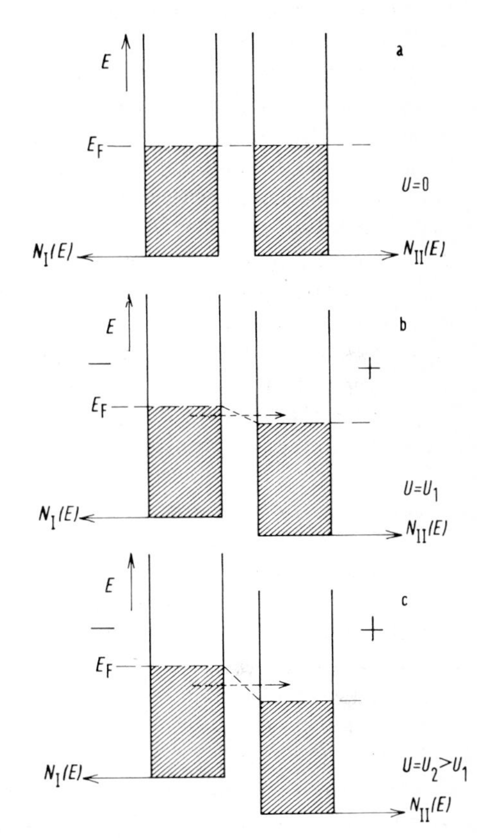

Fig. 28. Tunneling current between two normally conducting metals. For the sake of simplicity the occupation is shown for $T = 0$. There are no states in the region of E_F in the insulator.

are indicated by hatching. The temperature $T = 0$ has been chosen for simplicity. The energies of the electrons as negative particles are lower on the positive side of the voltage than on the negative. So more electrons can now tunnel from left to right than from right to left and an electron current flows (arrow in Figs. 28b and c). Since we have assumed the density of states to be constant the number of electrons that can tunnel from left to right increases in proportion to the voltage U. The net tunneling current I is, therefore, also proportional to the voltage U applied (Fig. 30, broken line[1]). It must be empha-

1 We are ignoring the effect of the applied voltage on the energetic height of the barrier in this treatment.

sized here that we are only considering tunneling processes at constant energy[1]; that is transitions which are in a horizontal direction in our representation.

The current-voltage characteristics of such a tunneling arrangement change if one or both sides are in the superconducting state. This is immediately obvious if we remember that in the superconducting state there is an energy gap in the term scheme of single electrons and that this fundamentally affects the density of states in the vicinity of the Fermi energy.

Figure 29 is a representation of the first case of normal conductor against superconductor after the manner of Fig. 28. We will again choose the simplifying case $T = 0$. The associated current-voltage characteristic is shown in Fig. 30 as Curve *2*. No tunneling current can flow until $U = \Delta/e$ is reached because the electrons of the normal conductor do not find any empty states in the superconductor. The tunneling current starts with a vertical tangent at $U = \Delta/e$. This steep rise is a result of the high density of empty states in the superconductor. At even higher voltages the curve is displaced to the tunneling characteristic for two normal conductors (Curve *1*). At finite temperatures the occupa-

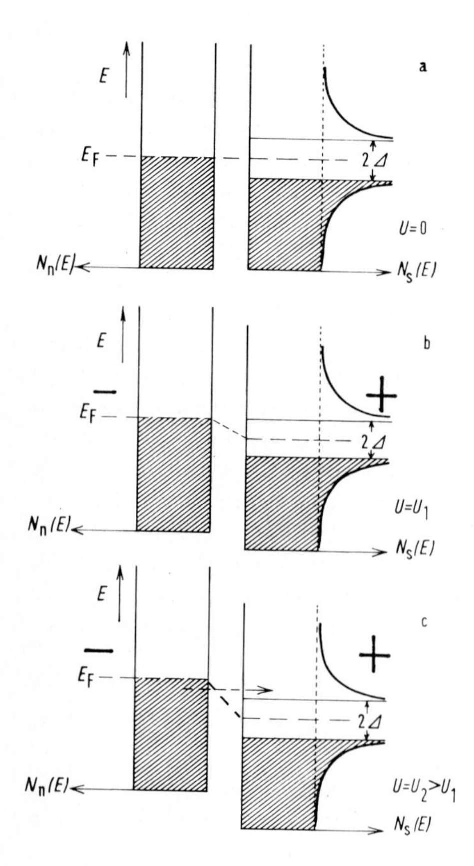

Fig. 29. Tunneling current between a normal conductor and a superconductor. $T = 0$ K.

1 During a tunneling process an electron can, for example, absorb or emit a phonon. Such processes, which are known as "phonon-assisted tunneling", are rare and will not be considered for the moment.

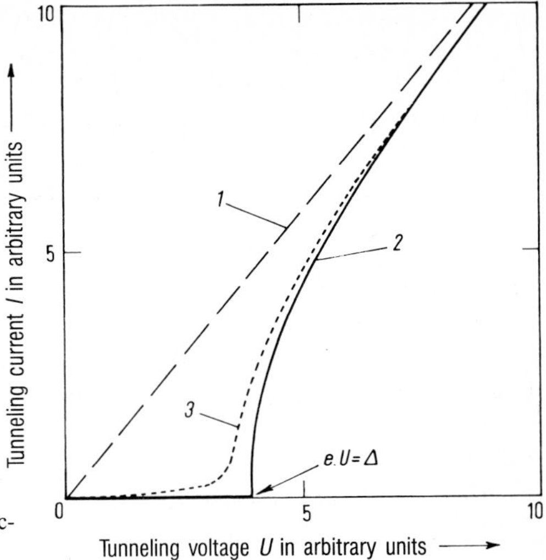

Fig. 30. Current-potential characteristics for a tunnel junction. 1: normal conductor/normal conductor, Fig. 28; 2: normal conductor/superconductor, $T = 0$ K, Fig. 29; 3: normal conductor/superconductor, $0 < T \ll T_c$.

tion of normal conductors is somewhat blurred and there are accordingly some single electrons of the superconductor above the energy gap, which is somewhat smaller too (see Fig. 20). We then have a characteristic such as is schematically represented by Curve *3*.

It is simple to determine the energy gap from such characteristic curves. If the density of states of the normal conductor is known it is possible to use the variation of the function $I(U)$ to provide quantitative evidence of the course of the density of states of unpaired electrons in the superconductor. When $T = 0$ and $N_n(E) = \text{const}$ the density of states of the single electrons $N_s(E)$ is given directly by the derivative dI/dU.

It must now be stated that the tunneling characteristic is independent of the direction of the voltage. If the voltage is reversed the empty and occupied states merely exchange roles. If the superconductor is made negative then at $U = \Delta/e$ the many single electrons can tunnel into the free states of the normal conductor. For the rule must be obeyed that because of the Pauli exclusion principle in the tunneling process electrons must not only be available in the starting state (e.g. left) but there must also be empty states for them in the final state (here right).

Before we treat case 2, that is a tunnel junction between two superconductors, we must consider the relationships between the important quantities. We have already established that the probability of a tunneling process depends on the height and thickness of the barrier. In the small energy region near the Fermi energy, which we are considering here, we can regard this probability as constant (independent of the energy). We name the probability D. The number of electrons tunneling in unit time from left to right at energy E is proportional to the number of states *occupied* on the right $N_I(E)F(E)$ and to the number of *empty* places on the right. This number is determined by the voltage U by $N_{II}(E + eU)(1 - F(E + eU))$ [1]. For we only wish to observe the tunneling process at

1 $F(E + eU)$ is the probability of the occupation of the states at $E + eU$, so that $1 - F(E + eU)$ is the probability of a free state at $E + eU$.

constant energy. The energy ε is calculated for what follows from the Fermi energy. We have, in a small energy interval dE at energy E, the small contribution to the tunneling current from left to right:

$$dI_{l\to r} \propto DN_{I}(\epsilon)\,F(\epsilon)\,N_{II}(\epsilon + eU)\,(1 - F(\epsilon + eU))\,d\epsilon\,. \tag{3-14}$$

The total tunneling current $I_{l\to r}$ is calculated by integration over all energies.

$$I_{l\to r} \propto D \int_{-\infty}^{+\infty} N_{I}(\epsilon)\,F(\epsilon)\,N_{II}(\epsilon + eU)\,(1 - F(\epsilon + eU))\,d\epsilon\,. \tag{3-15}$$

Because we measure the energy from E_F it is necessary to integrate from $-\infty$ to $+\infty$. The tunneling current from right to left $I_{r\to l}$ is obtained in the same way.

$$I_{r\to l} \propto D \int_{-\infty}^{+\infty} N_{II}(\epsilon + eU)\,F(\epsilon + eU)\,N_{I}(\epsilon)\,(1 - F(\epsilon))\,d\epsilon\,. \tag{3-16}$$

The net tunneling current is the difference between the two:

$$I = I_{l\to r} - I_{r\to l} \propto D \int_{-\infty}^{+\infty} N_{I}(\epsilon)\,N_{II}\,(\epsilon + eU)\,\{F(\epsilon) - F(\epsilon + eU)\}\,d\epsilon\,. \tag{3-17}$$

where according to (1–3):

$$F(\epsilon) = \frac{1}{e^{\epsilon/k_B T} + 1} \quad \text{with} \quad \epsilon = E - E_F\,. \tag{3-18}$$

We made this brief derivation because, as balance equation, it provides us, in such a simple manner, with the quantitative relationships. For instance, for the conditions in Fig. 29 we can assume that $N_n(\varepsilon) = \text{const}$ and use the relationship from Section 2.2 for $N_s(\varepsilon)$

$$N_s(\epsilon) = N_n(0)\,\frac{|\epsilon|}{\sqrt{\epsilon^2 - \Delta^2}}\,. \tag{3-19}$$

We can calculate the characteristics of the tunnel process. The case shown in Fig. 28 of a tunnel junction between two normal conductors is a particularly simple exercise[1].

---

1 In the case dealt with in Fig. 28 when $\varepsilon > 0$ the bracket $(F(\varepsilon) - F(\varepsilon + eU))$ equals zero, since both Fermi functions equal zero. Where it is assumed that $eU > 0$ independent of the sign of the electronic charge. The bracket is also zero for $\varepsilon < eU$, since then both Fermi functions equal 1. The bracket only equals 1 in the range $-eU < \varepsilon < 0$. We can place both the densities of state before the integral as they are assumed to be constant. We get:

$$I \propto DN_{I}N_{II} \int_{-eU}^{0} 1 \times d\epsilon \propto U$$

(Curve 1 in Fig. 30).

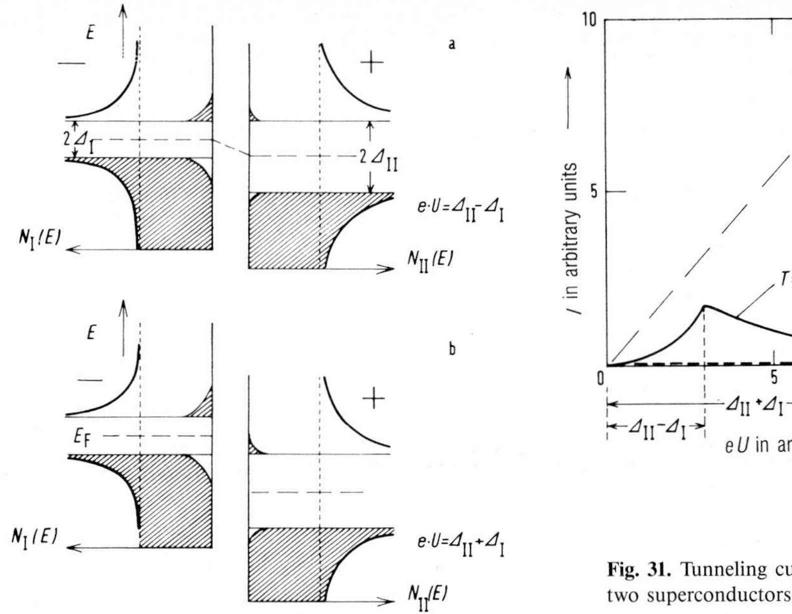

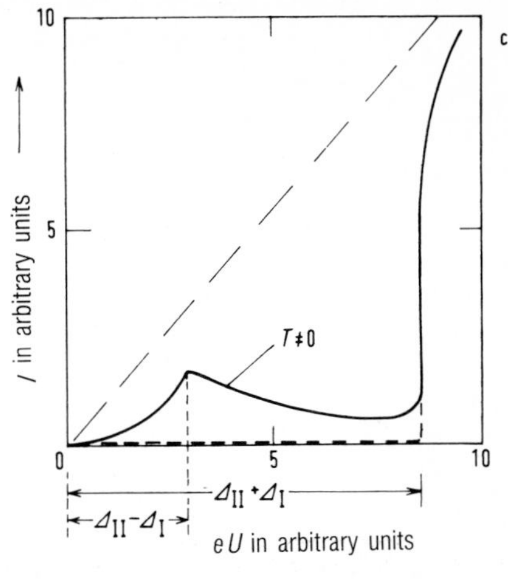

Fig. 31. Tunneling current between two superconductors $0 < T < T_c$.

Now let us consider the case of a tunnel junction between two superconductors. Such a junction is illustrated in the usual manner in Fig. 31a, b. Here, however, the portion of the diagram representing $U = 0$ has been omitted. Part c of the diagram is a schematic representation of the current voltage characteristics.

Since, in this case, there is a fundamental difference between the characteristics at $T = 0$ and finite temperatures we have used the occupation at $T \neq 0$. (The broken line in Fig. 31c would be obtained for $T = 0$). A maximum tunneling current is reached at $eU = \Delta_{II} - \Delta_I$, because now all the single electrons of superconductor I can tunnel to the right and there is a particularly high concentration of empty states. The current then falls as the potential is increased further, because the density of empty states in II falls. A particularly steep rise in I occurs at $eU = \Delta_{II} + \Delta_I$. The effect here results from the particularly high density of both occupied and empty states. This particularly steep rise constitutes the advantage of using two superconductors in measurement technology.

A wealth of results concerning the energy gap have been won using tunneling experiments. Usually two thin sputtered layers were employed with an oxide layer as barrier. In the case of substances that are difficult to evaporate, for example niobium, tantalum and such, it is possible to employ bulk samples. The barrier can also consist of an oxide layer here. But it can also be produced by sputtering with an insulator. The use of bulk samples is also necessary when single crystals are employed to determine the energy gap for a particular crystal orientation.

Figure 32 illustrates a design for sputtered or condensed layers; it is expedient to keep the area for the tunneling current small ($< 1\ mm^2$), so that the tunneling current is not large and to decrease the probability of holes in the oxide layer.

Figure 33 illustrates the results for a tunnel junction $Al - Al_2O_3 - Pb$ at 4 different temperatures [54]. Aluminium is normally conducting for Curves *1, 2* and *3*. Only Curve

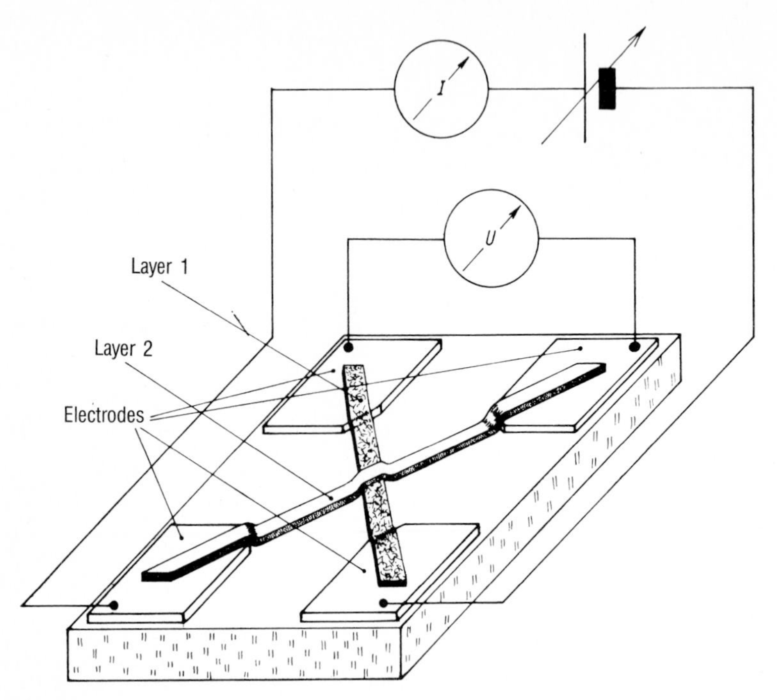

Fig. 32. Tunnel junction made up of two layers. Layer 1 was oxidized before deposition of layer 2. The layer thicknesses have been greatly exaggerated for clarity. They are usually less than 10^{-3} mm thick. Thicknesses of ca. 3 nm are usual for the oxide barriers.

4 represents an experiment with two superconducting, whereby, however, the measurement temperature 1.05 K was only slightly below the transition temperature for Al (1.2 K). However, the new shape of the characteristics Curve *4* is apparent.

Figure 34 illustrates the characteristics of a typical junction between two superconductors, namely Nb – Nb oxide – Sn at 3.4 K [55]. Nb (T_c = 9.3 K) and Sn (T_c = 3.7 K) are both superconducting at this temperature. The continuous curve is that recorded in the experiment. The points on it are calculated according to Eq. (3–17) for suitable values of the energy gap ($2\Delta_{Sn} = 0.74 \times 10^{-3}$ eV; $2\Delta_{Nb} = 2.98 \times 10^{-3}$ eV).

Finally Fig. 35 illustrates the temperature dependence of the energy gap, using tantalum as an example [56]. The circles give the experimental results and the solid line the dependence of $\Delta(T)/\Delta_0$ predicted by the BCS theory. The agreement is excellent.

Table 5 (p. 55) lists the values of $2\Delta_0$ for some superconductors. The BCS theory, being a general theory which does not initially employ any specific parameters of superconductors, yields:

$$2\Delta_0 = 3.5 k_B T_c \, .$$

The table which only lists a small proportion of the many results reveals that a series of metals confirm the BCS theory very well. Deviations such as that of niobium could easily be the result of surface impurities. Such metals are difficult to prepare with a completely clean surface.

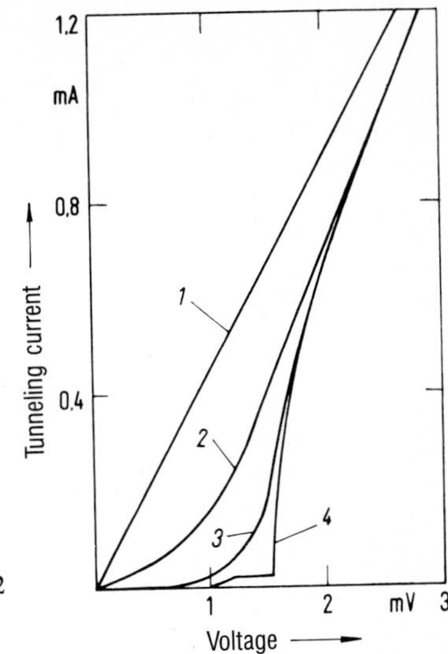

Fig. 33. Current-voltage characteristics of a tunnel junction Al-Al_2O_3-Pb, curves recorded at 1: T = 10 K; 2: T = 4.2 K; 3: T = 1.64 K; 4: T = 1.05 K. The aluminium is superconducting too at 1.05 K. The steep increase at $eU = \Delta_I + \Delta_{II}$ can readily be seen. Transition temperatures: Pb 7.2 K; Al 1.2 K (after [54]).

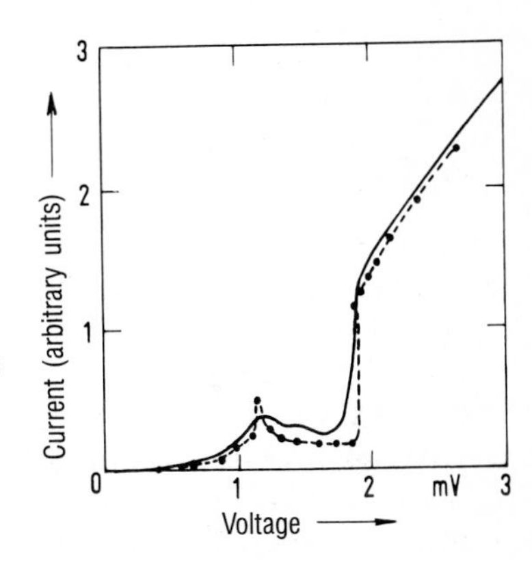

Fig. 34. Current-voltage characteristic of a niobium-insulator-tin tunnel junction at T = 3.38 K. The solid curve is the recording obtained in the experiment. The points have been calculated from Eq. (3-17) for $2\Delta_{Sn} = 0.74 \times 10^{-3}$ eV and $2\Delta_{Nb} = 2.98 \times 10^{-3}$ eV (after [55]).

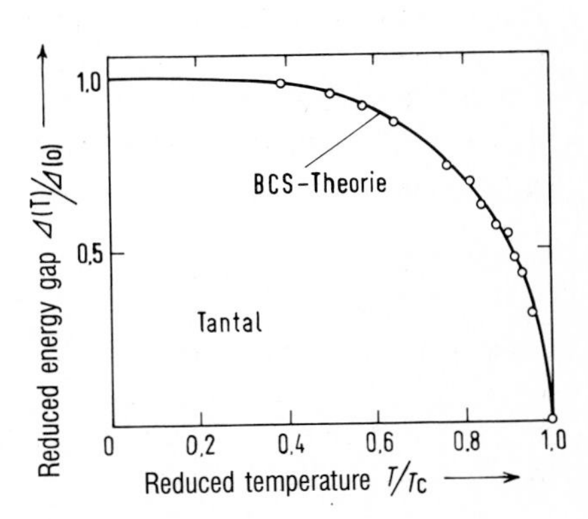

Fig. 35. Temperature dependence of the energy gap of tantalum $\Delta(0) = 1.3 \times 10^{-3}$ eV (after [56]).

The superconductors Pb and Hg exhibit considerable deviations from the values predicted by the BCS theory. These deviations can be understood in terms of the particularly strong electron-phonon interactions of these metals.

We must now mention two further questions with respect to the energy gap, which we will meet and treat in more detail later (see Sections 8.2 and 8.3).

1. The energy gap can be of different sizes in different orientations of a crystal. The superconductor is then described as being anisotropic. This anisotropy can be the reason for differing results in different experiments. Depending on the sputtering conditions specific crystal orientations can be preferred in the layers. Determinations on single crystals in various orientations reveal the magnitude of the anisotropy.

2. The energy gap can be greatly altered by very small concentrations of impurities possessing an atomic angular momentum and, hence, a magnetic moment (paramagnetic impurities). Superconductors that no longer possess an energy gap can be produced in this way (gapless superconductors), but they are still superconductors, since they still possess a pair correlation.

In the discussions of tunneling experiments with superconductors until now we have only considered the unpaired electrons. We have interpreted the observations in terms of the densities of the one-electron levels in the superconducting state. Cooper pairs and their binding energies were not discussed at all. Indeed, this was not possible, because in this picture for single electrons – they are often referred to as "quasi particles" or "excitations" – the total interaction is expressed in the change in the densities of one-electron states. Thus, for example, a pair must be broken up at $T = 0$, i.e. an energy of 2Δ is required to remove an electron from the superconductor. It is only if the voltage is sufficient to bring this process about that an electron can actually leave the superconductor.

As we already mentioned in Section 2.2, the picture of the single electron is very practical for all problems which involve energy exchange with the superconductor. In the case of tunneling experiments it is particularly useful for the description of the tunnel in terms of single electrons at constant energy.

Since, however, our understanding of the superconducting state is irrevocably involved with the concept of Cooper pairs it would be pleasant to have a picture where these Cooper pairs were plainly visible. We will discuss such a picture briefly now. Let us choose the case of a tunneling junction between two superconductors (Fig. 31) and take $T = 0$ for simplicity's sake. The superconductors are now characterized by the report of a state for the Cooper pairs and of the states available for the single particles produced on breaking up a Cooper pair (Fig. 36)[1]. In equilibrium without an external voltage the Cooper pair states are to be found at the same level. The system arranges itself so if an exchange of particles is allowed. When a voltage is applied then we expect, if $T = 0$ according to Fig. 31, no tunneling current for $eU < \Delta_I + \Delta_{II}$ and at $eU = \Delta_I + \Delta_{II}$ a very rapid increase in I. Figures 36b and c illustrate the situation for the two possible polarities of voltage $U = (\Delta_I + \Delta_{II})/e$. We now have to explain the occurrence of a tunneling current with the break-up of the Cooper pairs at this voltage. The applied voltage

1 The fact that *one* scheme is employed to represent the states for pairs, that is for collective states, and the states for single particles causes a certain amount of difficulty in this picture.

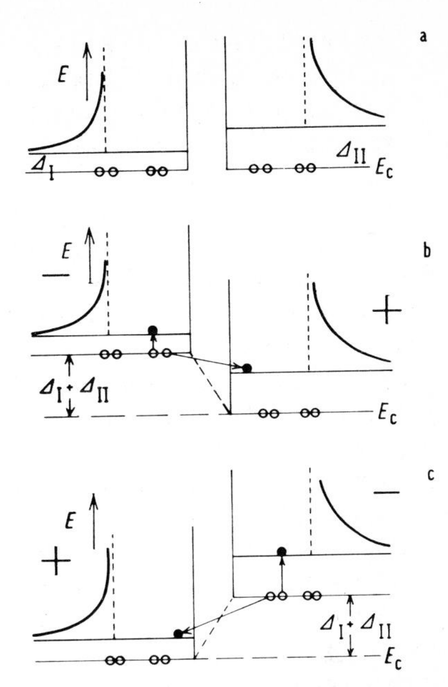

Fig. 36. Representation of the tunnel effect between two superconductors with respect to the Cooper pairs and the "excited" single electrons. Cooper pairs o o, single electrons (excitations) • .

must be at least large enough to be able to cause a pair to break up into an electron in superconductor I and an electron in superconductor II. This is just the process we wish to describe, namely the passage of *one* particle through the insulating layer. The first possibility of such a process occurring is as the voltage U is increased when $U = (\Delta_I + \Delta_{II})/e$. If the voltage lowers superconductor II with respect to I, then at $U = (\Delta_I + \Delta_{II})/e$ a pair can be broken up in I whereby one single electron will be in the lowest state of I and the other single electron tunnels to the lowest state of II. This process takes place at constant energy. The excitation energy of an electron can be supplied by the transfer of the other electron in the field of the external voltage. The final states that the two electrons can take up are decisive for the voltages required. For all voltages $U < (\Delta_I + \Delta_{II})/e$ there is no possibility that the electrons of a pair will split up into two one-electron states in I and II with conservation of energy. If superconductor II is made negative then a pair will again be able to break up when $|U| = (\Delta_I + \Delta_{II})/e$, but this time in II so that they occupy one-electron states in I and II [1]. On account of the great density of states for single electrons at $E = \Delta_i$ the tunneling current increases very rapidly at $|U| = (\Delta_I + \Delta_{II})/e$. The argumentation here is the same as in the other model. The number of these processes is proportional to the number of possible final states.

If we wish to retain this model for finite temperatures and obtain at least the qualitative behavior of the current-voltage characteristic from it, then it is necessary to include the thermal equilibrium of the single electrons present in the initial model.

Knowledge concerning the energy gap would also be very valuble in the case of the new superconductors. For this reason great efforts have been made to determine this important parameter. Here, in addition to infrared spectroscopy, which is described briefly in

1 Here the condition $I(U) = -I(-U)$, that applies to all tunnel junctions, is fulfilled.

Section 3.3.1, it has mainly been point contacts that have been studied (see Section 9.5.4, Fig. 175e). The results obtained so far are extremely confusing. While for classical superconductors the value of $2\Delta_0/k_B T_c$ closely approaches the 3.5 predicted by the BCS theory (Table 5, Section 3.3.1), values ranging from 2 to ca. 10 have been observed in the case of the oxide superconductors [57].

One basic difficulty involved in the new superconductors is that they possess a very small coherence length ξ_{Co} (mean distance for the pair correlations). So in tunneling experiments the observations only refer to the outermost atomic sites of the sample. In addition these substances possess a marked layer structure (see Section 8.7.4) and are, therefore, particularly highly anisotropic. Finally the superconductivity of oxides is sensitively dependent on the oxygen content. In particular the outer atomic sites of the sample, therefore, may possess differing properties depending on the previous treatment. All these difficulties have made it impossible until now to obtain clear information concerning the energy gap. But the experimental results so far available make it certain that one or even several energy gaps are present.

While we have until now concentrated on the tunneling of single electrons the diagram in Fig. 36 raises the question as to whether Cooper pairs can also tunnel through a sufficiently thin insulating layer. This is, in fact, the case. The phenomena this brings about constitute another excellent confirmation of the BCS theory and will be treated in the next section.

We will meet another method of measuring the energy gap in Chapter 4 (Section 4.2) during the treatment of the specific heat of the superconducting state. In principle every physical phenomenon that depends on the density of the states of single electrons can be employed to determine Δ. With light absorption, ultrasonic attenuation and one-electron tunneling we have discussed three of these possibilities which have provided us with a large proportion of the results and whose fundamentals are also very clear.

3.4 Phase coherence – Josephson effects

All Cooper pairs occur in the same quantum mechanical state. This macroscopic state is, as we maintained in Section 2.2, the reason for the unusual properties of the state of superconductivity. The Cooper pairs must, for they all have the same state, agree in all physical determinants. This particularly applies to the phase of the Cooper pair system which is very well defined. This strict phase correlation applies over very large (practically unlimited) distances as far as we know today.

We obtained the first evidence of the strict phase correlation of all Cooper pairs on observing the elementary flux quantum through a superconducting ring. The size of this elementary flux quantum $\Phi_0 = h/2e$ reveals that on the transition of a superconducting ring between two states, which differ by one flux quantum through the ring, *all* Cooper pairs must change their phase along the closed path around the ring by 2π.

We will now discuss some experiments that demonstrate the strict phase correlation of Cooper pairs in a particularly impressive manner. The impetus for these experiments was provided by a theoretical investigation by Josephson in 1962 [58]. In this work Josephson demonstrated that a passage of Cooper pairs is also to be expected through the insulating layer in tunneling experiments if the thickness is only small enough, namely 1 to 2 nm. He also predicted some remarkable and interesting phenomena that ought to occur in connection with this pair tunneling[1]. All the predictions of Josephson were later confirmed experimentally in a very convincing manner. The Josephson effects – as this complex of phenomena is now known – are not only of fundamental importance for an understanding of superconductivity they also offer some very interesting possibilities in the field of measurement technology (see Section 9.5.3 and 9.5.4).

The diagram in Fig. 36 suggests that Cooper pairs might also tunnel through a thin enough barrier[2]. This would mean, however, that a current carried by Cooper pairs could flow through a very thin layer of insulator. The current voltage characteristics of such a Josephson junction of two identical superconductors is illustrated schematically in Fig. 37. The novel feature in comparison with the characteristics, such as are illustrated in Fig. 30, lies in the existence of a supercurrent at $U_s = 0$, the Josephson direct current. The direction of this current depends on the polarity of U_0 in the external circuit (inset Fig. 37). The Josephson current can be increased to a maximum value by increasing U_0. If U_0 is increased further an electrical voltage appears at the junction, $U_s \neq 0$. The current jumps to a point on the characteristic curve which is determined by the resistance R of the external circuit.

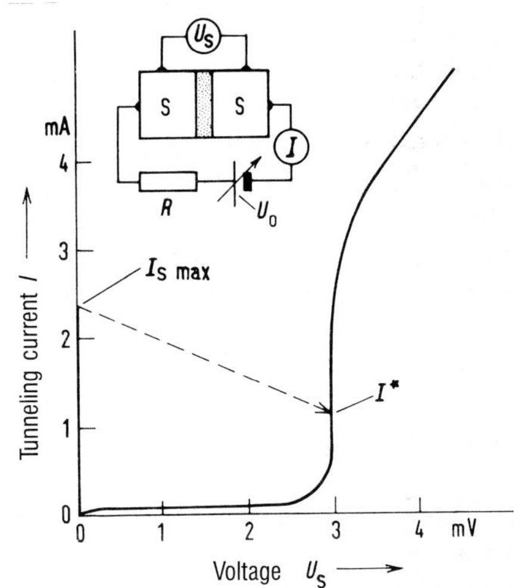

Fig. 37. Current-voltage characteristic of a Josephson tunnel junction between identical superconductors. The resistance R determines the behavior of the junction when instability starts.

Josephson made the particularly astonishing prediction concerning this state of the junction when $U_s \neq 0$, that under these conditions in the junction a high frequency alternating current would be set up, whose frequency is determined by the voltage U_s at the junction according to

$$\nu_J = \frac{2eU_s}{h}. \tag{3-20}$$

---

1 B.D. Josephson: Nobel prize for physics 1973.

2 There is naturally some difficulty in imagining this tunneling of a Cooper pair, for this is made up of two electrons with considerable opposing momenta. We must become accustomed to regarding Cooper pairs as new particles with charge $2e$, mass $2m$ and zero momentum in the current-free state. We pointed out in Section 2.2 that it only depends on the strict correlation of two electrons.

Shortly after its theoretical prediction this Josephson alternating current was also confirmed, first indirectly and then very directly.

All Josephson effects, those involving direct current and those involving alternating current, are dependent, as we will see, on the phase correlations between the Cooper pairs and, thus, confirm the strict phase correlations in the state of superconductivity. In addition to their importance for superconductivity the consideration and experimental study of these effects is itself fascinating on account of their physical excellence.

3.4.1 The Josephson direct current

The prediction that a supercurrent can flow through a thin insulating layer is evident from Fig. 36. The great importance of this prediction first becomes evident when we consider the effect of an external magnetic field parallel to the insulating layer on this Josephson direct current. For it is found that the maximum value of the Josephson direct current depends, in a very characteristic manner, on the intensity of the magnetic field. The dependence of $I_{s\,max}$ on the magnetic field is illustrated in Fig. 38 [59]. Here the maximum supercurrent through a Josephson junction is plotted as a function of an external magnetic field. A quantitative analysis reveals that the zero value of the current occurs at just those magnetic fields which represent an integral multiple of the elementary flux quantum for the tunnel arrangement. In addition the dependence of the maxima of the Josephson current can be seen to be very similar to the interference figure that is produced, for instance, when a wave passes through a slit.

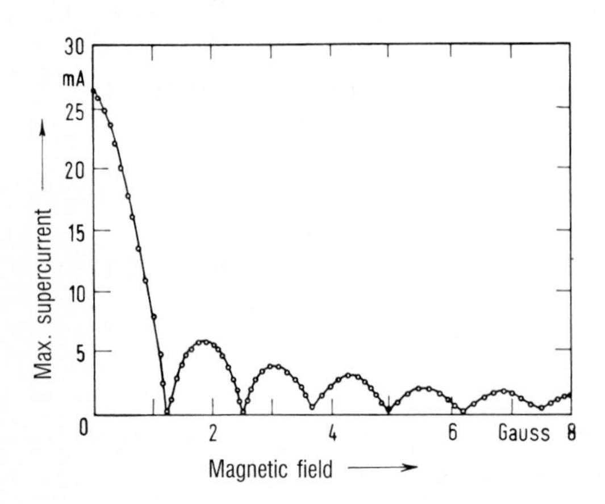

Fig. 38. Dependence of the maximum Josephson current on a magnetic field parallel to the insulating layer of a tunnel junction Sn – SnO – Sn (after [59]) (1 G = 10^{-4} T).

How can we understand these experimental results and the internal relationships they reflect? We must go a little further in order to be able to answer this question. Let us next consider a Josephson junction without an external magnetic field. We can characterize such a junction in the following manner. In superconductors I and II we have a Cooper pair system, with a strict phase correlation. Let us for simplicity assume that our junction is made up of two identical superconductors, i.e. the Cooper pair systems 1 and 2 are the same. If both systems are completely isolated each of them is in a well-defined quantum mechanical state. Since the systems are identical we may assume that they will take up the same states under the same conditions. We can describe this state, apart from a constant, by means of the energy E[1]. Every quantum mechanical state with energy E possesses an "internal frequency" $\nu = E/h$. So we can compare the Cooper pair systems with swinging pendulums. This analogy can be taken a long way and can provide a certain understanding of the processes taking place in the tunnel junction.

If we now bring the superconductors into contact via a very thin insulating layer, they can, as we have said, exchange Cooper pairs. The two systems are linked by the exchange of Cooper pairs. However, since the probability of a pair tunneling through even a very thin layer is very low, the linkage is very weak.

The special features of this tunnel arrangement lie in this weak linkage of two Cooper pair systems. Such weak coupling can also be achieved in other ways. For example, if you touch the fine tip of a superconductor onto a superconducting surface a transition zone is obtained with a very small cross section. This too represents a weak link between the two superconductors. The insulating layer of homogeneous thickness is the simplest model for a quantitative description. We will, therefore, limit ourselves to the treatment of such tunnel junctions.

In general the linkage effects a change in the states of both systems with time as a result of exchange processes. Here the phase difference between the two systems is of decisive importance for the size and direction of such exchange. Josephson postulated that for a tunnel junction:

$$I_s = I_{smax} \sin(\varphi_2 - \varphi_1) \tag{3-21}$$

where $I_{s\,max}$ is the maximum Josephson direct current through the junction. Its size depends solely on the properties of the insulating layer (e.g. the thickness and height of the barrier). φ_1 and φ_2 are the phases of the two Cooper pair systems left and right of the barrier. We can also write Equation (3-21) for the current density:

$$j_s = j_{smax} \sin(\varphi_2 - \varphi_1) \,. \tag{3-22}$$

The direct current I_s is then found from the integration of the current density over the total area A of the junction. In the case of a homogeneous current density:

$$I_s = j_s A \,. \tag{3-23}$$

We will now clarify the predictions of Equation (3-21), for which, on account of its importance, a simple derivation is given in Appendix A, in terms of the analogy between the Cooper pair systems and pendulums. The linkage of the pendulums can occur in any desired manner, e.g. by means of a weak spring. The linkage leads to each pendulum "sensing" the oscillations of the other. If we lift pendulum 1 and release it we have as starting condition pendulum 1 with oscillation energy and pendulum 2 without energy since it is at rest and practically not displaced to the side. Experience shows us that pen-

1 Since here we are taking the energy differences as the physically relevant quantities the choice of constants is without importance.

dulum 2 will be "excited" by the oscillations of pendulum 1. In our terminology this means: The linkage causes the exchange of oscillation energy. In the example both our pendulums have the same frequency and all the oscillation energy is exchanged from 1 to 2. The energy exchange then begins in the opposite direction, that is from 2 to 1, pendulum 2 now excites pendulum 1. The decisive point in our considerations now lies in the phase relationship between the oscillations. The energy exchanged goes from 1 to 2 if the oscillation of 1 is *in front* of 2 by a phase difference of $\Delta\varphi$ where $0 < \Delta\varphi < \pi$ [1]. The maximum energy flow occurs at $\Delta\varphi = \pi/2$. If, on the other hand, pendulum 2 is running in front of pendulum 1 then the energy will pass from 2 to 1. We can express this in two ways with respect to the phase difference. We can say that the phase difference $\Delta\varphi$ is negative in this case. But this is the same as saying that we assume a positive phase difference, but this lies between π and 2π or, in general, between $(2n + 1)\pi$ and $(2n + 2)\pi$.

It is not easy in the mechanical model to set up any desired phase difference to observe the associated energy flow. Only the limiting cases are simple. For $\Delta\varphi$ equals 0, π, 2π... we have no energy flow and the same amplitude for each pendulum. In one case the pendulums swing in phase in the other exactly in antiphase. Both are stationary states if the pendulums have the same energy, i.e. the same maximum amplitude. If, on the other hand, we have phase differences of $(2n + 1)\pi/2$ then the energy exchange will be at its maximum, with exchange from 1 to 2 if n is even and from 2 to 1 if n is uneven.

The energy flow of our mechanical model corresponds to the exchange of Cooper pairs through the tunnel junction. The simplest mechanical model has, however, the disadvantage that the energy of the stimulating pendulum is used up in the exchange. An analogous effect (reduction of the density of Cooper pairs on one side of the junction) does not occur with the Cooper pairs because our tunnel junction is connected to a battery. The battery can supply electrons via the external circuit and these can become Cooper pairs in the superconductor. We can easily remove this disadvantage of our model by incorporating in the pendulums an internal mechanism that makes up the energy that flows away from them and absorbs the energy that flows to them, in other words by maintaining the pendulums at constant amplitude. It is not very difficult to invent such a mechanism [2].

For us it is important that we detect an energy flow that is dependent on the phase difference of the oscillations of the two pendulums. The behavior of our tunnel junction is analogous. The exchange of the Cooper pairs depends on the phase difference between the internal oscillations of systems 1 and 2. This is the prediction of Equation (3-21).

---

1 In the same manner the energy will flow from 1 to 2 if:

$$2n\pi < \Delta\varphi < (2n + 1)\pi .$$

The maximum current is obtained when

$$\Delta\varphi = (2n + 1)\frac{\pi}{2} .$$

2 It is naturally questionable whether such a mechanical model is more vivid than the equations that describe the system. What we find vivid depends on what objects we are particularly famillliar with. For instance, the electrical specialist would certainly replace the mechanical pendulums by two weakly coupled oscillating circuits, for which, in fact, some experimental conditions are more readily realized.

A supercurrent flows from 2 to 1 if the phase differences $\varphi_2 - \varphi_1$ lie between 0 and π (or $n2\pi$ and $(2n + 1)\pi$). When $\Delta\varphi = 0$ and $\Delta\varphi = n\pi$ no net Cooper pair current flows. Correspondingly for phase differences of $\Delta\varphi = \varphi_2 - \varphi_1$ between π and 2π (and the corresponding multiples) a Cooper pair current will flow from 1 to 2. The level of this supercurrent is determined by Equation (3–21).

It is necessary to emphasize that this analogy to the mechanical model does not constitute a proof of Equation (3–21). The intention was merely to demonstrate that a relationship of type (3–21) is not so unusual as it might at first appear. We only have to take seriously the fact that our Cooper pairs have a well defined phase.

Now let us return to the influence of a magnetic field. A quantitative treatment is necessary here. We first have to find out whether the phase relationship of our system of Cooper pairs can be affected at all by a magnetic field in the insulating layer. Then we have to try to describe the changes in phase caused by the magnetic field quantitatively.

In order to understand that the magnetic field can affect the phase difference between Cooper pairs we must recall the quantum mechanical rule that was required in Section 3.2, namely that the description of the motion of a charged particle in a magnetic field requires two momentum-like quantities, namely the momentum $\vec{p}$ of the wave function of the particle and the momentum $m\vec{v}$. The two quantities are related by the Equation (3–7):

$$m\vec{v} = \vec{p} - q\vec{A} . \tag{3–7}$$

m, $\vec{v}$ and q are the mass, the velocity and the charge of the particle. $\vec{p}$ is the "canonical" momentum that determines the wavelength $\lambda = h/|\vec{p}|$ and appears under quantum conditions. $\vec{A}$ is known as the vector potential which is related to the magnetic field $\vec{B}$ via $\text{curl}\,\vec{A} = \vec{B}$. For our Cooper pairs the equation is:

$$2m\vec{v} = \vec{p} + 2e\vec{A} . \tag{3–24}$$

On account of the negative charge of the electrons we write $\vec{p} + 2e\vec{A}$. But the sign of the charge is not decisive for the following treatment.

If we hold one of the three quantities constant, e.g. the canonical momentum $\vec{p}$, then the kinetic momentum $2m\vec{v}$, which is, after all, related to the current density j_s, must vary with $\vec{A}$. Then if we keep the kinetic momentum $2m\vec{v}$ constant then $\vec{p}$ and, hence, the wavelength $\lambda = h|\vec{p}|$ will depend on $\vec{A}$.

This possibility of altering the wavelength of a charged particle by a vector potential has been demonstrated quite independently of superconductivity in beautiful experiments with electron waves in vacuo by Möllenstedt et al. [60]. Let us consider the fundamentals of these experiments briefly, because they can contribute to our understanding of the situation at a Josephson junction.

Möllenstedt et al. divided a beam of electrons [1)] into two parts which were passed on opposite sides of a tiny coil (diameter ca. 20 μm) and then superimposed to form an interference system. They observed the normal interference pattern for a double slit. The

1 As in optical interference certain conditions of beam geometry, namely coherence geometry, must be fulfilled in order to obtain good interference figures.

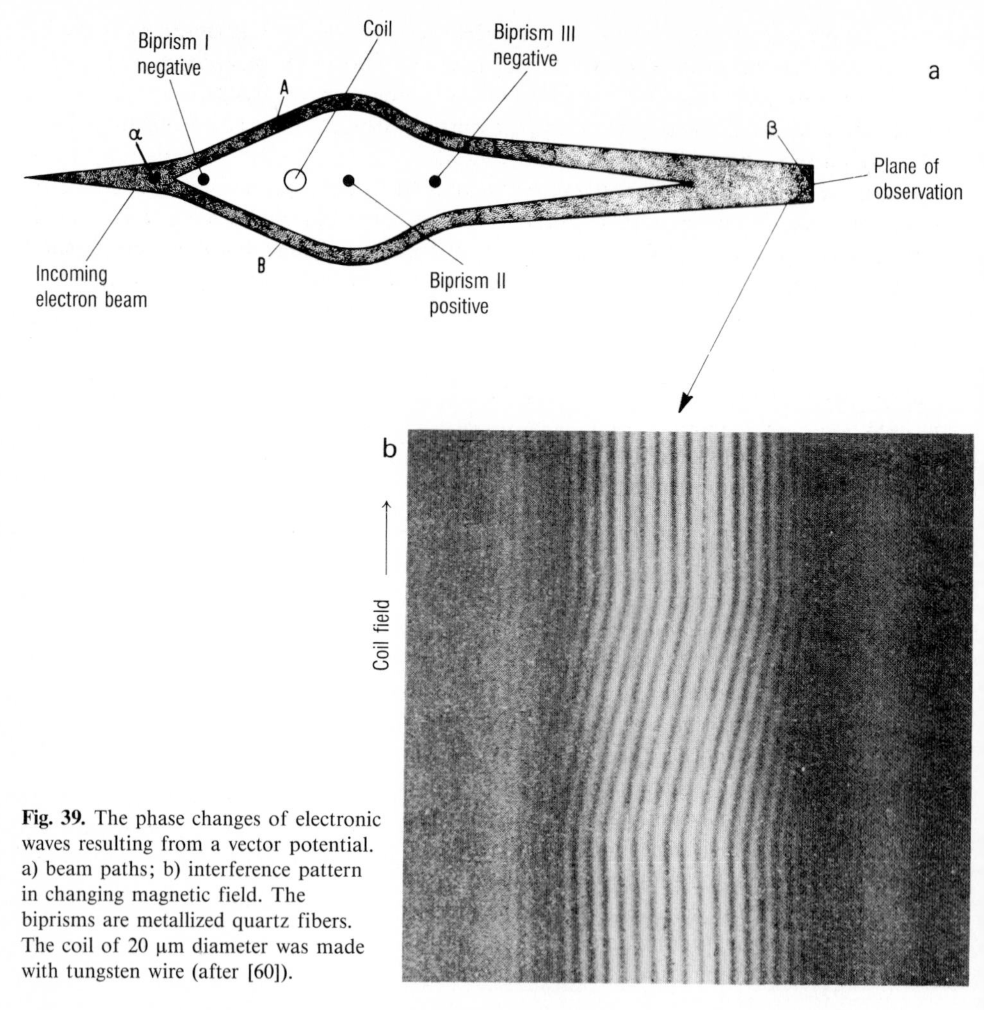

Fig. 39. The phase changes of electronic waves resulting from a vector potential. a) beam paths; b) interference pattern in changing magnetic field. The biprisms are metallized quartz fibers. The coil of 20 μm diameter was made with tungsten wire (after [60]).

decisive experiment as far as our considerations are concerned was that the interference pattern was recorded when various currents were passed through the coil. It was found that there was a displacement of the pattern as the magnetic field was changed. This means that the phase difference of the two split beams was altered by the magnetic field produced by the coil. Figure 39 illustrates the experiment schematically (a) and the record of the pattern (b). Here the film was moved vertically as the magnetic field was changed. The displacement of the pattern is clearly visible[1]. It amounts to ca. three complete periods in Fig. 40b, i.e. the phase difference between the split beams was changed by the magnetic field by about $3 \times 2\pi$ in this experiment.

---

1 The fact that there was a change of $\vec{B}$ in time during the record of the line system is not decisive. Each value of $\vec{B}$ is associated with a particular time-independent constant interference system. A $\vec{B}$ in the coil which is constant with time also alters the phase relationship of the two split beams via the associated vector potential $\vec{A}$.

The quantitative analysis reveals a phase displacement of 2π (displacement of the interference system by a full period) when the magnetic flux Φ in the coil is altered by $\Delta\Phi = h/e$. This result can readily be understood from Equation (3–7). In the experiment in Fig. 39 we have no magnetic field along the path of the electrons. The magnetic field of a long coil can be limited as well as desired to the inside of the coil. In this experiment the returning lines of force were kept away from the loops formed by the electrons by means of yokes of magnetic material. There is also no additional electric field when $\vec{B}$ in the coil is kept constant. We are, therefore, allowed to conclude that $m\vec{v}$ = const, since the magnetic field of the coil does not exert any force on the electrons [1]. So that:

$$\vec{p} = \text{const} - e\vec{A}\,. \tag{3–25}$$

Now let us calculate the phase difference $\Delta\varphi = \varphi_2 - \varphi_1$ between the two halves of the split beam on account of the magnetic field $\vec{B}$ in the coil. The phase difference between two points 1 and 2 along the path of the wave is given by:

$$\Delta\varphi_{21} = 2\pi \int_1^2 \frac{|\mathrm{d}\vec{l}|}{\lambda} = \frac{2\pi}{h} \int_1^2 \vec{p}\,\mathrm{d}\vec{l} \tag{3–26}$$

or

$$\Delta\varphi_{21} = \frac{2\pi e}{h} \int_1^2 \vec{A}\,\mathrm{d}\vec{l}\,. \tag{3–27}$$

As stated for (3–24) the sign is unimportant for this treatment. Let us assume that the phase difference between the two halves of the split beam is zero at point α in Fig. 39; we are now in a position to calculate the phase difference at point β. For beam components A and B it is:

$$\Delta\varphi^{\mathrm{A}}_{\beta\alpha} = \frac{2\pi e}{h} \int_\alpha^\beta \vec{A}\,\mathrm{d}\vec{l}_{\mathrm{A}} \tag{3–28}$$

and

$$\Delta\varphi^{\mathrm{B}}_{\beta\alpha} = \frac{2\pi e}{h} \int_\alpha^\beta \vec{A}\,\mathrm{d}\vec{l}_{\mathrm{B}}\,. \tag{3–29}$$

From this the phase difference between split beams A and B at point β is:

$$\Delta\varphi_{\mathrm{AB}}(\beta) = \Delta\varphi^{\mathrm{A}}_{\beta\alpha} - \Delta\varphi^{\mathrm{B}}_{\beta\alpha} = \frac{2\pi e}{h}\left\{\int_\alpha^\beta \vec{A}\,\mathrm{d}\vec{l}_{\mathrm{A}} - \int_\alpha^\beta \vec{A}\,\mathrm{d}\vec{l}_{\mathrm{B}}\right\}. \tag{3–30}$$

1 The electrostatic forces produced by the biprism are kept constant during the experiment.

By reversing the integration path for split beam B it is found that:

$$\Delta\varphi_{AB}(\beta) = \frac{2\pi e}{h}\left\{\int_{\alpha}^{\beta} \vec{A}\,d\vec{l}_A + \int_{\alpha}^{\beta} \vec{A}\,d\vec{l}_B\right\} = \frac{2\pi e}{h}\oint_{C} \vec{A}\,d\vec{l}\,. \tag{3-31}$$

The integral of $\vec{A}$ over a closed path C can be converted into an area integral, of an area encompassed by C (Stokes' theorem). Whith it we get:

$$\Delta\varphi_{AB}(\beta) = \frac{2\pi e}{h}\oint_{C} \vec{A}\,d\vec{l} = \frac{2\pi e}{h}\int_{S} \operatorname{curl}\vec{A}\,d\vec{a} = \frac{2\pi e}{h}\int \vec{B}\,d\vec{a} =$$

$$= \frac{2\pi e}{h}\,\Phi_S\,. \tag{3-32}$$

The phase difference is proportional to the magnetic flux that is surrounded by the two split beams A and B. For a flux h/e or a whole multiple thereof the phase difference will be just 2π or $n2\pi$. But this just means that changing the magnetic flux by h/e causes a displacement of the interference system by a whole period, because this change brings about a phase difference between the two split beams of 2π [1].

We have discussed this interference experiment with electron waves in such great detail because fundamentally similar considerations can be applied to the calculation of the phase changes of Cooper pair waves in a Josephson junction.

Figure 40 illustrates a Josephson junction schematically. The Josephson direct current is flowing in the x axis. The height and width of the junction are represented by a and b. The two superconductors, which we will regard for the sake of simplicity as being the same, are separated at $x = 0$ by the plane tunneling layer (oxide layer) with the very small thickness d. The magnetic field $\vec{B}$ lies in the y direction and decays in a surface layer of the superconductors because of screening (see Section 5.1.2). The superconducting screening currents flow in this layer. In this geometry they only possess a z component. We can substitute a stepwise function for the decaying field in the calculation by assuming a constant field $\vec{B}$ to a "penetration depth" λ_{eff}" and a field of zero deeper in the superconductor. Since the penetration depth λ is much greater than d ($d \approx$ 1–2 nm; $\lambda \approx$ 30–50 nm), we ignore d with respect to λ without any great error and allow the superconductor to start at $x = 0$ (Fig. 40) [2].

We now ask what is the phase difference between the two systems of Cooper pairs to the left and to the right of the plane $x = 0$. In the presence of a magnetic field this phase

1 Because F. London had to assume that the supercurrent was carried by single electrons he came with the requirement $\Delta\varphi = n2\pi$ for a complete circulation of the wave function for the superconducting electrons in a ring, which corresponds in our quantization conditions in Section 3.2 to an elementary flux quantum of $\Phi^{L}_{0} = h/e$.

2 In the correct description it must be ensured that as $d \to 0$ the height of the barrier is allowed to increase so that the tunneling probability stays constant.

difference will be a function of the positions we occupy on the *y-z* plane. If we know the phase difference as a function of position we also know the current density j_s of the Josephson direct current and we can by integration over the *y-z* surface determine the total Josephson current through the junction. We will assume that the magnetic field of the Josephson current is negligible with respect to the external field. This assumption is generally very well fulfilled on account of the smallness of the Josephson current.

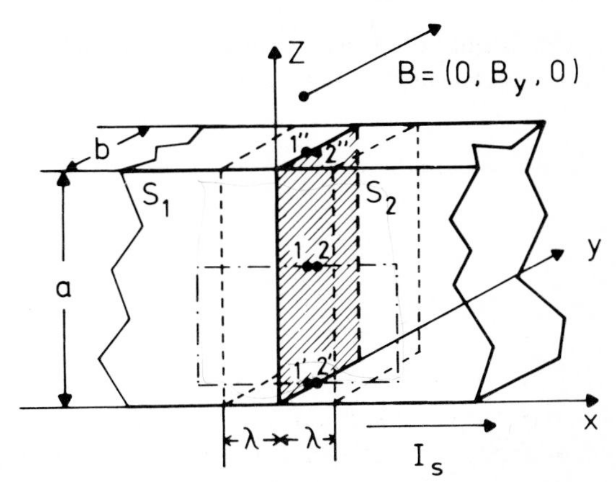

Fig. 40. Tunnel junction in magnetic field. The depth of penetration λ is a few times ten nm.

In order to find out the effect of the magnetic field we must start out from Equation (3–7). So we require the vector potential $\vec{A}$ that delivers the assumed magnetic field via curl $\vec{A} = \vec{B}$. You can readily convince yourself on the basis of the calculation rule given in Section 3.2, page 52, footnote, that $\vec{A} = \{0,0, -xB_y\}$ describes the field $\vec{B} = \{0, B_y, 0\}$. This representation holds for the region $|x| < \lambda$ that is where we have assumed the field is homogeneous. With the present geometry $\vec{A}$ maintains in the field-free interior of the superconductor the value that it had on the edge of the region penetrated by $\vec{B}$[1]. Since we are interested in the contribution of the magnetic field, or better the vector potential of the assumed magnetic field, to the phase difference we can assume that the phase differences not associated with the magnetic field are constant and set them equal to $\Delta\varphi_0$ for, e.g., points 1′ and 2′ at $z = 0$.

In order to determine the difference between the phase differences at points 1′ and 2′ and points 1 and 2 we must, as in the interference experiment with an electron beam described above, integrate the momentum $\vec{p}$ over a path from 1′ to 1 and from 2′ to 2. According to (3–26) we get:

$$\Delta\varphi_{11'} = \frac{2\pi}{h} \int_{1'}^{1} \vec{p}\,\mathrm{d}\vec{l} \tag{3–33}$$

$$\Delta\varphi_{22'} = \frac{2\pi}{h} \int_{2'}^{2} \vec{p}\,\mathrm{d}\vec{l}\,. \tag{3–34}$$

---

1 It is a condition here that $\oint_C \vec{A}\mathrm{d}\vec{l} = \Phi$ for the closed path C (flux through the area enclosed by C). This condition is fulfilled if we assume a value of $\vec{A}$ in the field-free interior of the superconductor equal to the value at the edge of this region.

We can choose our integration path in such a manner that the term $m\vec{v}$ occurring in $\vec{p}$ does not make any contribution. Since this term, which is proportional to the current density of the screening, only possesses a component in the z direction, we choose our path through the screening layer in a direction perpendicular to the z direction. The vector potential does not make any contribution along this path, since, in the form chosen, it does not possess an x component. In the field- and current-free interior of the superconductor we integrate along straight lines parallel to the z axis. Here the vector potential has, in the form chosen – also referred to as the calibration – the value $-\lambda B_y$ in superconductor S_1 and the value $+\lambda B_y$ in superconductor S_2. We, thus, get only *one* contribution for each of the two paths from 1′ to 1 and from 2′ to 2 (see dash-dotted line in Fig. 40):

$$\Delta\varphi_{11'} = \frac{4\pi e}{h} \lambda B_y z \tag{3-35}$$

$$\Delta\varphi_{22'} = \frac{4\pi e}{h} \lambda B_y z \, . \tag{3-36}$$

Since we are considering Cooper pairs here the charge must be taken as $2e$.
Finally for the whole phase difference produced by the magnetic field we find:

$$\Delta\varphi_{21} - \Delta\varphi_0 = \frac{4\pi e}{h} 2\lambda z B_y \tag{3-37}$$

$$\Delta\varphi_{21}(z) = \Delta\varphi_0 + 2\pi \frac{\Phi(z)}{\Phi_0} \tag{3-38}$$

where $\Phi(z)$ is the flux which passes through the area $A(z) = 2\lambda z$. Φ_0 is the flux quantum (Section 3.2). With our choice of $\vec{B}\{0, B_y, 0\}$ the phase difference does not depend on y[1)].

The current density

$$j_s(z) = j_{smax} \sin\left\{\Delta\varphi_0 + \frac{2\pi\, 2e}{h} \Phi(z)\right\} \tag{3-39}$$

is, therefore, dependent on z. In particular the maximal phase difference caused by the magnetic field will be:

$$\Delta\varphi_{max}(\vec{A}) = \frac{2\pi\, 2e}{h} \Phi(a) \, . \tag{3-40}$$

---

1 The choice of the vector potential was not obligatory. We could also represent $\vec{B} = \{0, B_y, 0\}$ by $\vec{A} = \{zB_y, 0, 0\}$. The 3rd possibility, $\vec{A} = \{1/2zB_y, 0, -1/2xB_y\}$ is merely a combination of the two basically different representations. The vector potential $\vec{A} = \{zB_y, 0, 0\}$ also yields the phase difference $\Delta\varphi_{21}$. We must note that with this representation of the vector potential $A_x = 0$ must be employed in the field-free interior of the superconductor. This follows from the requirement that $\int_c \vec{A}\, d\vec{l} = \Phi_A$. The paths parallel to the x axis then provide the contribution to the phase difference.

If this makes $\Phi(a) = \Phi_0 = h/2e$, then the magnetic field causes a phase difference between the two points 1″ and 2″ at $z = a$ which is 2π, that is a whole period, greater than the phase difference between points 1′ and 2′.

Let us now increase the voltage U_0 in the presence of the magnetic field (Fig. 37); the system will then so adjust its phases that it can conduct the maximum supercurrent corresponding to the particular magnetic field. When $\Phi(a) = h/2e$ it is not possible for a Josephson direct current to flow, since the phase over the height a of the tunnel junction will be altered by 2π independently of $\Delta\varphi_0$. This means that the integral of the current density across the junction is equal to zero, because according to Equation (3–22) the current density has equal components from right to left and from left to right. The distribution of the current density for this case is illustrated in Fig. 41b. The current density distributions for $\Phi(a) = 1/2\ \Phi_0$ and for $\Phi(a) = 3/2\ \Phi_0$ are illustrated in Figs. 41a and 41c. One can see from these three examples how the dependence illustrated in Fig. 38 comes about.

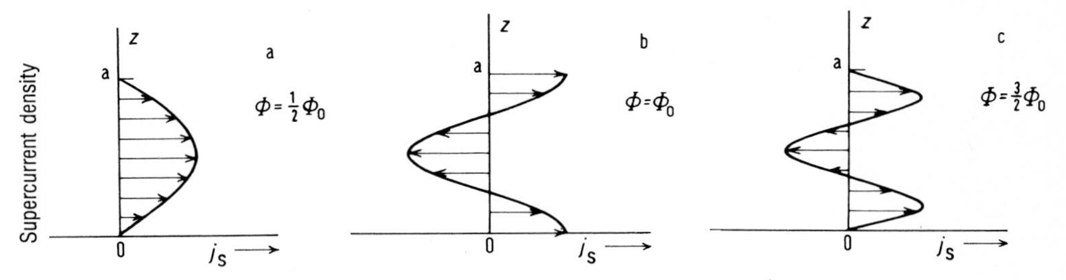

Fig. 41. Variation of the Josephson supercurrent along a junction with magnetic field.

An exact calculation yields the maximum supercurrent as a function of the magnetic flux through the tunnel junction.

$$I_{s\max}(B) = I_{s\max}(0)\,\frac{\sin\dfrac{\pi\Phi(a)}{\Phi_0}}{\dfrac{\pi\Phi(a)}{\Phi_0}}\,. \qquad (3\text{–}41)$$

Here $\Phi(a)$ is the magnetic flux produced in the whole tunnel unit by field $\vec{B}$ and Φ_0 is the elementary flux quantum ($\Phi_0 = h/2e = 2 \times 10^{-11}$ T cm^2).

Thus, the Josephson direct current is zero for those fields where $\Phi(a) = n\Phi_0$. In the example of Fig. 38 the zero points are separated by 1.25 gauss ($= 1.25 \times 10^{-4}$ T) along the $\vec{B}$ axis. This yields an effective area of the contacts A (in the nomenclature of Fig. 40: $A = 2\lambda a$) of ca. 1.6×10^{-7} cm^2 [1]. The great sensitivity to external fields could well be the reason that this supercurrent was only detected experimentally after it had been predicted theoretically by Josephson.

1 The depth of penetration λ in this case is ca. 10^{-6} cm.

Before we turn to a very interesting modification of this experiment, namely the behavior of a Josephson double junction, we ought to consider the formal analogy to the corresponding optical interference system. Let a parallel beam of light[1] pass through a very slightly wedge-shaped cuvette, so that we can arrange, say by altering the air pressure in the cuvette with respect to that outside, that a phase difference occurs between the waves at the thin and the thick end of the cuvette. Beyond the cuvette we focus the beam at the focal point of a lens, i.e. we produce an image of the light source. The intensity of this image now depends on the maximum phase difference that occurs on passage through the prism. If the phase difference is just 2π or a whole multiple of 2π the intensity of the image will be exactly zero. In the usual simple language we say that across the light beam two waves are always to be found with a phase difference of π and, thus, cancel each other out. As a function of the maximum phase difference a variation in intensity of the image occurs which is the square of the curve in Fig. 38. It is very fascinating to investigate how far the analogy between the Josephson direct current effects and the interference phenomena of light can be taken.

The analogies become particularly close for an arrangement of two Josephson tunnel junctions. Figure 42a illustrates such a double junction schematically and Fig. 42b gives an impression of how such a double junction can be realized. The area enclosed by the two superconductors is A. The effective areas a of the two Josephson junctions are very much smaller than A. The maximum Josephson direct current flows when there is no magnetic field through the device, if the two junctions are identical this is exactly double the maximum current of one junction. If we now apply a magnetic field $\vec{B}$ perpendicular to area A this magnetic field will first be screened by superconducting ring currents flowing through the Josephson junctions. But since these currents can only be very small on account of the insulating layers the magnetic field is able to penetrate to the interior of the ring and create a magnetic flux through A. The importance of the tunnel junction in this device is merely that it allows the penetration of the flux into the ring (Fig. 42a) even at very low fields. We can again ask what the phase difference is at two points 1 and 2 on each side of the barrier. Since the area A is very much greater than the effective area a of the junctions we can initially, at low magnetic field, ignore the phase changes taking place within the two junctions, i.e. our expression for the phase will only contain the magnetic flux Φ_A through the area A.

If we make the phase differences of the Cooper pair systems in the two superconductors in junction $\alpha(z = 0)$ equal to $\Delta\varphi_0$ and take account of the experiences with the single junction when the phase difference increased with increasing z like $\Phi(z)$ then we obtain in analogous manner to a single junction for β[2]:

$$\Delta\varphi_\beta = \Delta\varphi_0 + \frac{4\pi e}{h}\,\Phi_A\,. \tag{3-42}$$

We now calculate the total current by adding the two currents through the two junctions. To this we always make the initial assumption that the current density across an

1 The beam must fulfil the coherence conditions to produce good interference figures.
2 In the calculation the vector potential must be integrated along a path around the total area A.

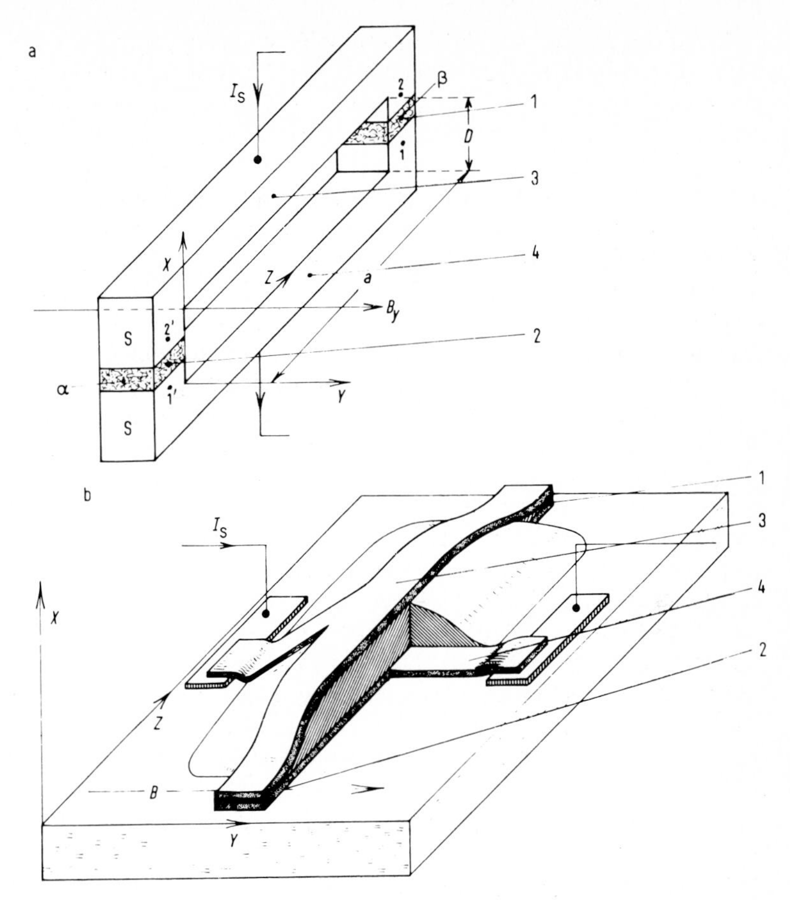

Fig. 42. Arrangement with two parallel Josephson junctions (a: schematic, b: possible realization). 1 and 2: insulating layers (thickness very much exaggerated) – 3 and 4: superconducting layers (thickness also very much exaggerated).

individual junction is constant, since the phase difference can be neglected on account of the small size of the area a. The currents at the two junctions α and β are:

$$\alpha\colon I_s = I_{smax} \sin \Delta\varphi_0 \tag{3-43}$$

$$\beta\colon I_s = I_{smax} \sin \left\{ \Delta\varphi_0 + \frac{2\pi\, 2e}{h}\, \Phi_A \right\}. \tag{3-44}$$

And the total current is thus:

$$I_{stot} = I_{smax} \left\{ \sin \Delta\varphi_0 + \sin \left(\Delta\varphi_0 + \frac{2\pi\, 2e}{h}\, \Phi_A \right) \right\}. \tag{3-45}$$

Since the phase difference $\Delta\varphi_0$ has not yet been fixed, we will substitute $\Delta\varphi_0$ by:

$$\Delta\varphi_0^* = \Delta\varphi_0 + \frac{\pi\, 2e}{h}\,\Phi_A\,. \tag{3-46}$$

We then obtain for the total current the simple expression:

$$I_{stot} = 2I_{smax}\sin\Delta\varphi_0^*\cos\pi\,\frac{2e}{h}\,\Phi_A\,. \tag{3-47}$$

If we now increase the voltage U_0 in the external circuit as for a single junction, then $\Delta\varphi_0{}^*$ will change in such a manner that the maximum supercurrent can flow under the conditions obtaining. But $\sin\Delta\varphi_0{}^*$ cannot exceed 1. We have, therefore, a periodic variation of I_{stot}. The maxima are at:

$$\Phi_A = n\,\frac{h}{2e} = n\,\Phi_0 \tag{3-48}$$

and the zero values at:

$$\Phi_A = (2n+1)\,\Phi_0/2\,. \tag{3-49}$$

If we now take account too of the changes in phase difference in each separate junction then the known current variation of a single junction is superimposed (3-41).
So that:

$$I_{stot} = 2I_{smax}\,\frac{\sin\pi\dfrac{\Phi_a}{\Phi_0}}{\pi\,\dfrac{\Phi_a}{\Phi_0}}\times\cos\frac{\pi\,\Phi_A}{\Phi_0}\,. \tag{3-50}$$

This dependence of the Josephson direct current through a double junction is completely equivalent to the interference structure produced by a light wave passing through a double slit.

Since the area A can be made very much greater than the effective area of a single junction the structure with respect to the magnetic field is very much finer. Figure 43 illustrates the variation of the Josephson current through a double junction as a function of the external magnetic field [61]. The length of the period is ca. 40 mG (= 4×10^{-8} T) here. This corresponds to an effective area A of ca. 5×10^{-4} cm^2 [1]. For an area of 1 cm^2 a change in the field of 2×10^7 Gauss is sufficient to create a flux quantum through S and, thus, for the structure to pass through one cycle. For this reason such dou-

1 Such almost ideal dependence of I_s on the magnetic field is only obtained when $LI_{smax}\ll\Phi_0$ (L = the coefficient of self-induction of the ring with the two junctions), i.e. when the screening of the ring is very small.

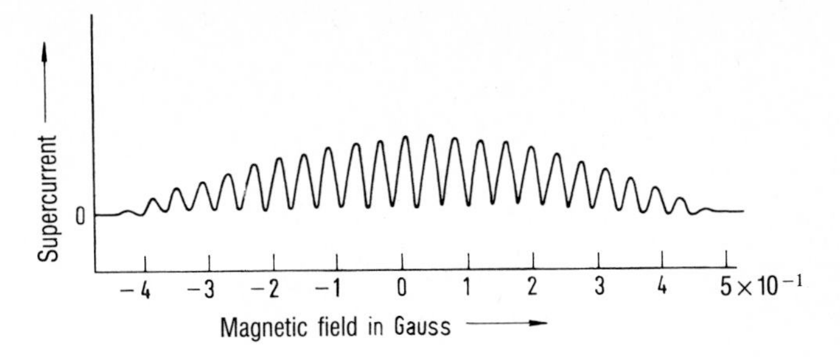

Fig. 43. Josephson current through a double junction as a function of the magnetic field (recorded curve) (1 G = 10^{-4} T). The maximum supercurrent corresponds to ca. 1 mA. The displacement of the origin was the result of an interfering field (after [61]).

ble junctions, which are also known as "quantum interferometers", are very sensitive instruments for the measurment of magnetic fields (see Section 9.5.4).

Multiple junctions can also be employed in a complete analogy to the interference phenomena with light waves. As the number of parallel connected junctions increases the maxima become sharper. This allows a further increase in the sensitivity to magnetic fields [62].

The experiments with two junctions have been carried out with areas up to $A \approx 3\ \text{cm}^2$ and interference phenomena have always been observed, as expected for coherent waves. This proves the strict phase correlation of Cooper pair waves over macroscopic distances.

3.4.2 The Josephson alternating current

The occurrence of a high frequency alternating current in a Josephson junction, to which a voltage $U_s \neq 0$ has been applied is also understandable from the basic equation (3–21), whose decisive importance has already been demonstrated in the Josephson direct current experiments. Equation (3–21) relates a supercurrent through the junction with the phase difference between the two Cooper pair systems left and right of the insulating layer. If there is now an electrical voltage U_s across the junction, this means an energy difference of $\Delta E = 2eU_s$ between the two Cooper pair systems. When a Cooper pair passes through the insulating layer from the negative to the positive side it can take up just this amount of energy as a result of the voltage.

However, according to the rules of quantum mechanics the energy difference between the two Cooper pair systems means a difference $\Delta\nu$ in the internal frequencies of the systems:

$$\Delta\nu = \frac{2e\,U_s}{h}\,. \tag{3–51}$$

When the two systems oscillate with different frequencies that stay constant with time then the phase between the two systems changes linearly with time. As can readily be demonstrated with two pendulums of different frequencies it follows that:

$$\Delta\varphi = 2\pi\,\Delta\nu\,t = 2\pi\,\frac{2e\,U_s}{h}\,t\,. \tag{3-52}$$

Here we have made the assumption that $\Delta\varphi$ equals zero at time $t = 0$. This assumption is not a restriction of any sort because only differences are involved. It must be specifically stated that Equation (3-52) is independent of the absolute energy levels of the two systems of Cooper pairs. It is only the difference in energy that is of importance.

If, however, we have a phase difference $\Delta\varphi$ that increases with time then it is immediately clear from Equation (3-21) that an alternating current must flow in the junction. This is governed by:

$$I_s = I_{smax}\,\sin\left(2\pi\,\frac{2e\,U_s}{h}\right)t\,. \tag{3-53}$$

Thus, the frequency of this alternating current is determined by the voltage across the junction. The importance of this for measurement technology should be stated immediately. Frequency determinations are today the most accurate determinations that can be made, so the Josephson alternating current offers the opportunity of measuring voltages (namely U_s) particularly accurately.

A potential difference of 1 mV at the junction produces a frequency of this alternating current of

$$\nu_J = \frac{2e\,U_s}{h} = 4.85 \times 10^{11}\ \mathrm{sec}^{-1}\,. \tag{3-54}$$

This frequency corresponds to that of electromagnetic radiation in vacuum with a wavelength of ca. 600 μm, that is a very long-wave infrared radiation.

The difficulty of direct detection lies less in the low power of this radiation than in the fact that it is difficult to couple out the high frequency power from the tiny tunnel junction to a corresponding high frequency conductor. The first confirmation of the Josephson alternating current was, therefore, made indirectly [63]. If such a junction is introduced into the high frequency alternating field of an oscillating microwave cavity, then equidistant steps of the current-voltage characteristic curve are observed. Their separation on the axis ΔU_s corresponds to the relationship:

$$\frac{2e\,\Delta U_s}{h} = \nu_M \tag{3-55}$$

where ν_M is the frequency of the high frequency field. These steps result from the superposition of the Josephson alternating current and the microwave field. Whenever the frequency of the Josephson alternating current is an integer multiple of the microwave frequency the superposition yields an additional Josephson direct current, which causes the characteristic to acquire a staircase structure.

Another indirect confirmation of the existence of a Josephson alternating current was found for junctions in a small static magnetic field. Here, without the use of external radiation, equidistant steps are observed in the characteristic at low voltages U_s [64]. These steps correspond to the resonances of the junction. The tunnel arrangement corresponds to a cavity. At suitable voltages U_s and a suitable field $\vec{B}$ the current density variations of the Josephson alternating current fit exactly with the vibrational modes of the junction. The current is particularly high when this resonance occurs.

A first, more direct demonstration of the Josephson alternating current was achieved in 1965 by Giaever [65]. As already mentioned the main difficulty of direct detection, say with normal radio frequency instrumentation, lies in the coupling out of the power from the tiny tunnel junction. Giaever assumed that a second tunnel device, immediately attached to the Josephson junction, would provide particularly favorable coupling (Fig. 44).

Here the demonstration of the output took place at the second tunnel junction via the change in the characteristic of the one-electron tunneling current by means of an applied high frequency field, namely that high frequency field generated by the Josephson junction. It had been demonstrated several years previously that a high frequency field produced a structure in the characteristic of the one-electron tunneling current [66]. This effect of a high frequency field is readily understandable on the basis of the discussion in Section 3.3.3, e.g. Fig. 31. The electrons are able to interact with the high frequency field by taking up or emitting the vibrational quanta of the field with energy $E = h\nu_M$. This means that a photon assisted tunneling process can begin at a voltage of $U_s = \Delta_I + \Delta_{II} - h\nu_M$. If an electron takes up several photons in the tunneling process, which can occur at high photon densities – or put another way – at high intensities of the high frequency field, then a structure appears in the current voltage characteristic with the characteristic interval of the voltage U_s:

$$\Delta U_s = \frac{h\,\nu_M}{e} \tag{3-56}$$

where h is Planck's constant, ν_M is the frequency of the high frequency field, e is the elementary charge. Note that here e, the elementary charge of a single electron, appears.

A typical one-electron characteristic is obtained for the tunnel junction at junction 2–3 when it is measured in the usual way without a voltage at junction 1–2 (Curve 1 in

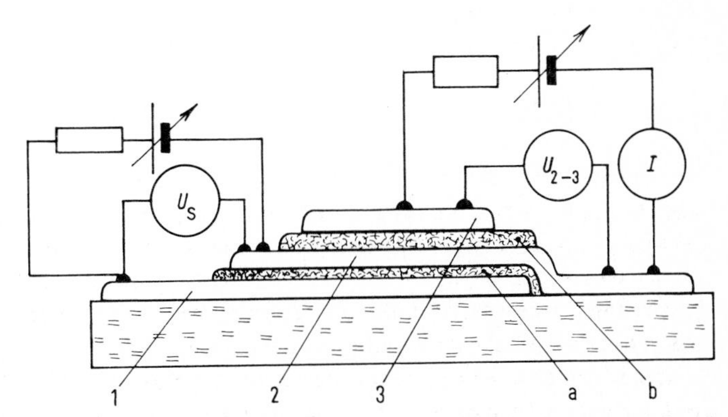

Fig. 44. Arrangement for detection of the Josephson alternating current according the Giaever. 1, 2 and 3 are Sn layers, *a* and *b* are oxide layers. The thicknesses of *a* and *b* are chosen so that 1 and 2 form a Josephson junction and so that no Josephson currents are possible between 2 and 3 (after [65]).

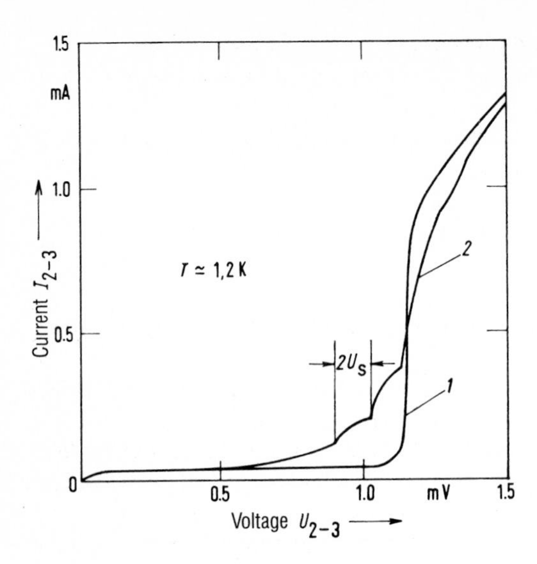

Fig. 45. Characteristic of the junction 2–3 in Fig. 44. Curve 1 without applied voltage at junction 1–2. Curve 2 with 0.055 mV across junction 1–2.

Fig. 45). The decisive experiment consists in applying a small potential U_s to Josephson junction 1–2. If the expected high frequency current is then produced at this junction then because of the relatively good coupling with junction 2–3 it should produce the structure in the tunneling characteristics. Giaever was able to observe the expected effects. Curve 2 in Fig. 45 reproduces such a characteristic. Here a potential U_s of 0.055 mV was applied to junction 1–2, the generator of the high frequency field. The frequency of the Josephson alternating current is then:

$$\nu_J = \frac{2e\,U_s}{h} \tag{3-57}$$

and the structure in the characteristic of junction 2–3 should be at voltage intervals of

$$\Delta U_{2\text{-}3} = \frac{h\,\nu_J}{e} = 2\,U_s\,. \tag{3-58}$$

The curve illustrated in Fig. 45 yields $\Delta U_{23} = 0.11$ mV which was observed by Giaever.

An American and a Russian group have succeeded in demonstrating the Josephson alternating current by coupling out the power in a high frequency conductor. The Americans [67] were able to detect the high frequency output of a Josephson junction by placing the tunnel junction in a tuned cavity resonator and working in the resonance mode of the junction by suitable choice of magnetic field. It was still necessary to reach extremely high sensitivity. An output of ca. 10^{-11} watts was detected, with a detection sensitivity of 10^{-16} watts (!). In a review article in "Scientific American" [68] the scientists D.N. Langenberg, Douglass J. Scalapino and Barry N. Taylor illustrated this detection sensitivity as follows. The detectable output corresponded approximately to the light falling on the human eye from a 100 watt bulb at a distance of ca. 500 km. This investigation was a magnificent example of experimental performance. The Russian group I.K. Yanson, V.M. Svistunov and J.M. Dmitrenko [69] were able to detect a power output of ca. 10^{-13} watt from a Josephson junction.

In all cases the relationship between the frequency of the Josephson alternating current and the voltage at the junction was found to be

$$\nu_J = \frac{2e\,U_s}{h}\,. \tag{3-59}$$

This is another convincing demonstration of the importance of electron pairs in superconductivity.

The measurement accuracy was increased so much by the American group that they were able to make an excellent precision determination of $2e/h$ [70].

All the experiments described in Sections 3.2, 3.3 and 3.4 were carried out after 1960. They have confirmed the basic premises of the microscopic theory in an impressive manner. Both the importance of the correlation into electron pairs and the macroscopic occupation of one quantum mechanical state by the electron pairs, the Cooper pairs, are, thereby, confirmed.

It has also proved possible to demonstrate Josephson effects in the new superconductors. Here the contacts between the grains of the sintered oxides have frequently been employed as tunnel barriers or better as sites of ”weak linkage“ [71]. Individual contacts between two grains of the superconducting phase with high T_c have also been studied [72].

4. Thermodynamics and the thermal properties of superconducting states

In Chapters 2 and 3 we have attempted to present the fundamental ideas of the microscopic theory of superconductivity and to demonstrate their correctness with experiments. The historical development occurred by a quite different route. F. and H. London (1935) [73] developed a phenomenological description of superconductivity long before the proposal of an adequate microscopic theory. Even earlier in 1924 Keeson [74] had attempted to apply the very general laws of thermodynamics [1] to superconductivity. The difficulty here was that it was not yet possible to regard the superconducting state as *one* new thermodynamic phase. In those days practically the only property of superconductivity known was the infinitely good electrical conductivity. For a material with the electrical resistivity zero, however, there are different states for the same temperature $T < T_c$ and the same external field $\vec{B}$ depending on how the material is brought into the field. If the field is applied when $T < T_c$ a persistent current will be induced which shields the interior of the sample from the magnetic field. If, however, the field is applied at $T > T_c$ and the sample is cooled to $T < T_c$ in the field then, so long as we only know the property $R = 0$, there is no reason for shielding of any sort.

So one could bring a substance with only the property $R = 0$ into quite different states for the same external variables T and $\vec{B}$ depending on the previous history. This meant that there was not *one* well-defined superconducting phase when the variables were fixed but a continuous multiplicity of superconducting phases with any desired shielding current. The existence of such a multiplicity of superconducting states was so incomprehensible that people spoke even before 1933 – that is without any experimental evidence – of *one* superconducting state [75].

The discovery of the Meissner-Ochsenfeld effect in 1933 brought about clarification. Meissner and Ochsenfeld [76] discovered that when superconductivity sets in a magnetic field is excluded from the interior of the superconductor no matter what the experimental arrangement. That is if a sample is cooled in the presence of an external field this field is excluded from the interior of the sample with the onset of superconductivity even though here no induction process can be made attributable for setting up the persistent current. This was the experimental confirmation of the existence of *one* superconducting phase. It is not possible to derive this behavior from the property $R = 0$, for it is a new property of the superconducting state. It will be discussed in detail in Chapter 5 [2]. We will first begin with some remarks concerning the thermodynamic treatment of physical systems.

---

1 Thermodynamic statements are valid for all properties of a physical system. The combination of thermodynamics and thermal properties appeared appropriate but in no way means that the thermodynamic description applies preferentially to the thermal properties.

2 In the case of type 2 superconductors (see Section 5.2), where the field penetrates the superconductor, the thermodynamic equilibrium state is also unequivocally determined by the values of T and $\vec{B}$.

Introductory remarks

The decisive characteristic of the thermodynamic treatment of a macroscopic physical system lies in the reduction of the gigantic number of independent coordinates of the individual particles to a few macroscopic variables describing the system. Thus, the thermodynamic treatment of a perfect gas does not employ the $3N$ spacial and $3N$ momentum coordinates of the N atoms that make up the gas, but describes the behavior in terms of variables such as temperature T, volume V, number of particles N and so on.

Questions are put concerning the macroscopic behavior of the system, such as about the stability of individual phases, such as solid, liquid and gaseous, when one variable is altered and the rest maintained constant. The thermodynamic equilibrium states under the chosen conditions play an important role. This takes place somewhat as follows. A system, e.g. a liquid-vapor mixture, is defined by setting its temperature T, volume V and total number of particles N. We wish to know the number of atoms N_1 in the vapor phase in thermodynamic equilibrium, i.e. in the state which is set up under the given conditions when all the other quantities are free to exchange. There is then, in the example given, a particular state that the system takes up. To do this the particles must be able to pass freely between the liquid and the vapor phases and, in addition, there must be heat exchange possible with a heat reservoir.

It should be mentioned here that thermodynamic equilibrium is very often not or only very slowly achieved, because the free exchange of one or other of the quantities is not possible under the particular experimental conditions. The question of thermodynamic equilibrium is completely independent of the possibility of realizing it.

All such questions are encompassed by specifying a suitable thermodynamic function, the Gibbs function, also known as the thermodynamic potential. These Gibbs functions are made up from the variables in such a way that a particular set of independent variables form a particular Gibbs function. If the Gibbs function is known then the system is completely determined thermodynamically.

The difficulty lies in choosing the correct Gibbs function for a system. It is first important to find a set of independent variables. This is not always simple. Known sets of variables are e.g.: temperature T, volume V and number of particles N, or temperature T, pressure p and number of particles N. Other variables are naturally included when the system has the opportunity to vary in other ways, e.g. when affected by a magnetic or electrical field. The behavior in a magnetic field, in particular, is of decisive importance for the treatment of a superconductor.

If the Gibbs function for a set of variables is known then the equilibria are laid down by the extremum of the Gibbs function [1]. Two phases of a system are in equilibrium if their Gibbs functions have the same value. This means that, in principle, we can answer the question concerning the stability of a phase. If the equilibrium state is defined by a minimum of the Gibbs function, then phase I will be unstable with respect to phase II if the Gibbs function of phase I is greater than that of phase II.

---

1 The extremum for the entropy S as a Gibbs function is very well-known. Systems under conditions where the entropy is Gibbs function develop in the direction of increasing entropy. The equilibrium state corresponds to a state of maximal entropy, always, of course, under the given conditions. The extremum requirements for equilibrium states for the other Gibbs functions are always attributable to these experiences laid down by the 2nd law of thermodynamics.

The Gibbs functions are characterized by the fact that the differentials of the independent variables appear in their differentials. Here are some well-known examples by way of illustration.

The internal energy U is given by:

$$dU = TdS + \delta A \,. \tag{4-1}$$

If only the work of compression $\delta A^{V} = -pdV$ is considered then:

$$dU = TdS - pdV \,. \tag{4-2}$$

where U is the Gibbs function for the variables S and V. The negative sign of pdV is necessary because we wish to define that all energy which is *given* to a system shall be taken as positive. When the volume is reduced, that is $dV < 0$, work is carried out on the system[1]. Other variables can naturally be involved too, e.g. the number of particles N. In all the examples that follow we will keep the number of particles constant, so that we can ignore this variable.

The free energy F is given by

$$F = U - TS \tag{4-3}$$

and, hence,

$$dF = dU - TdS - SdT = -SdT - pdV \,. \tag{4-4}$$

Thus, F is the Gibbs function for the variables T and V.

It is frequently easier to alter the pressure p at will than the volume V and, thus, use p as the independent variable. The Gibbs function for the variables T and p is the free enthalpy G where:

$$G = U - TS + pV \tag{4-5}$$

and:

$$dG = -SdT + Vdp \,. \tag{4-6}$$

We will use this free enthalpy in the following consideration of the superconducting phase, since T and p are very practical variables. They are relatively easy to choose at will.

1 Another convention is usually employed in applied thermodynamics. Here the work done by the system is taken as positive being the technologically interesting work contribution.

4.1 The stability of the superconducting state

Already in Section 2.2 (page 37) we demonstrated that the superconducting state must be unstable at high currents. When the kinetic energy of the Cooper pairs is so great that it is sufficient to overcome the energy of binding, then the process of pair breaking occurs. This means that a critical current density j_c may not be exceeded without destroying the superconductivity.

From the existence of a critical current density it follows directly that there must also be a critical magnetic field $\vec{B}_c$.[1] For if we apply a magnetic field $\vec{B}$ to a superconductor this will begin to carry a persistent current. As the magnetic field is increased this current will also increase. When the critical current density is reached the superconducting state must become unstable. The details of the behavior of a superconductor in a magnetic field are dealt with in Chapter 5. The decisive fact here is that an external magnetic field $\vec{B}$ can destroy the superconductivity on exceeding a critical value.

In the thermodynamic treatment of superconducting phases[2] we must, therefore, include a variable that takes account of the behavior in the magnetic field. We will choose the magnetic field $\vec{B}$ as a further independent variable. The Gibbs function associated with T, p and $\vec{B}$ is[3]:

$$G = U - TS + pV - \vec{m}\vec{B} \tag{4-7}$$

where $\vec{m}$ is the magnetic moment of the superconductor[4]. Since $\vec{m}$ and $\vec{B}$ are always parallel or antiparallel and we wish to take account of this in the sign of $\vec{m}$ we can disregard the vector character in this treatment.

Since for the variation of the internal energy U we have[5]:

$$\mathrm{d}U = T\mathrm{d}S - p\mathrm{d}V + B\mathrm{d}m \,. \tag{4-8}$$

(See, for example, Becker: "Theorie der Wärme", p. 7, Springer, Heidelberg (1961), then:

$$\mathrm{d}G = -S\mathrm{d}T + V\mathrm{d}p - m\mathrm{d}B \,. \tag{4-9}$$

---

1 We will use the quantity $\vec{B}$ throughout as the quantity describing the size of the magnetic field and call it, for simplicity, the "magnetic field" and not the "magnetic flux density". Since all magnetic fields occurring here, those in the superconductor too, are generated by macroscopic currents it is unnecessary to distinguish between $\vec{H}$ and $\vec{B}$.

2 The terms "superconducting state" and "superconducting phase" are used here with the same meaning.

3 We remind the reader that the number of particles is maintained constant.

4 Since here the magnetic moment $\vec{m}$ always appears in association with $\vec{B}$ there is no danger of confusing it with the mass m.

5 The differential work $\delta A^{\mathrm{m}} = B\mathrm{d}m$ corresponds to the work of compression $\delta A^{\mathrm{V}} = -p\mathrm{d}V$, where B stands for pressure p and the magnetic moment m for the negative volume $-V$. If we consider a paramagnetic substance, for example, then $\vec{m}$ is parallel to $\vec{B}$. If $\vec{m}$ is increased, i.e. $\mathrm{d}\vec{m} > 0$ work is done on the system, that becomes apparent as heat of magnetization.

As required, the independent variables T, p and B appear in the differential in our choice of Gibbs function $G(T,p,B)$.

We, therefore, have the starting point for a thermodynamic treatment of the superconducting phase. First let us consider the superconducting phase without an external field. The term mB then disappears. We also wish to maintain the superconductor under constant pressure. Only the temperature is to be altered. Experiment shows us that the superconducting phase is stable at temperatures below T_c. Thus, at $T < T_c$ the Gibbs function G_s of the superconductor must be smaller than G_n, the Gibbs function of the normal conductor. When $T = T_c$ then $G_s = G_n$.

Since the density of Cooper pairs and the energy gap 2Δ increase with decreasing temperature, we would expect that the difference between the Gibbs functions $G_n - G_s$ would also increase with decreasing temperature. This difference is a direct measure of the stability of the superconducting state.

We can determine $G_n - G_s$ as a function of the temperature. To do this we exploit the fact that an external magnetic field of sufficient strength makes the superconducting phase unstable. The reason for this is an increase in G_s with increasing field B, which causes G_s to rise above a critical value larger than G_n and, thus, become unstable. The Gibbs function G_n for the normal-conducting phase is practically independent of the magnetic field, since the magnetic moments produced in the normal conductor are very small[1].

In order to determine $G_n - G_s$ as a function of T it is, therefore, necessary to determine the critical field B_c at various temperatures. The term $\int_0^{B_c} m\,\mathrm{d}B$ then yields $G_n - G_s$ at the particular temperature. For:

$$
\begin{aligned}
G_s(B) - G_s(0) &= -\int_0^B m\,\mathrm{d}B \\
G_n(B) - G_n(0) &= 0 \\
G_n(B_c) - G_s(B_c) &= 0\,.
\end{aligned}
\tag{4-10}
$$

The Gibbs functions G_s and G_n are exactly equal at B_c. From these three equations it follows that:

$$
G_n(T) - G_s(T) = -\int_0^{B_c(T)} m\,\mathrm{d}B\,. \tag{4-11}
$$

Since the magnetic moment of a superconductor in equilibrium is always antiparallel to B, it follows that it must be

$$
G_n(T) > G_s(T) \quad \text{for all} \quad T < T_c\,.
$$

---

1 This statement is no longer true if we consider substances which remain superconducting at very high fields ($B \approx 50$ T $= 500$ kG).

And now to obtain somewhat more quantitive information concerning the difference between the Gibbs functions of superconducting and normalconducting phases we must consider the magnetic moment m as a function of B. Since our experiment only served to determine the expression $G_n - G_s$ we can choose the simplest experimental conditions without any loss of generality. These are:

a) homogeneous field B

b) long, thin wire parallel to the field.

This shape of sample and the field geometry give very simple relationships because the homogeneous field is practically not affected by the appearance of a magnetic moment in the wire and the magnetic moment m in this case is simply related to the homogeneous magnetization M by

$$m = MV \tag{4-12}$$

(V = volume of sample in the state considered). Under these conditions we have[1]:

$$G_n - G_s = -V_s \int_0^{B_c} M \mathrm{d}B . \tag{4-13}$$

For further conclusions it is necessary to know M as a function of B. Here, as will be demonstrated in Chapter 5, it is necessary to distinguish between various types of superconductor. However, for the sake of thermodynamic simplicity we will limit ourselves to superconductors of the 1st type, in order not to lose track in a maze of equations. The treatment of superconductors of the 2nd type is basically the same.

In a superconductor of the first type the magnetic field is, apart from a very thin screening layer with a thickness of only a few times 10^{-6} cm, excluded from the interior of the superconductor (Meissner-Ochsenfeld effect). Macroscopic samples of such superconductors behave externally as ideal diamagnets with a susceptibility of $\chi = -1$[2]. The magnetization M is to a good approximation:

$$M = \chi \frac{B}{\mu_0} = -\frac{B}{\mu_0} . \tag{4-14}$$

(μ_0 is a constant of the system of units used here, $\mu_0 = 4\pi \times 10^{-7}$ Vs/Am)

This yields a very simple expression for $G_n - G_s$, namely:

$$G_n - G_s = \frac{V_s}{\mu_0} \int_0^{B_c} B \mathrm{d}B = V_s \frac{B_c^2}{2\mu_0} . \tag{4-15}$$

---

1 This equation is only approximately valid. In fact V_s is also a function of T, p and B and should be left under the integral. Since the change of V_s with B is only very small, Equation (4-13) is a very good approximation.

2 We know that this behavior is the result of macroscopic persistent currents flowing on the surface. For an observer outside the superconductor it is of no consequence whether this diamagnetism results from surface currents or atomic currents.

We are now in the position of being able to quantify $G_n - G_s$, when we have determined the critical field B_c as a function of T in our simple system.

Figure 46 reproduces the critical field B_c as a function of T for some superconductors of type I. These results are represented to a very good approximation by the expression:

$$B_c(T) = B_c(0)\,(1 - (T/T_c)^2)\,. \tag{4-16}$$

We will now investigate the conclusions that can be made from this temperature dependence concerning the other variables of the superconductor [1]. Here we will exploit the decisive advantage of the thermodynamic approach, which is, that all important quantities can be obtained as a result of simple differentiation of the Gibbs function.

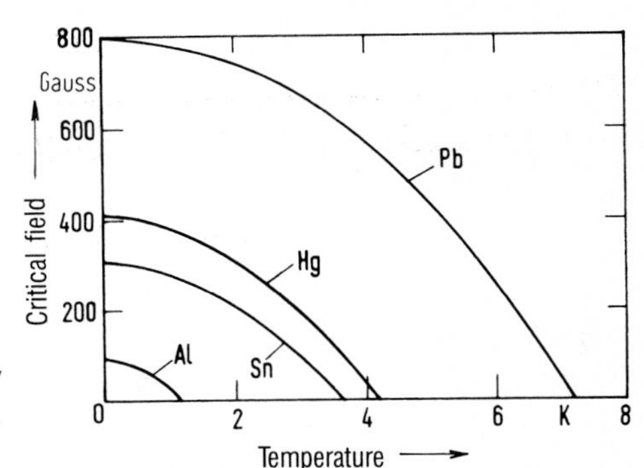

Fig. 46. Critical magnetic field as a function of T for some type I superconductors (1 G = 10^{-4} T).

Thus, the entropy S is obtained from G via

$$S = -\left(\frac{\partial G}{\partial T}\right)_{B,p} \tag{4-17}$$

which becomes without magnetic field:

$$S_n - S_s = -V\frac{B_c}{\mu_0}\frac{\partial B_c}{\partial T}\,. \tag{4-18}$$

Here, since it is so small, we have neglected the variation of V with T. We will discuss this dependence of V separately in Section 4.3. The justification for our approximation will also become clear there.

Let us now consider our temperature dependence of the critical field B_c, as illustrated in Fig. 46 and well approximated in Equation (4-16); we can immediately write down the general dependence of $S_n - S_s$.

---

1 It is possible to set up a general relationship in the form

$$B_c(T) = B_c(0)\,(1 + \sum_{n=2}^{N} a_n\,(T/T_c)^n)\,,$$

that is as a polynomial of the Nth degree in T/T_c.

When $T \to T_c$ then $B_c \to 0$. So $S_n - S_s$ also tend towards zero at $T \to T_c$. The entropies of the normal and superconducting phases are also equal at T_c (without magnetic field). Since according to the 2nd law of thermodynamics the entropy change ΔS in a reversible process is proportional to the heat involved δQ we have learnt an important consequence: There is no heat of transition at T_c in the absence of a magnetic field. The transition here is one of 2nd or higher order. This result forms the basis of the phenomenological theory of Ginsburg and Landau which we will meet in the treatment of superconductors of the type II. This theory, which starts from the assumption that the transition from the normal to the superconducting state at T_c is a 2nd order transition, achieved great importance in past years particularly in connection with the application of superconductivity.

The entropy difference $S_n - S_s$ is positive at temperatures $T - T_c$, for $dB_c/dT < 0$ over the whole range $0 < T < T_c$. According to the third law of thermodynamics $S_n - S_s$ must also approach zero as $T \to 0$[1]. But this means that $S_n - S_s$ must go through a maximum. This fact will be of importance for the behavior of the specific heat.

Another important conclusion is that there is a finite entropy difference $S_n - S_s$ for all temperatures in the range $0 < T < T_c$, i.e. the transitions at these temperatures are first order with a finite heat of transition.

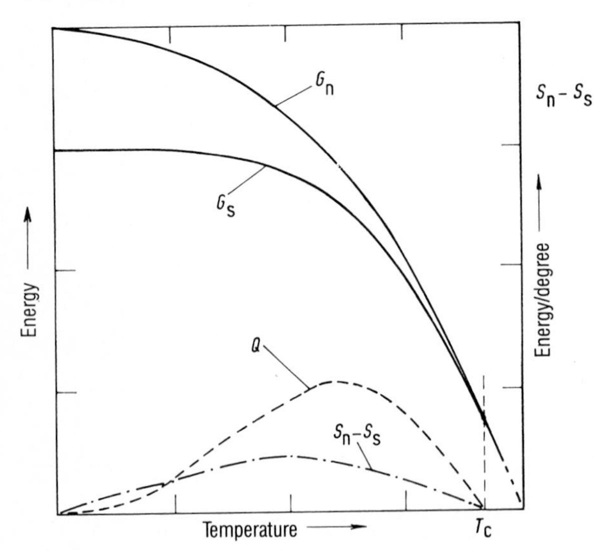

Fig. 47. Gibbs functions G_n and G_s, entropy difference and heat of transition as a function of temperature. Figures for Sn: $T_c = 3.72$ K; $(G_n - G_s)_{T=0} = 5 \times 10^{-3}$ Ws; $(S_n - S_s)_{max} = 2.28 \times 10^{-3}$ Ws/K; $Q_{max} = 5 \times 10^{-3}$ Ws.

Figure 47 reproduces the temperature dependence of G_n, G_s and $S_n - S_s$ with arbitrary units for the ordinate. Both of the curves $G_n(T)$ and $G_s(T)$ must not only have the same value at $T = T_c$ but also, as we have seen, the same slope. That is the characteristic of a phase transition of greater than 1st order. But as we will see directly the transition to superconductivity is a 2nd order phase transition, since at T_c the 2nd differentials of the Gibbs functions G_n and G_s are different. L.D. Landau[2] has produced a sound theory for the treatment of 2nd order phase transitions [77].

1 Our Equation (4-16) fulfils this condition.

2 L.D. Landau, Russian physicist, 22.10.1908–2.4.1968, Nobel prize 1962.

The results obtained here from thermodynamics are in very good agreement – as must be demanded – with our microscopic picture. It follows from $S_n - S_s$ that the superconducting phase has a lower entropy than the normal phase. We can remind ourselves, without going into details, that the entropy is a measure of the "disorder" of a physical system, so it follows from $S_n > S_s$ that the degree of ordering of the superconducting phase is greater than that of the normal phase. This greater degree of ordering can readily be seen in the correlation of single electrons into Cooper pairs and of the Cooper pairs with each other. This correlation means extra order in our system.

When $T \to T_c$ the density of Cooper pairs and the energy gap fall continuously towards zero. We cannot, therefore, expect there to be a heat of transition at T_c. Rather our microscopic picture reveals that we must expect a transition of second or higher order. On the other hand, we have below T_c a finite density of Cooper pairs. A persistent current is set up when a magnetic field is applied, but the number of Cooper pairs remains practically constant until the critical field is reached[1)]. The superconductivity then collapses with the destruction of all Cooper pairs. This requires a finite energy of transition. Heat, namely $(S_n - S_s)T$ is consumed on the transition from the superconducting to the normally conducting state, i.e. it must be supplied to the system if isothermal conditions are to be maintained. If, however, the S → N transition is carried out without heat exchange ($\delta Q = 0$), i.e. adiabatically, then the sample will be cooled since heat will be withdrawn from the remaining degrees of freedom. We have in the superconductor a substance which can be cooled by adiabatic magnetization [78]. Since substantially more effective cooling methods have now been developed this possibility has no special practical importance.

A second differentiation of the Gibbs function with respect to T yields the specific heat, here at constant pressure and constant B. The questions concerning the specific heat will be discussed in the next section.

4.2 The specific heat

The specific heat c is quite generally defined by the relationship:

$$\Delta Q = c\, m' \, \Delta T \, . \tag{4-19}$$

ΔQ is the heat, required to raise a mass of substance m' by a temperature ΔT. We distinguish between various types of specific heats depending on the conditions under which the heat is supplied. For instance, if we maintain the pressure constant we obtain the specific heat c_p.

Since for our considerations we employ the Gibbs function for p as an independent variable, we can use this Gibbs function very directly to obtain the specific heat c_p. For:

$$-T(\partial^2 G/\partial T^2)_{p,B} = T(\partial S/\partial T)_{p,B} = c_p \, . \tag{4-20}$$

1 This statement applies to superconductors of type I.

So we get from

$$G_n - G_s = V\frac{B_c^2}{2\mu_0} \tag{4-15}$$

for $c_n - c_s$

$$c_n - c_s = -\frac{VT}{\mu_0}\left\{\left(\frac{\partial B_c}{\partial T}\right)^2 + B_c\frac{\partial^2 B_c}{\partial T^2}\right\} \tag{4-21}$$

or [1]

$$c_s - c_n = \frac{VT}{\mu_0}\left\{\left(\frac{\partial B_c}{\partial T}\right)^2 + B_c\frac{\partial^2 B_c}{\partial T^2}\right\}. \tag{4-22}$$

We have again regarded the volume V as being constant. It should be noted here that we have used the specific volume, that is the volume per gram, and that the specific heat is in terms of the gram as amount of substance.

It can be seen from Equation (4-22) that when $T = T_c$ $c_s > c_n$. The critical field $B_c = 0$ at T_c that is $c_s - c_n > 0$. The specific heat makes a jump at T_c which is given by:

$$(c_s - c_n)_{T=T_c} = \frac{VT_c}{\mu_0}\left(\frac{\partial B_c}{\partial T}\right)^2_{T=T_c}. \tag{4-23}$$

This important relationship is known in the literature as "Rutgers formula" [79]. It relates a thermal quantity, namely the jump in specific heat, with the critical magnetic field. This relationship is fulfilled very well by a range of superconductors. Table 6 lists some values for $(c_s - c_n)_{T=T_c}$, obtained from caloric determinations and from determinations of the critical field.

Since $\mathrm{d}^2B_c/\mathrm{d}T^2 < 0$ and $\mathrm{d}B_c/\mathrm{d}T$ becomes ever smaller as the temperature is reduced we have a temperature $0 < T < T_c$, where $c_s = c_n$. The point of intersection of $c_s(T)$ and $c_n(T)$ must be at the temperature where $S_n - S_s$ is maximal (Fig. 47).

Figure 48 represents the specific heat of tin as a function of temperature [80]. The solid curve was observed without external magnetic field. At fields higher than the critical field $B > B_c$ it is possible to determine the specific heat c_n of the normalconducting state even when $T < T_c$ (broken line in Fig. 48).

This specific heat of a normal conductor can be divided into two parts, namely the contribution of the conducting electrons c_{nE} and the contribution of the lattice vibrations c_{nG}.

---

1 This result was derived by Keesom [74] as early as 1924, but without any sure basis for the existence of a superconducting phase in the thermodynamic sense.

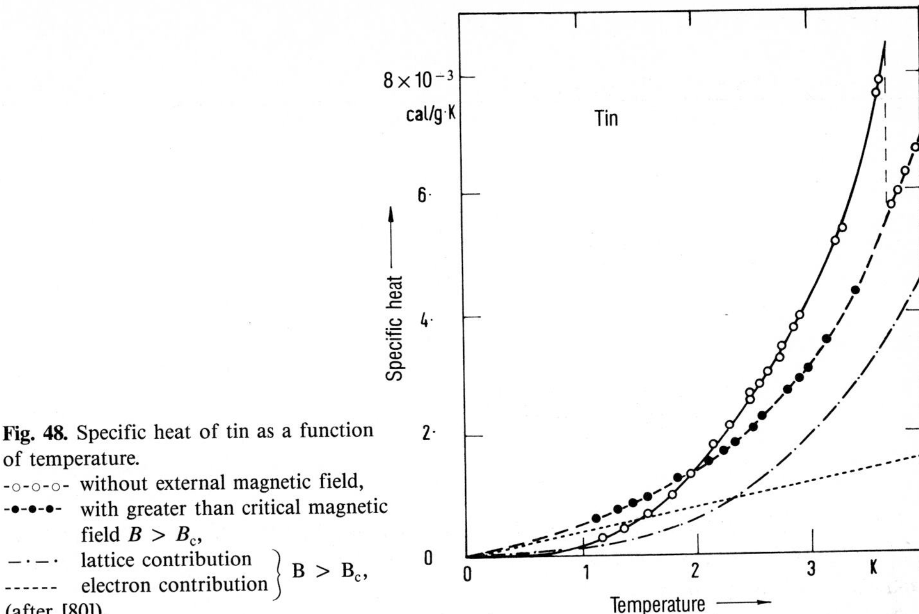

Fig. 48. Specific heat of tin as a function of temperature.
-o-o-o- without external magnetic field,
-•-•-•- with greater than critical magnetic field $B > B_c$,
— · — · lattice contribution } $B > B_c$,
------ electron contribution }
(after [80]).

Table 6. Values of $(c_s - c_n)T_c$ from calorimetric and magnetic data-test of the "Rutger" formula.

Element	T_c in K	$(c_s - c_n)$ calor. det. in 10^{-3} W · s (mol · K)	$(c_s - c_n)$ magnet. det. in 10^{-3} W · s (mol · K)
Sn [a]	3.72	10.6	10.6
In [a]	3.40	9.75	9.62
Tl [b]	2.39	6.2	6.15
Ta [a]	4.39	41.5	41.6
Pb [b]	7.2	52.6	41.8

a) Mapother, D. E.: IBM Journal *6*, 77 (1962)
b) Shoenberg, D.: "Superconductivity". Cambridge University Press 1952.

To a good approximation:

$$c_{nE} = \gamma T \tag{4-24}$$

$$c_{nG} = \alpha (T/\Theta_D)^3 \tag{4-25}$$

γ and α are constants [1], Θ_D = Debye temperature.

1 γ is generally known as the Sommerfeld constant of the electron system.

The constant γ is proportional to the density of states of the electrons at the Fermi energy (see Section 2.1). In general:

$$\gamma = \frac{2}{3} \pi^2 k_B^2 N(E_F) \tag{4-26}$$

where k_B = Boltzmann's constant, value: 1.38×10^{-23} W s/K, $N(E_F)$ = density of states in $(\mathrm{Ws})^{-1}\ \mathrm{mol}^{-1}$.

Since the density of states is a decisive variable for superconductivity the possibility of determining γ and, hence, $N(E_F)$ will be discussed here. Experience has shown that the contribution of the lattice to the specific heat remains practically unchanged on entering the superconducting state. We can, therefore, ascribe the difference $c_s - c_n$ completely to the electron system. Therefore:

$$c_{sE} - c_{nE} = \frac{VT}{\mu_0} \left\{ \left(\frac{\partial B_c}{\partial T} \right)^2 + B_c \frac{\partial^2 B_c}{\partial T^2} \right\}. \tag{4-27}$$

On the basis of Fig. 48 and in view of the higher order of the superconducting state we may assume that the specific heat of the electrons in the superconducting state approaches zero more rapidly as $T = 0$ is approached than is the case for the normal state. We express this by $c_{sE} \propto T^{1+a}$ with $a > 0$. This means, however, that c_{sE}/T approaches zero as $T \rightarrow 0$. So that for sufficiently small temperatures when c_{sE}/T can be neglected with respect to c_{nE} we get:

$$-\frac{c_{nE}}{T} = \frac{V}{\mu_0} \left\{ \left(\frac{\partial B_c}{\partial T} \right)^2 + B_c \frac{\partial^2 B_c}{\partial T^2} \right\}. \tag{4-28}$$

However, the expression c_{nE}/T is just γ. We also assume that $\left(\frac{\partial B_c}{\partial T} \right)^2 \ll B_c \frac{\partial^2 B_c}{\partial T^2}$ which is always true if the temperature is sufficiently low, so we get:

$$\gamma = -\frac{V}{\mu_0} B_c \frac{\partial^2 B_c}{\partial T^2} \tag{4-29}$$

and with:

$$B_c = B_c(0) \left(1 - \left(\frac{T}{T_c} \right)^2 \right) \tag{4-16}$$

then

$$\gamma = \frac{V}{\mu_0} 2 \frac{B_c^2(0)}{T_c^2}. \tag{4-30}$$

If for B_c the polynomial is employed instead of the simple quadratic temperature dependence then only the factor a_2 appears on the right-hand side (see Section 4.1).

The observation of B_c at sufficiently low temperatures is, hence, independent of the specific temperature dependence of the critical field and suitable for the determination of γ, thus providing information concerning $N(E_F)$.

The temperature dependence in the superconducting state is very well approximated by a parabola of the third order

$$c_s(T) \propto T^3 . \tag{4-31}$$

This dependence yields the parabolic law (4–16) for the temperature dependence of B_c. This is only mentioned as an illustration of the fact that the thermodynamic relationships connect the different variables very closely.

We will now turn our attention to what temperature dependence for the specific heat in the superconducting state can be deduced from our fundamental understanding of this state. At temperatures close to T_c the Cooper-pair density and the energy gap change with temperature. In this range we cannot expect to obtain simple relationships from our qualitative microscopic picture. At very low temperatures, on the other hand, the energy gap ought to be almost independent of T. The supply of energy to the electron system is then fundamentally related to the break-up of Cooper pairs[1]. This requires excitation over the energy gap. Since the probability of excitation should decrease according to an exponential function of type $\exp(-A/k_B T)$ (A is a constant, basically the energy of excitation), we would expect for our model, at low temperatures, a basically exponential fall in c_{sE}, the electronic part of the specific heat.

For $T \to 0$ the BCS theory predicts:

$$C_{sE} = 9.17\gamma T_c \exp\left(-\frac{1.5T_c}{T}\right) . \tag{4-32}$$

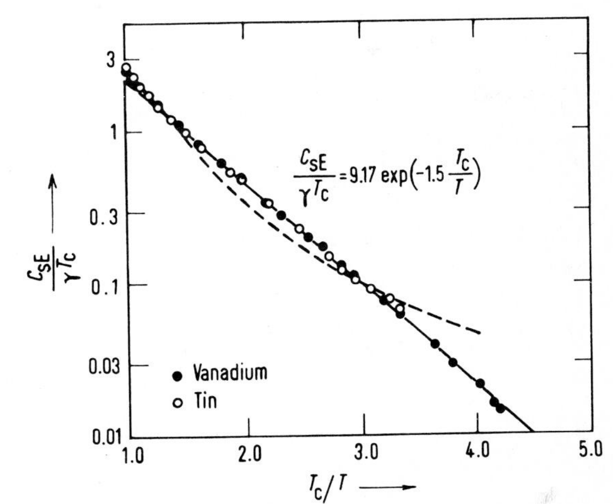

Fig. 49. Specific heat of tin and vanadium. The straight line that has been drawn in corresponds to the relationship predicted by the BCS theory (after [81]). The broken curve represents the T^3 law for c_s (4–31).

1 The density of the unpaired free electrons has then become so small that their energy uptake can be neglected.

Figure 49 illustrates an example of this exponential dependence. With regard to Equation (4–32) $C_{sE}/\gamma T_c$ is plotted against T_c/T. The straight line that has been drawn in corresponds to Equation (4–32). Even before the BCS theory had been proposed exponential temperature dependences had been observed at very low temperatures, thus confirming the assumption of an energy gap in the excitation spectrum by Daunt and Mendelssohn [50].

The exponential dependence of C_{sE} on temperature is a further very direct confirmation of the existence of an energy gap.

While resistance measurements are very unsuitable for providing evidence concerning the proportion of superconducting phase in a sample – the first continuous superconducting path for the current immediately short circuits the rest of the sample –, the heat capacity is determined by the whole mass of the superconducting phase. It was, therefore, important in the case of the new superconductors, which often occur as mixtures of phases, to determine the specific heat. The contribution $c_{nE} = \gamma T$ can be employed to determine the proportion of normally conducting phase(s).

The BCS theory predicts a value of 1.43 for the "jump" in specific heat $(c_{sE} - c_{nE})/c_{nE}$. It is difficult to determine this quantity in the case of the new superconductors, because the contribution of the lattice to the heat capacity is still large at the high transition temperatures [82, 83, 84].

4.3 The influence of pressure on the superconducting state

In the previous section we discussed the difference between the free enthalpies of normal and superconducting states at constant pressure as a function of temperature. To do this we made three assumptions in the quantitative discussion in order to simplify the formulae.

1. An external magnetic field is completely excluded from the superconducting material in the superconducting state – a perfect Meissner-Ochsenfeld effect is assumed. This assumption, which is very well fulfilled for macroscopic samples of type I superconductors in the whole range of their stability, gave us the relationship between the external field and the magnetization.

2. In order to derive the magnetization M from the magnetic moment m, we must know the volume occupied by the sample. Here we employed an approximation by ignoring the dependence of the volume on the external magnetic field. Experience shows that this dependence is very small.

3. Finally we took the temperature dependence of the critical field B_c from experiment and employed the very good analytical approximation of parabolic dependence $B_c(T) = B_c(0)(1 - (T/T_c)^2)$.

Constant pressure was assumed for all this. In this section we ask what is the influence of pressure on superconductivity. For instance, Sizoo and Onnes had already shown in 1925 that it had an effect on the transition temperature T_c [85]. Figure 50 illustrates the change in the transition temperature of tin [86] with increasing pressure in the absence of an external magnetic field. The transition temperature decreases as the pressure is increased. The effect is not very large. In the case of tin, for instance, a pressure of ca. 2000 bar [1)] is required to change T_c by 0.1 K.

The behavior illustrated in Fig. 50 is typical for many superconductors. But there are also some other substances (e.g. Ti, Zr, V, La, U etc.), whose transition temperatures are raised with increasing pressure [87, 88].

A change in T_c with pressure when the magnetic field is zero must be associated with a change in the critical field B_c with pressure. Figure 51 illustrates the effect of pressure on B_c for cadmium [89]. We have chosen the investigation of cadmium as an example because it is a typical example of excellence in the art of experimentation. The very small temperature differences involved and the simultaneous employment of high pressures place great strains on experimental technique.

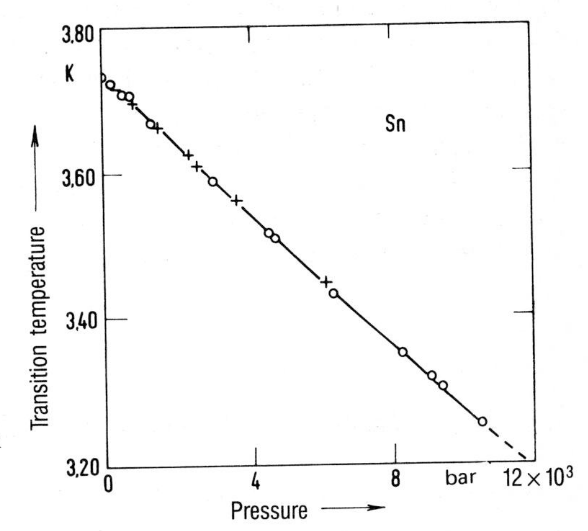

Fig. 50. Pressure dependence of the transition temperature of tin (after [86]).

The pressure was produced in what is called an "ice bomb" [90]. A steel container is filled completely with water. When it is cooled below the freezing point a pressure of ca. 1800 bar is produced on account of the volume being kept constant. When water freezes at constant pressure its volume increases by ca. 10% [2)]. If the liquid solid phase transition is forced to take place at constant volume then the ice must be formed under a pressure which is sufficient to effect a ca. 10% reduction in the volume of the ice. Many interesting results concerning behavior under pressure have been obtained by means of this ice bomb technique.

In the investigations of Cd the ice bomb was cooled to ca. 6×10^{-2} K by means of a magnetic cooling process, adiabatic demagnetization, and the critical field B_c measured down to this temperature. In the case of Cd the value of B_c decreases when the pressure is increased, as would be expected from the decrease in T_c with increasing pressure when $B = 0$.

A reduction in the critical field under pressure means that the difference $G_n - G_s$ becomes smaller with increasing pressure at constant B and T. If we take the volumes of

---

1 $1 \text{ bar} = 10^5 \text{ Pa} = 10^5 \text{ N/m}^2 \approx 1 \text{ kp/cm}^2$

2 Icebergs float with about 10% of their volume out of the water.

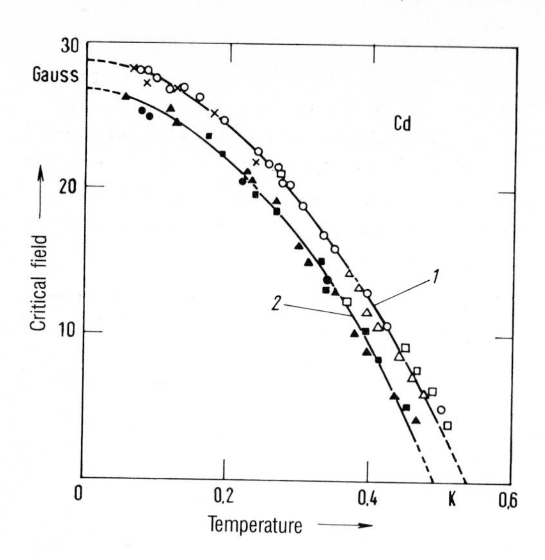

Fig. 51. Temperature dependence of the critical field (1 G = 10^{-4} T) of cadmium at normal pressure (curve 1) and at 1550 bar (curve 2 after [89]). The original curves contained more result points than are reproduced (after [91]).

the normal conductor V_n and of the superconductor V_s as functions of our independent variables T, p and B, then we can calculate the p dependence of $G_n - G_s$. However, the volume changes $\Delta V/V$ are so small, namely only a few times 10^{-8}. The experimental difficulties of measuring these changes with some certainty are considerable. The determinations of B_c under pressure are somewhat simpler. Figure 51 reproduces an example. These results can be used to determine the volume change on phase transition by use of the Gibbs function.

In general:

$$(\mathrm{d}G/\mathrm{d}p)_{T,B} = V\,. \tag{4-33}$$

On phase transition we get[1]:

$$(V_n - V_s)_{B=B_c} = V_s(B_c)\,\frac{\partial}{\partial p}\left(\frac{B_c^2}{2\mu_0}\right) = V_s(B_c)\,\frac{B_c}{\mu_0}\,\frac{\partial B_c}{\partial p}\,. \tag{4-34}$$

This is analogous to the expression that we derived in Section 4.1 for the entropy difference. We see from this expression that $V_n - V_s$ disappears at T_c, because B_c equals zero. Further we see that $V_s > V_n$ when $T < T_c$ and – as is usually the case – B_c falls with p that is when $\partial B_c/\partial p < 0$. A second differentiation with respect to p or T delivers the difference between the compressibilities $\varkappa_s$ and $\varkappa_n$ or the coefficients of thermal expansion α_s and α_n in the superconducting and normal states.

The volume changes on transition from the superconducting to the normal states have been determined using extremely sensitive measuring methods. Here it is primarily the

1 The volume change on transition from superconducting to normal state takes place at $B = B_c$ and is reached on the coexistence curve by a differential change in p. For this determination we do not need to know $\mathrm{d}V_s/\mathrm{d}p$ since this term is unimportant for the transition.

length of a rod or strip-shaped sample that has been measured. Figure 52 reproduces the results obtained for tin [91]. As expected $l_s > l_n$ for all temperatures $T < T_c$ where T_c is the transition temperature without applied magnetic field.

The sample expands on transition to the superconducting state. This is in agreement with the predictions of the Gibbs function where we assumed – and this should be stated clearly once again – an ideal exclusion of the magnetic field. It should be noted that the measurement of the change in length of a rod 10 cm long in the region of T_c requires a sensitivity of length measurement of ca. 10^{-8} cm.

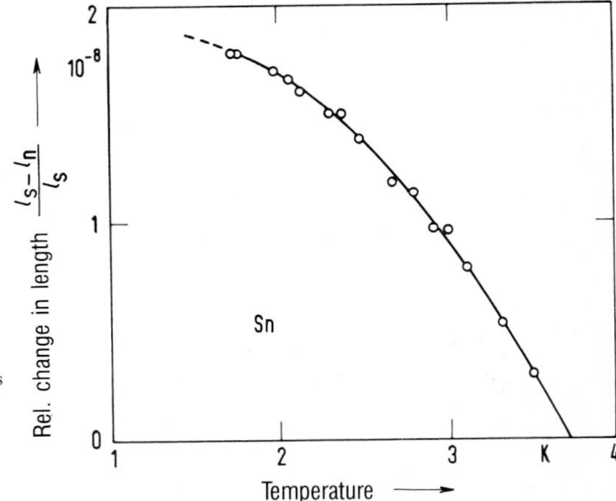

Fig. 52. Relative change in length $(l_s - l_n)l_s$ of a tin rod on transition to the superconducting state.

The investigations of the pressure dependence of the transition temperature have greatly increased in importance since the development of the microscopic theory, of the BCS theory. The application of pressure from all sides makes it possible to vary the lattice constant of a substance continuously. The lattice constant is, however, an important parameter for quantum states of both the electrons and the lattice vibrations, that is the phonons. Since both quantities are of fundamental importance for superconductivity it is possible to obtain new results for the quantitative improvement of the microscopic theory from pressure experiments. We saw in Section 1.2 that our knowledge concerning the quantitative relationships between metal parameters and superconductivity is very incomplete. The BCS theory provides the following relationship for the transition temperature T_c (see Section 3.1):

$$T_c \propto \Theta_D \exp\left(-\frac{1}{N(E_F)\,V^*}\right). \tag{3-1}$$

We differentiate with respect to p and get:

$$\frac{\partial T_c}{\partial p} = \frac{\partial \Theta_D}{\partial p} \exp\left(-\frac{1}{NV^*}\right) + \Theta_D \exp\left(-\frac{1}{NV^*}\right)\left(\frac{1}{NV^*}\right)^2 \frac{\partial (NV^*)}{\partial p}. \tag{4-35}$$

And to determine the relative change in T_c we divide by T_c

$$\frac{1}{T_c}\frac{\partial T_c}{\partial p} = \frac{1}{\Theta_D}\frac{\partial \Theta_D}{\partial p} + \left(\frac{1}{NV^*}\right)^2 \frac{\partial (NV^*)}{\partial p}. \tag{4-36}$$

From the determination of the dependence of T_c on pressure and of the Debye temperature Θ it is possible to obtain information concerning the pressure dependence of $N(E_F)\,V^*$ (density of states times interaction parameter) [88].

We have, until now, considered the effect of pressure within *one* crystalline phase. It is often possible to produce new crystalline phases, new modifications, under pressure. It is naturally to be expected that the superconductivity properties alter as a result of such a phase change. The high pressure modification is simply a new material.

Those substances, where the phase stable at normal pressures does not exhibit superconductivity, but where the high pressure phase is a superconductor, are of interest. In recent years a whole range of such superconducting high-pressure phases have been found amongst the semiconductors and the transition region between metals and semiconductors. These elements are listed in Table 1 (see Section 1.2) and marked by dark shading in Fig. 11.

The typical semiconductor germanium will be discussed here as an example. At a pressure of ca. 100 kbar germanium is converted into a metallic phase with a tetragonal structure like that of metallic tin [92]. At equilibrium pressure this high pressure phase of germanium becomes superconducting at 5.4 K [93].

Figure 53 illustrates another example of such a transition to a high pressure phase [94]. Tin is also a superconductor in its normal pressure phase. As the pressure is increased T_c for this phase falls. The transition takes place at somewhat over 100 kbar. At the equilibrium pressure the high pressure phase has an appreciably higher transition temperature of 5.3 K. We will discuss the correlation between high pressure phases and condensed films in metastable states in Section 8.5.4.

The change in the transition temperature under hydrostatic pressure can be employed for the construction of a superconducting manometer. Lead is particularly suitable, because it does not exhibit any phase changes up to about 160 kbar and is also insensitive to lattice defects. So a lead wire can be placed in the pressure cell and the pressure in the cell at low temperatures can be determined from its transition temperature T_c. A cal-

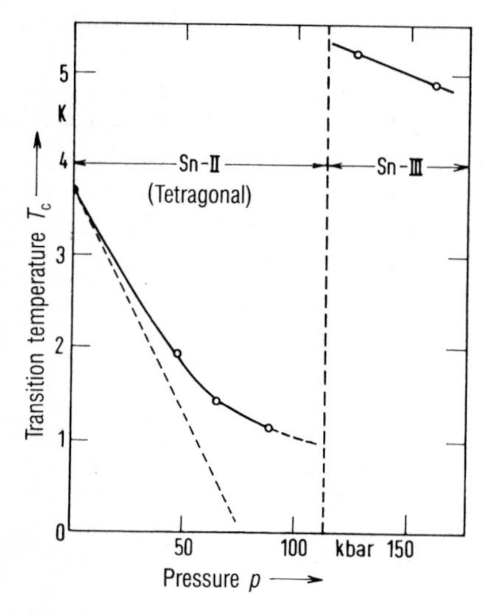

Fig. 53. Transition temperature of tin under all-round pressure. A high pressure modification having a higher transition temperature is stable above ca. 100 kbar (after [94]).

ibration of the Pb manometer up to ca. 160 kbar has been carried out by Eichler and Wittig [95]. The lack of sensitivity of T_c to lattice defects is very important because a certain plastic deformation always occurs at high pressures – particularly when they are applied at low temperatures – and this can lead to the production of lattice defects.

The new oxides of the system Ba-La-Cu-O were investigated very early under hydrostatic pressure [96]. This resulted in an unexpectedly high increase of T_c under pressure. Since hydrostatic pressure reduces the size of the lattice constant, a "chemical" reduction of the lattice constant, such as is achieved by the incorporation of smaller ions, could also cause an increase in T_c. Following this consideration C.W. Chu at al. decided to replace La by the smaller ion Y. In this manner they discovered the system Y-Ba-Cu-O and samples with values for T_c higher than 80 K [19][1].

The results of the pressure experiment on $YBa_2Cu_3O_7$ are somewhat contradictory [97, 98]. But it seems certain that the pressure effect for the $YBa_2Cu_3O_7$ system is very much smaller than that for samples of the Ba-La-Cu-O system [99].

4.4 Thermal conductivity

If you maintain a temperature difference ΔT along a rod – we choose to use a simple geometry – of length l then energy flows in the form of heat from the hot to the cold end. The thermal conductivity λ_w is a matter constant and is defined by the following equation:

$$\frac{\Delta Q}{\Delta t} = \lambda_w \frac{A}{l} \Delta T \tag{4-37}$$

where $\Delta Q/\Delta t$ = thermal energy per unit time, A and l are the cross section and the length of the rod, ΔT = temperature difference[2].

The transport of heat through a metal is carried out by both the conduction electrons and the lattice vibrations. The contribution of the electrons is generally appreciably greater than that of the lattice[3].

In this case we can very readily predict from our model of the superconducting state how the thermal conductivity should behave at temperatures below T_c. Below T_c more and more electrons are correlated to Cooper pairs and, thus, taken away from energy exchange. So the contribution of the electrons to the thermal conductivity will get smaller and smaller below T_c. Therefore, we would expect the thermal conductivity of the superconducting state to be less than that of the normal state, as far as it is mainly dependent on electrons.

---

1 Chinese scientists in Peking who had independently discovered the system Y-Ba-Cu-O answered the author's question as to how they had thought of Y, simply by saying: "Since the replacement of Ba by Sr increases T_c the obvious thing for us seemed to be to replace La by Y".

2 A linear variation of T is assumed along the length of the rod.

3 In insulators heat can only be transferred by the lattice vibrations since there are no free electrons available.

Figure 54 illustrates the behavior for tin and mercury [100]. We will not discuss the dependence of the thermal conductivity of the normal state on temperature. It is only of importance here that the thermal conductivity in the superconducting state is, as expected, lower than in the normal state. When the temperature is low enough so that in the superconducting state there are practically no free electrons since almost all of these have condensed to Cooper pairs, then the behavior observed for a superconductor approaches that of a completely insulating crystal. The electron system is practically uncoupled from the thermal surroundings. If superconductivity is destroyed by a magnetic field greater than the critical value the metal regains the much greater thermal conductivity of the electron system. A superconductor can, therefore, be employed as a switch for currents of heat. When the magnetic field is above the critical level the thermal conductivity is very good, the switch is – by analogy with an electrical circuit – closed. The thermal conductivity is very much less when there is no field, the switch is open (see also Section 9.5.2).

This behavior is characteristic for pure metals. In the case of alloys and very disordered metals, on the other hand, the situation is appreciably more complex. If foreign atoms are incorporated into a metallic lattice, this causes the free paths of the electrons to be reduced, because impacts occur with these "impurities". The associated increase in the electrical resistance is observed as a residual resistance at low temperatures (Fig. 2). The obstruction of the movement of the electrons also yields an additional thermal resistance.

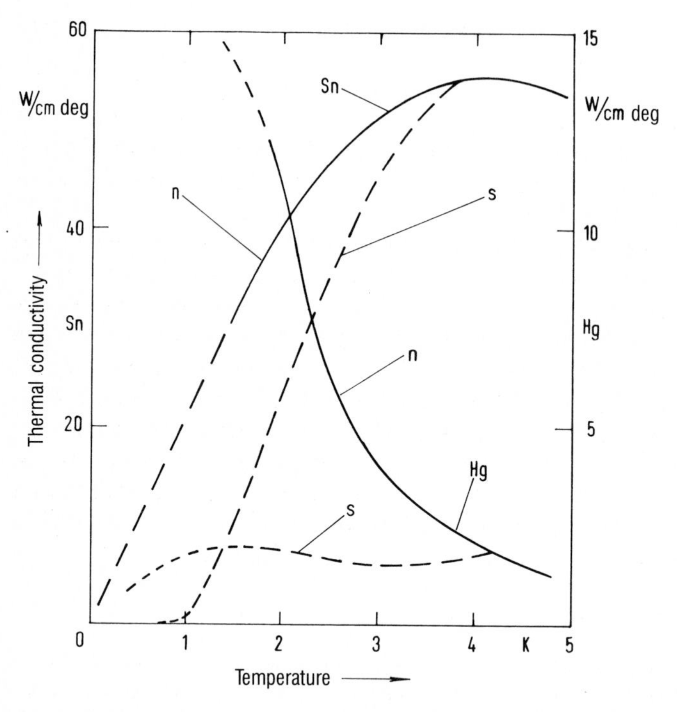

Fig. 54. Thermal conductivity of pure tin and mercury (after [100]). Left-hand scale for tin.

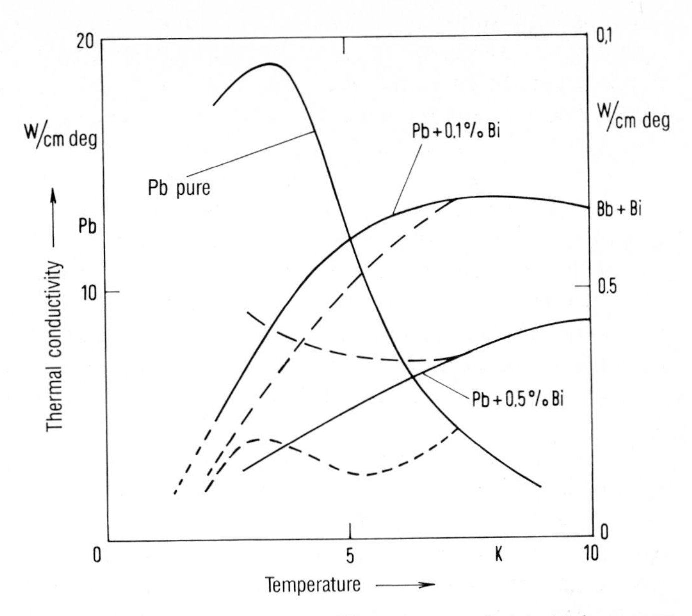

Fig. 55. Thermal conductivity of lead and lead-bismuth alloys. Solid curves normal state, broken curves superconducting state (after [101]).

In contrast to the electrons the lattice vibrations, the phonons, are much less impeded in their transmission by the atomic defects[1]. This means that the contribution of the phonons to the thermal conductivity is very much less affected on incorporation of such defects. Thus, the thermal conductivity via phonons can become very much greater than that via electrons. The thermal conductivities in the superconducting and normal states differ only very little. The thermal conductivity of a lead-bismuth alloy containing 0.1 % Bi is illustrated in Fig. 55 as an example [101].

Finally, if the phonon portion of the thermal conductivity is appreciably greater than the electronic part, which is true for some alloy systems, then the thermal conductivity in the superconducting state can actually be greater than it is in the normal state. An example of this behavior is included in Fig. 55 where the thermal conductivity is illustrated of a lead-bismuth alloy containing 0.5 % Bi. This behavior can be understood when account is taken of the fact that electron and phonon systems also interact with each other, i.e. that scattering processes take place between electrons and phonons. The whole temperature dependence of the electrical resistance depends on such scattering processes of electrons at lattice vibrations. As the temperature is increased the vibrations of the lattice increase, i.e. the numbers of phonons increase. Hence, the amount of scattering of electrons and the electrical resistance increase with increasing temperature.

1 The scattering of a wave at an obstacle becomes considerable when the wavelength and the dimensions of the obstacle are comparable. At the momentum corresponding to the Fermi energy the electron waves have a wavelength of a few times 0.1 nm and are, therefore, very heavily scattered by atomic obstacles, whereas the long lattice waves pass the same obstacles with much less interaction.

In the case of those substances where the lattice vibrations make an appreciable contribution to the thermal conductivity we have to consider electron-phonon scattering from the point of view of the phonons. These scattering processes inhibit the propagation of the phonons and, hence, the thermal conductivity of the phonon system. If in the superconducting state the electrons are decoupled then these scattering processes no longer occur. The thermal conductivity of the phonon system will be larger. In this way states can be reached where the thermal conductivity is greater in the superconducting state than in the normally conducting state. This explains the results for Pb-Bi alloys in Fig. 55.

The thermal conductivity in what is known as the intermediate state constitutes a special case. In the intermediate state, which, as we will see in Section 5.1.4, can be stabilized in an external magnetic field if the sample geometry is suitable, we have the normal and superconducting states present next to each other in a sample. The thermal conductivity can then, in addition to all other processes, be inhibited by a scattering at the phase boundary between n- and s-conducting regions. A maximum is observed in the thermal resistance on passing through the intermediate state [101].

In an introduction such as this we cannot go into this seemly complex situation any further. Qualitatively – and that is what is important here – we are able to understand thermal conductivity in the superconducting state very well on the basis of our fundamental model of this state.

5. Superconductors in magnetic fields

It was believed for a long time that the only characteristic property of the state of superconductivity was that it possessed no detectable resistance for direct currents. This meant that, in the thermodynamic treatment of this state, there was the difficulty that a substance, that is only characterized by the condition $R = 0$, has quite different states for the same independent variables T, p and B, depending on the experimental conditions during the phase transition.

This situation, which was briefly discussed at the beginning of Chapter 4, is illustrated again in Fig. 56. The B-T plane is divided into two sections by the $B_c(T)$ curve[1]. Points (T, B), where the superconducting state is thermodynamically stable, are only to be found in the hatched area. Let us now consider two points α and β in the stability ranges of the normally conducting and superconducting states respectively. We start from α and take our sample to point β by route 1. Here the transition takes place at zero field. Field B_β is switched on at T_β. The induction process starts currents, which flow as persistent currents on account of the condition $R = 0$. They prevent the magnetic field penetrating the interior of the sample. So we have at point T_β, B_β a superconductor without a magnetic field in its interior and we can understand the result on the basis of the requirement that the resistance $R = 0$.

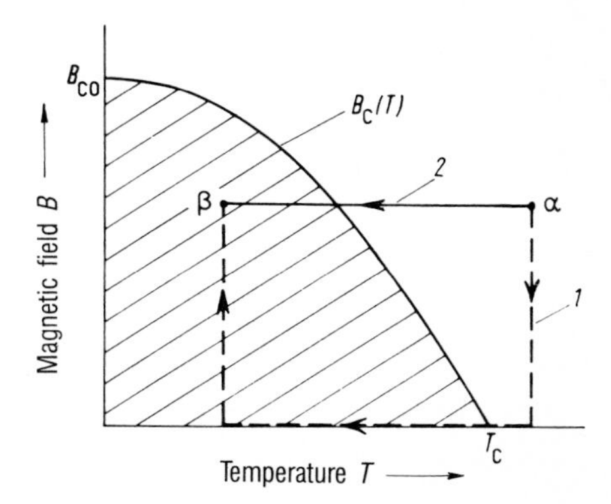

Fig. 56. Phase diagram of a superconductor. Region of stability of the superconducting phase hatched.

Now let us bring the sample from point α to point β by route 2. Here the transition takes place at field B_β. We see no reason for the setting-up of currents and must expect that at point (T_β, B_β) we produce this time a superconductor which is penetrated by field B_β and does not possess any shielding current. Hence, we would apparently have produced a superconductor with different properties under the same conditions T_β and B_β. We would have no right to speak of *one* superconducting state.

An experiment carried out in 1924 by Onnes seemed to confirm this complicated behavior unequivocally. Onnes [102] brought a lead sphere into the region of superconductivity by route 2 and then switched off the external magnetic field. This resulted, as must be expected for the condition $R = 0$, in a persistent current, which imparted magnetic

1 We will keep the variable p constant for the whole of this treatment.

moment to the sphere. The observation was correct but the fact that a hollow sphere had been used in order to require less liquid helium for cooling had been overlooked. In the case of a hollow sphere it is possible for a ring-shaped superconducting region to be produced during cooling which then holds a flux constant through its area. The hollow sphere can, therefore, act like a superconducting ring (Fig. 5, see also Section 7.1).

The treatment of the superconducting state as a new thermodynamic phase was, nevertheless, so enticing that in 1933 Gorter [103] developed a thermodynamic theory of superconductivity in full knowledge of its hypothetical character. A few months later Meissner and Ochsenfeld [76] were able to show in their famous experiments that besides the property $R = 0$ the superconducting state also possess the property of always – independently of the experimental conditions – excluding a magnetic field from its interior. The thermodynamic treatment was, therefore, justified. But, in addition to this, a quite new side of superconductivity was opened up, one that was of decisive importance for the development of our understanding of superconductivity and leading up to the microscopic theory.

The exclusion effect can, like the property $R = 0$, be demonstrated vividly with the "floating magnet" (see Section 1.1). In Section 1.1 we lowered the permanent magnet towards the superconducting lead dish, thereby setting up a persistent current by means of the associated induction effect, in order to demonstrate the property $R = 0$. To demonstrate the Meissner-Ochsenfeld effect we place the permanent magnet in the lead dish at $T > T_c$ (Fig. 57 left) and then cool it down. On transition to superconductivity field exclusion commences, the magnet is repelled by the diamagnetic superconductor and rises to float at equilibrium height (Fig. 57 right). In the limit of ideal exclusion of the magnetic field the same floating height will be achieved as in Fig. 7.

We will now treat the behavior of superconductors in magnetic fields in detail. Because of its importance many books treat the magnetic side of superconductors first. However, with this treatment, which also reflects the historical development of the subject, the

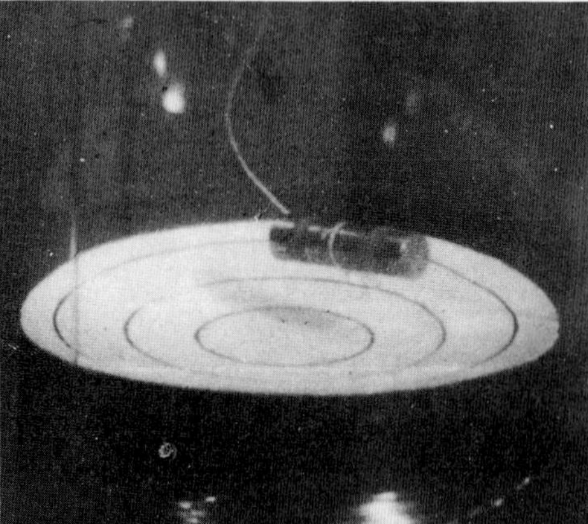

Fig. 57. "Floating magnet" to illustrate the Meissner-Ochsenfeld effect.
Left: starting point (point α in Fig. 56)
Right: equilibrium position for $T < T_c$ (point β in Fig. 56)
(My thanks are due to Frau Stremme for producing the pictures.)

close internal relationships between type I and type II superconductors is often blurred. In order to demonstrate this properly the magnetic behavior will now be treated as a whole.

Here it is important for the clear distinction of the various possibilities that we adopt an unambiguous terminology from the start. We will make a clear distinction between three concepts.

1. We always speak of a Meissner phase when the magnetic field is excluded from the superconductor apart from a thin surface layer.

2. "Type I superconductors" exhibit this exclusion effect up to a value of the magnetic field corresponding to the thermodynamically critical field B_{cth} which is determined by the relationship:

$$G_n - G_s = \frac{1}{2\mu_0} V_s B_{cth}^2 . \qquad (4-15)$$

3. "Type II superconductors" exhibit the exclusion effect at sufficiently low fields $B < B_{c1}$ with $B_{c1} < B_{cth}$ but for fields $B_{c1} < B < B_{c2}$ where $B_{c2} > B_{cth}$ go into a state which is knonw as a "mixed state" or also as a Shubnikov phase. So type II superconductors also have a Meissner phase, but it is limited to fields $B < B_{c1}$. We will discuss the meaning of B_{c1} and B_{c2} in detail.

5.1 Type I superconductors

5.1.1 Field exclusion

Figure 58 illustrates once more the Meissner-Ochsenfeld effect for a rod-shaped superconductor. If the length of the rod l is very great compared with its diameter, the magnetic field will only be somewhat distorted at the ends. We will have practically the same magnetic field B_e along the length of the rod as at great distances from the rod. It is possible in the case of especially simple bodies (ellipsoids with an axis parallel to the field) to express the field distortion in the form of a number, the demagnetization factor n_M. A long rod with its axis parallel to the field has a demagnetization factor $n_M = 0$, i.e. the field at the surface does not require to be corrected, it is identical with the external field B_e. This constitutes the specially simple nature of this sample geometry [1)]

On account of its importance we will describe the exclusion effect in the Meissner phase in yet another manner. The screening current, which completely compensates the external field in the interior of the rod-shaped sample (Fig. 58), endows the rod with a magnetic moment $\vec{m}$. We can speak purely formally of magnetization $\vec{M}$ if we set $\vec{M} = \vec{m}/V$, where V is the volume of the sample. This magnetization corresponds to that of an ideal diamagnet with a susceptibility ($\chi = -1$ [2)]).

---

1 We will discuss demagnetization factors for other sample geometries in Section 5.1.4.

2 The thin surface layer into which the magnetic field penetrates can be ignored in an integral consideration of the sample.

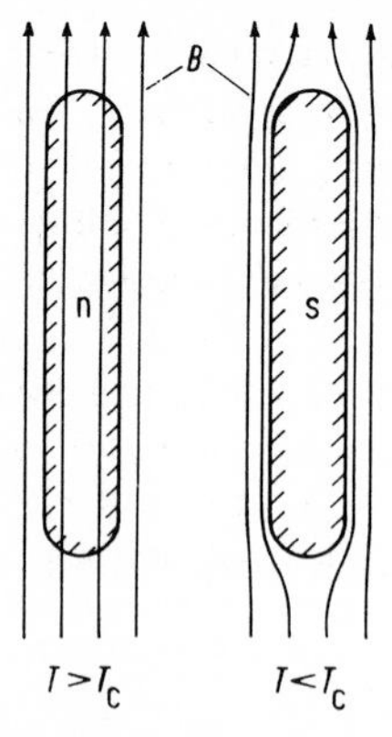

Fig. 58. Field exclusion by a rod-shaped sample. The sample is cooled in field B.

This magnetization M is shown as a function of the external field B_e for a long rod with axis parallel to the field (Fig. 59). The magnetization increases proportionally to the external field. Only when the critical field strength B_c is exceeded does the superconductivity break down. In the case of a "thick" superconductor, i.e. a superconductor with completely developed screening, the critical field B_c is identical with the thermodynamic field B_{cth}. For a "thin" superconductor (see Section 5.1.3), on the other hand, B_c is larger than B_{cth}. The area under the magnetization curve multiplied by the volume V of the sample yields, as we saw in Section 4.1, the difference of the thermodynamic potentials $G_n - G_s$ at field $B = 0$.

$$G_n - G_s = -V_s \int_0^{B_c} M \mathrm{d}B . \tag{4-13}$$

With complete exclusion we get, since $M = -B/\mu_0$,

$$G_n - G_s = \frac{1}{2\mu_0} V_s B_c^2 . \tag{4-15}$$

But it must be emphasized once again that the area under the magnetization curve $M(B)$ always yields the difference of the Gibbs functions $G_n - G_s$ even when there is a complicated dependence of the magnetization on the external field such as we will meet

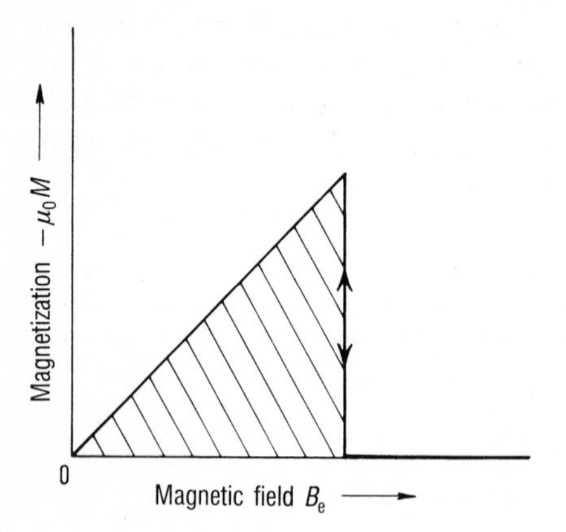

Fig. 59. Magnetization of a rod-shaped sample ($n_M \approx 0$) in a field parallel to the axis of the rod. With reversible transition the field will pass along the curve with increasing and decreasing field B.

for type II superconductors. The only condition is that the magnetization must be established reversibly, i.e. reached only via equilibrium states. Reversibility is achieved if the same magnetization curve is followed with both increasing and decreasing external fields. The conditions for reversibility no longer apply in the case of the "hard superconductors" which we will discuss separately on account of their technological importance (Chapter 7).

Finally Fig. 60 illustrates the magnetic field B_i as it could be observed in the interior of the sample, say in a thin channel parallel to the axis, plotted against the external field B_e. The screening current renders the internal field B_i practically zero until the critical field B_c is reached[1]. Then for all external fields $B_e > B_c$ we find that $B_i = B_e$ since the sample is normally conducting.

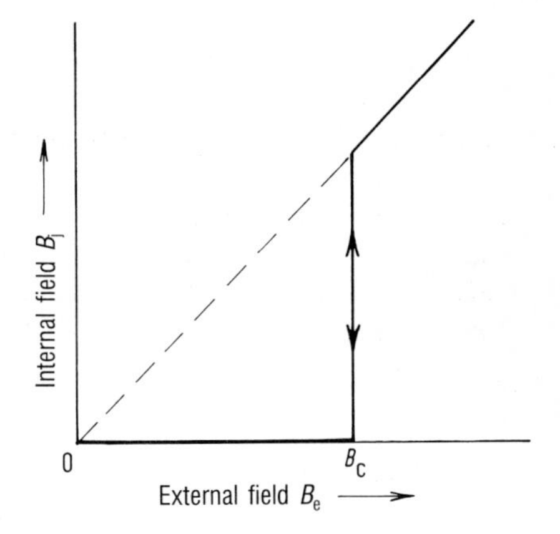

Fig. 60. Magnetic field in the interior of a rod-shaped sample ($n_M \approx 0$) with an external field parallel to the axis. With reversible changes the field will pass along the curve with increasing and decreasing field *B*.

The presentations in Figs. 59 and 60 both make the same statement. Both are frequently employed and are adapted for different experiments (measurement of *M* by induction or of B_i with a Hall sensor[2]). The behavior of *M* or B_i as a function of B_e makes the difference between type I and type II superconductors particularly clear (Section 5.2.1).

Measurement of the magnetization is a particularly simple method for the determination of the transition temperature T_c. The sample is placed in an induction coil and its self- inductivity is determined as a function of temperature by means of a weak alternating current. The self-inductivity suddenly becomes smaller as superconductivity occurs. This method of determination of the transition temperature has the advantage over determination of the electrical resistance in a current-voltage measurement (see Section 1.1) that the current has to flow over the whole surface to screen the volume of the sample. Inhomogeneous samples, where a single continuous current path can yield superconduc-

1 In this explanation we explicitly exploit the knowledge that macroscopic currents on the surface compensate the external field. For the description of field relationships in the interior of matter it is important to know the mechanism controlling the magnetic behavior. It is only for the variation of the external field that this is unimportant.

2 With modern sensors the Hall effect can be employed to measure magnetic fields very sensitively. Here a potential is observed that is set up perpendicular to the current and the magnetic field as a result of the Lorentz force. This Hall potential is proportional to the applied field.

tivity in current-voltage determinations, frequently do not exhibit complete screening, thus, allowing the recognition of inhomogeneities. However, if the superconducting phase is present in the form of a thin network (e.g. as a precipitate) then induction measurements will also reveal full superconductivity. Unequivocal evidence concerning the proportion of the volume that is superconducting can only be obtained from specific heat measurements. If a portion of the sample remains normally conducting, then from Equation (4–24) the specific heat of the electrons of this part of the sample will provide the contribution $c_{nE} = \gamma T$, which is readily observed.

5.1.2 The depth of penetration

The field exclusion from the interior of the superconductor cannot take place exactly up to the surface of the sample. For this would mean that the magnetic field at the surface had to jump discontinuously from the value B_e to the value zero. This would require an infinitely large current density at the surface. Such infinite current densities can naturally not occur. So the magnetic field must penetrate somewhat into the material. The screening currents flow in the thin surface layer across which the magnetic field falls to very low levels.

A quantitative treatment of the screening currents and the magnetic field in the surface layers is obtained from what are known as the London equations. Shortly after the discovery of the Meissner-Ochsenfeld effect F. and H. London [73] proposed a phenomenological theory of superconductivity – which is now generally referred to as the London theory. This theory was capable of describing a large number of observations and, what was almost more important, indicated where a microscopic theory must start. Here we will not discuss the London theory in detail but merely use its basic equations to describe the behavior of the screening layer quantitatively.

The two basic equations of the London theory encompass the characteristic electromagnetic properties of a superconductor. These equations are:

$$\frac{\mathrm{d}(\Lambda \vec{j}_s)}{\mathrm{d}t} = \vec{E} \tag{5-1}$$

$$\mathrm{curl}\,(\Lambda \vec{j}_s) = -\vec{B}\,. \tag{5-2}$$

The first equation describes a conductor with $R = 0$. In it the charges are accelerated evenly under the influence of an electrical field, i.e. the change of the current density with time is proportional to the electrical field $\vec{E}$. This is expressed by Equation (5–1). The constant Λ, that has already appeared for flux quantization (see Section 3.2), is given by:

$$\Lambda = \frac{m_s}{n_s e_s^2} \tag{5-3}$$

where m_s, n_s and e_s are the mass, the number per unit volume and charge of the carriers of the supercurrent. F. and H. London had to assume that the current was carried by single electrons. Today we know that paired electrons, Cooper pairs, carry the supercurrent.

They have the mass $2m$, the charge $2e$ and their number per unit volume is $n_s/2$ where n_s is the number of single electrons. Thus, the value of Λ is not changed by going from single electrons to Cooper pairs.

The second equation describes the Meissner-Ochsenfeld effect. It provides us with the decay of the magnetic field in the thin surface layer of the superconductor. We will take a closer look at this immediately.

The two London equations have the same importance for superconductors that Ohm's law has for normal conductors. For a normal conductor:

$$\vec{j} = \sigma \vec{E} \qquad (\sigma = \text{the electrical conductivity})\ .$$

The relationship between current density and electric or magnetic field[1] of a superconductor is given by the two London equations. Alongside these equations of state for normal and superconductivity the general Maxwell relationships, as expressed by the Maxwell equations, naturally apply too.

For the treatment of the surface layer we start from Equation (5-2). In order to make the calculations straightforward we will assume a simple geometry, namely a plane surface perpendicular to the x axis (Fig. 61) and the magnetic field $\vec{B}$ parallel to the z axis, so $\vec{B} = \{0,0,B_z\}$[2].

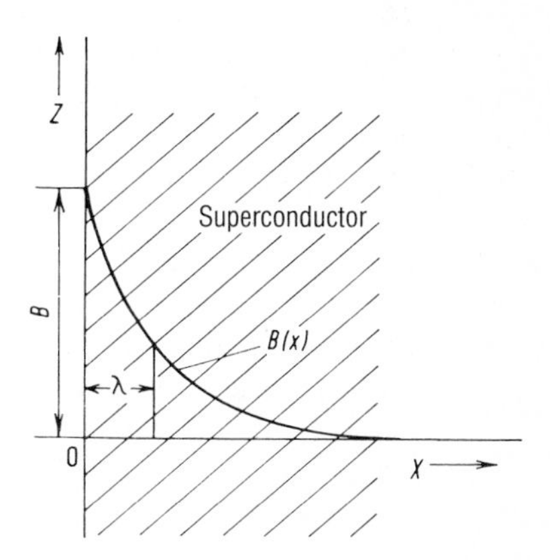

Fig. 61. Decrease of the magnetic field in the screening layer of a plane surface.

Then

$$\operatorname{curl}(\Lambda \vec{j}_s) = -\vec{B} \qquad \text{2nd London equation} \qquad (5\text{-}2)$$

$$\operatorname{curl} \vec{B} = \mu_0 \vec{j}_s \qquad \text{1st Maxwell equation}\ . \qquad (5\text{-}4)$$

---

1 Since a persistent current is set up in every induction process the current density must be related to the magnetic field itself and not just to its changes with time.

2 This geometry is also a good approximation for a rod with a circular cross section when the radius $r \gg \lambda$. The curvature of the surface is then unimportant. An exact solution must naturally consider the cylindrical symmetry of a rod with circular cross section.

We obtain the rotation from Equation (5–4):

$$\text{curl curl } \vec{B} = \text{curl } \mu_0 \vec{j}_s = -\frac{\mu_0}{\Lambda} \vec{B} .$$

This yields according to the general rules of vector calculus for our geometry:

$$\frac{d^2 B_z(x)}{dx^2} - \frac{\mu_0}{\Lambda} B_z(x) = 0 . \qquad (5-5)$$

This differential equation has a simple solution [1]:

$$B_z(x) = B_z(0)\, e^{-\frac{x}{\sqrt{\Lambda/\mu_0}}} . \qquad (5-6)$$

This dependence of the magnetic field is illustrated in Fig. 61. The magnetic field falls to the eth part over a distance $\lambda = \sqrt{\Lambda/\mu_0}$. This distance λ is known as the penetration depth [2]. Since it follows from the London theory it is often known as the London penetration depth λ_L. A rough value of λ_L is obtained from Equation (5–3) if the certainly incorrect assumption is made that one electron per atom with the mass of the free electron m_E contributes to the supercurrent. This approximation yields $\lambda_L = 26$ nm for tin, for instance. This value differs little from the measured value given in Table 8, p. 134.

The penetration depth λ is temperature-dependent. This follows from Equation (5–3) where n_s is a function of temperature. In the region of T_c the BCS theory predicts a number of Cooper pairs per unit volume n_C:

$$\frac{n_c(T)}{n_c(0)} \propto \left(1 - \frac{T}{T_c}\right) \qquad (5-7)$$

and hence of $\lambda(T)$:

$$\frac{\lambda(T)}{\lambda(0)} \propto \left(1 - \frac{T}{T_c}\right)^{-1/2} . \qquad (5-8)$$

The experimentally observed temperature dependence is very well approximated by the expression (see for example [II])

$$\frac{\lambda(T)}{\lambda(0)} = \left(1 - \left(\frac{T}{T_c}\right)^4\right)^{-1/2} . \qquad (5-9)$$

---

1 The second linear independent solution $B_z(x) = B_z(0)\, e^{+x/\sqrt{\Lambda/\mu_0}}$ is not suitable for our boundary conditions.

2 In more general decay curves $B_z(x)$, which are obtained from other theories, the depth of penetration λ can be defined according to the expression:

$$\lambda = \frac{1}{B_z(0)} \int_0^\infty B_z(x)\, dx .$$

When $T \to T_c$ this relationship agrees with the BCS theory [1]:

$$\frac{\lambda(T)}{\lambda(0)} \propto \left(1 - \frac{T}{T_c}\right)^{-1/2} \tag{5-8}$$

Figure 62 illustrates this temperature-dependence of the penetration depth schematically.

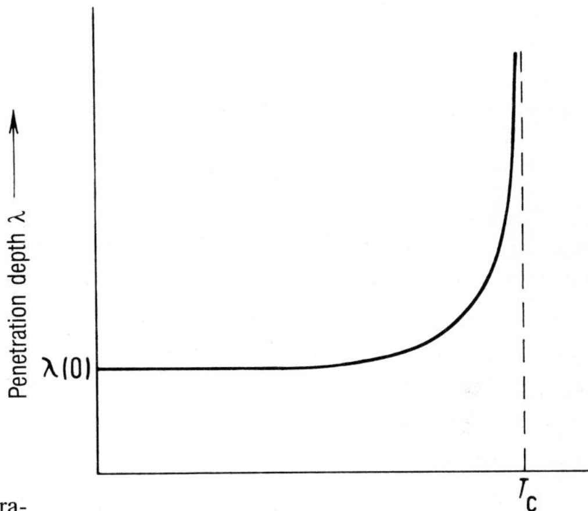

Fig. 62. Temperature dependence of the penetration depth (after [II]).

We have arrived at the important result that the London penetration depth increases very rapidly in the neighborhood of T_c. As $T \to T_c$ the magnetic field penetrates further and further into the superconductor.

Some values for λ are listed in Table 8 p. 134. The depth of penetration is the decisive constant of a superconductor within the framework of the London theory.

When determining the depth of penetration experimentally it is necessary, in principle, in all methods to investigate the effect of the thin screening layer on the diamagnetic properties. Various methods have been employed [II]. Only relative determinations are necessary for a study of the temperature dependence. For instance, the resonance frequency of a cavity made of superconducting material can be determined. The resonance frequency is sensitively dependent on the geometry of the cavity. If the depth of penetration varies with temperature then this amounts to a change in the geometry of the cavity and, hence, in its resonance frequency, from which it is possible to determine the change in λ [104].

Another method uses small samples. "Small" means samples where at least one dimension is comparable with λ. The basic idea behind these experiments is very clear. The smaller a sample the more the thin surface layer, into which the magnetic field penetrates,

---

1 We write $1 - (T/T_c)^4 = (1 - (T/T_c)^2)(1 + (T/T_c)^2)$ and substitute the second factor for $T \to T_c$ by 2. We employ the same process on $1 - (T/T_c)^2$ to get:
$1 - (T/T_c)^4 \approx 4(1 - (T/T_c))$.

determines the behavior of the whole sample. Thus, the larger will be the deviations from a complete geometrical screening effect. Shoenberg et al. [105] used the determination of the magnetization of colloidal mercury, that is of tiny spheres, to determine the temperature dependence of λ. Figure 63 illustrates the results in a manner which allows comparison with Equation (5–9). The solid line represents Equation (5–9). It can be seen here how well the analytical expression describes the results. It is also now possible to determine the penetration depth very accurately even for single crystals with the aid of SQUIDS (Section 9.5.4) [106].

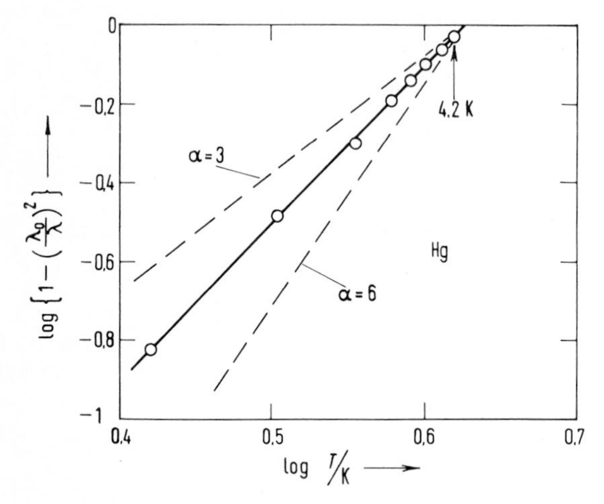

Fig. 63. Temperature dependence of the penetration depth for Hg. The solid curve represents the exponent $a = 4$ in the brackets of Eq. (5–9). The curves for $a = 3$ and $a = 6$ are included for comparison as dashed lines.

Thin layers possessing a thickness d comparable to λ can also be used as samples whose behavior in a magnetic field is basically dependent on the penetration depth. We will consider the behavior of such layers in more detail, not just because they can be used to determine the penetration depth [107], but also and particularly because we will require these considerations in our treatment of the "intermediate state" in Section 5.1.4.

5.1.3 Thin layers in a magnetic field parallel to the surface

The sample geometry, that we wish to consider now, is illustrated in Fig. 64. The sample is so large in directions y and z that we do not need to take account of edge effects in these directions. Differential Equation (5–5) also applies here for the size of the magnetic field in the interior of the sample:

$$\frac{d^2 B_z(x)}{dx^2} - \frac{1}{\lambda^2} B_z(x) = 0 \qquad \lambda = \sqrt{\frac{\Lambda}{\mu_0}} \,. \tag{5–5}$$

But the boundary conditions have now become:

$$B_z\left(\frac{d}{2}\right) = B_z\left(-\frac{d}{2}\right) = B_0 \,. \tag{5-10}$$

We require the two linear independent solutions in order to fulfil these boundary conditions:

$$B_1(x) = B_1\, e^{-\frac{x}{\lambda}}\,; \qquad B_2(x) = B_2\, e^{+\frac{x}{\lambda}}\,. \tag{5-11}$$

The general solution is a linear combination of the two independent solutions, whereby the boundary conditions determine the coefficients. For $x = d/2$ we get:

$$B_1\, e^{\frac{d}{2\lambda}} + B_2\, e^{-\frac{d}{2\lambda}} = B_0 \,. \tag{5-12}$$

Since the problem is symmetrical in x and $-x$ because of our choice of coordinate system, it must follow that $B_1 = B_2 = B^*$.
We get

$$B^*\left(e^{\frac{d}{2\lambda}} + e^{-\frac{d}{2\lambda}}\right) = B_0 \qquad B^* = \frac{B_0}{2\cosh\frac{d}{2\lambda}} \tag{5-13}$$

so that in the superconductor

$$B(x) = B_0\,\frac{\cosh\frac{x}{\lambda}}{\cosh\frac{d}{2\lambda}}\,. \tag{5-14}$$

This behavior is illustrated in Fig. 64. It can clearly be seen that as the thickness d becomes smaller the field variation becomes ever smaller because the screening layer can no longer be completed, when $d \ll \lambda$ there can only be a very small variation of the field over the thickness. The magnetic field penetrates the superconducting layer practically homogeneously.

The behavior of the critical field for such thin layers is a particularly important experimental result. The critical field is understood as being that field which must be applied to destroy the superconductivity or, expressed in another manner, to make G_s equal G_n. This critical field B_c becomes ever larger as the layer is decreased in thickness. For thicknesses $d \ll \lambda$ it can be more than a factor of 10 larger than the thermodynamic field B_{cth} which is observed when the screening layer is formed completely.

This extraordinary finding can be understood very readily. In Section 4.1 we saw that the free enthalpy of the superconducting state increases with the external field B_e. For:

$$G_s(B) = G_s(0) - \int_0^B m\,\mathrm{d}B \tag{4-10}$$

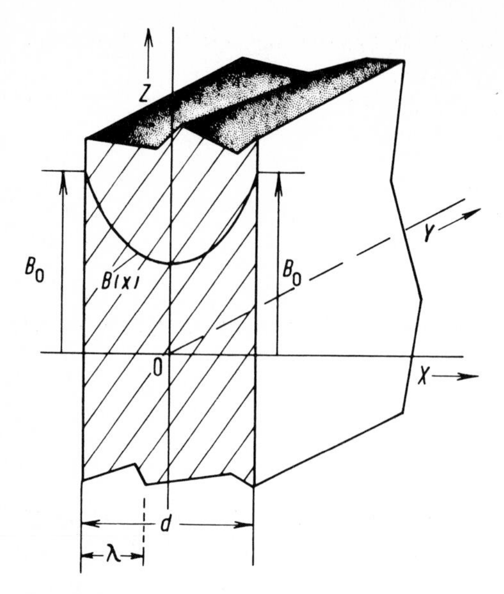

Fig. 64. Spatial variation of the magnetic field in a thin superconducting layer. For the assumed ratio $d/\lambda \approx 3$ the magnetic field only falls to about half of its value outside the film.

where m is the diamagnetic moment of the sample, that arises as a result of the screening current. When the screening layer is complete the current density j_s and its decrease in the interior of the superconductor is determined by the external field B_a independently of the macroscopic dimensions of the sample. In the case of samples which are small or comparable in size with the penetration depth in at least one dimension the relationship between external field and screening current will depend on the geometry of the sample. As the thickness of our layer is reduced the screening current will become ever smaller for a particular fixed value of B_a. That is the same as saying that the free enthalpy of the thin superconductor increases more slowly with B_a than does that of a "thick" one. So higher fields B_a are necessary to make G_s equal to G_n.

Qualitatively it is also very evident in Fig. 64. The reduction of the magnetic field on the inner side of the layer is a measure of the diamagnetic behavior of the sample. As the layer decreases in thickness the field reduction in the interior and, hence, the diamagnetism becomes ever smaller. In order to make G_s equal G_n by means of an external magnetic field it is necessary to apply ever larger fields as the layer thickness is decreased.

In order to comprehend these relationships, which are very important for an understanding of the behavior of type I superconductors in the magnetic field, quantitatively it is necessary to consider the current density of the screening current more closely. The screening current is obtained from the behavior of the magnetic field with the aid of the 1st Maxwell equation[1)]

$$\operatorname{curl} \vec{B} = \mu_0 \vec{j}_s \,. \tag{5-15}$$

1 We fix, as before, the magnetic field at the surface.

The current density only has a y component. This, like the magnetic field, falls from the surface to the interior of the conductor. In detail:
1. In the case of a complete screening layer (5.1.2) with B in the z direction (Fig. 61):

$$\vec{B}(x) = \vec{B}_0 \, e^{-\frac{x}{\lambda}} \tag{5-6}$$

$$\text{curl}\,\vec{B} = -\frac{\partial B_z}{\partial x}\,\vec{e}_y\,. \tag{5-16}$$

$\vec{e}_y$ is the unit vector in the y direction. The expression $-\vec{e}_y$ means a unit vector in the $-y$ direction. So that:

$$\mu_0 \vec{j}_s = \text{curl}\,\vec{B} = \frac{1}{\lambda} B_0 \, e^{-\frac{x}{\lambda}}\,\vec{e}_y\,. \tag{5-17}$$

This means: The screening current flows in the $+y$ direction. The current density at the surface is related to the external field B_e at the surface by the relationship[1]:

$$\sqrt{\mu_0 \Lambda}\; j_{s0} = B_0\,. \tag{5-18}$$

2. For the thin layer being considered in this section:

$$\mu_0 \vec{j}_s = -\frac{B_0}{\lambda}\,\frac{\sinh\dfrac{x}{\lambda}}{\cosh\dfrac{d}{2\lambda}}\,\vec{e}_y \tag{5-19}$$

$$\vec{j}_s = -\frac{B_0}{\sqrt{\mu_0 \Lambda}}\,\frac{\sinh\dfrac{x}{\lambda}}{\cosh\dfrac{d}{2\lambda}}\,\vec{e}_y\,. \tag{5-20}$$

Figure 65 illustrates the current distribution in a thin layer with a magnetic field B_0 on each side which is parallel to the layer and directed in the z direction.
On the surface at $x = d/2$ we have, for example:

$$\vec{j}_s(d/2) = -\frac{B_0}{\sqrt{\mu_0 \Lambda}}\,\frac{\sinh\dfrac{d}{2\lambda}}{\cosh\dfrac{d}{2\lambda}}\,\vec{e}_y\,. \tag{5-21}$$

As the thickness is reduced the current on the surface associated with a constant B_0 becomes ever smaller. Here it becomes very evident that as the thickness is reduced, what

---

1 Now we have decided on the directions we can write Equation (5-18) as an equation of moduli.

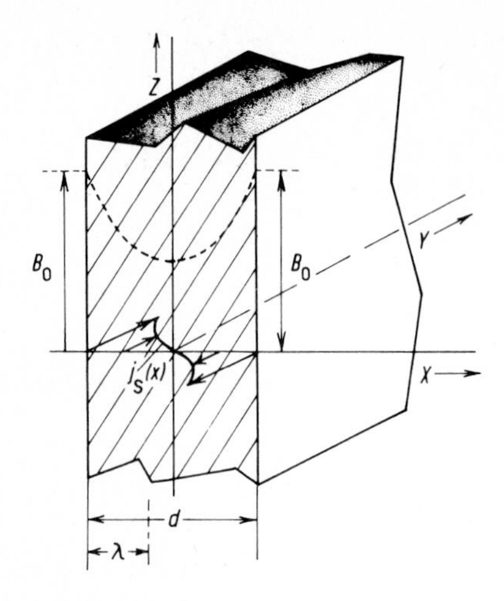

Fig. 65. Spatial variation of the density of the screening current $j_s(x)$ in a thin superconducting layer in a magnetic field parallel to the surface of the layer.

may be termed the reaction of the superconductor to the field becomes ever smaller. Ever larger values of B_0 are required to make the free enthalpy of the superconducting state equal to that of the normal state by means of the screening current.

We can now express the free enthalpies explicitly with the aid of the critical current density j_c at the surface of a superconductor. Thus, according to (5–18) and (4–15):

$$G_n - G_s = \frac{\Lambda}{2} j_c^2 V . \tag{5–22}$$

This is not a new result for "thick" superconductors with complete screening layers. In the case of thin superconductors we now have a quantity j_c that is not dependent on the geometry. It, therefore, corresponds to the thermodynamic critical field B_{cth}, while the actual critical field B_c, which is required to abolish the superconductivity, becomes ever larger with decreasing thickness.

Let us summarize once more the important result of this section. For superconductors, which are comparable in size with the penetration depth in at least one dimension, the external field B_e required to abolish superconductivity increases. This result is of great importance for what follows.

5.1.4 The intermediate state

In view of the results obtained in the previous section we must formulate a question. *Why is it possible to make the superconducting state unstable by means of a greater than critical field?* We would expect from the results in Section 5.1.3 that when the field reached the critical value the superconductor would decay into regions of very fine alternating superconducting and normalconducting phases parallel to the magnetic field. The superconducting regions could thereby be thin on account of the penetration depth and, thus, resist a larger magnetic field without becoming unstable.

Experience teaches us that this is not the case. Rather the superconducting state of a rod-shaped sample ($n_M = 0$) becomes unstable when the critical field is reached. From this we have to conclude that division into fine regions is energetically unfavorable. One simple assumption would make this understandable. Additional energy is assumed to be associated with every interface between normalconducting and superconducting regions. This prevents the breaking-up of a type I superconductor into many small regions. The energy that would be required for this would make this state less energetically favorable than the normalconducting state.

The interfacial energy determines the magnetic structure of what is known as the intermediate state. With an intermediate state we mean a situation in which a homogeneous type I superconductor is neither completely superconducting nor completely normalconducting. Such a situation can readily be produced by the geometrical shape of the sample.

Let us consider a superconducting sphere in a homogeneous external field B_e (Fig. 66). The screening current in a thin surface layer keeps the interior of the sphere free from field. This exclusion of the magnetic field leads directly, as can be seen in Fig. 66, to an intensification of the field on the surface of the sphere in the vicinity of its equator. The variation in field over the surface is given by the superimposition of the homogeneous external field B_e and the field produced by the screening current.

It is evident that the field intensification is a consequence of the geometry of the sample. For simple bodies this can be described in terms of a single number, the demagnetization factor. It can be shown that a body with an ellipsoid as external surface will have a homogeneous magnetization in a homogeneous external field. If a major axis of the ellipsoid is parallel to the external field then the magnetization is also parallel to the field[1]. This magnetization now changes the magnetic field that the sample experiences. We call the maximum value of this field the effective field B_{eff}.

This is:

$$B_{eff} = B_e - n_M \mu_0 M . \tag{5-23}$$

For a superconductor where the Meissner effect is total the magnetization M at an effective field B_{eff} is given by

$$M = -\frac{B_{eff}}{\mu_0} . \tag{5-24}$$

Whereby

$$B_{eff} = B_e + n_M B_{eff} \tag{5-25}$$

or:

$$B_{eff} = \frac{1}{1 - n_M} B_e . \tag{5-26}$$

1 It should be remembered that the magnetization of a superconductor is determined by the screening current at its surface.

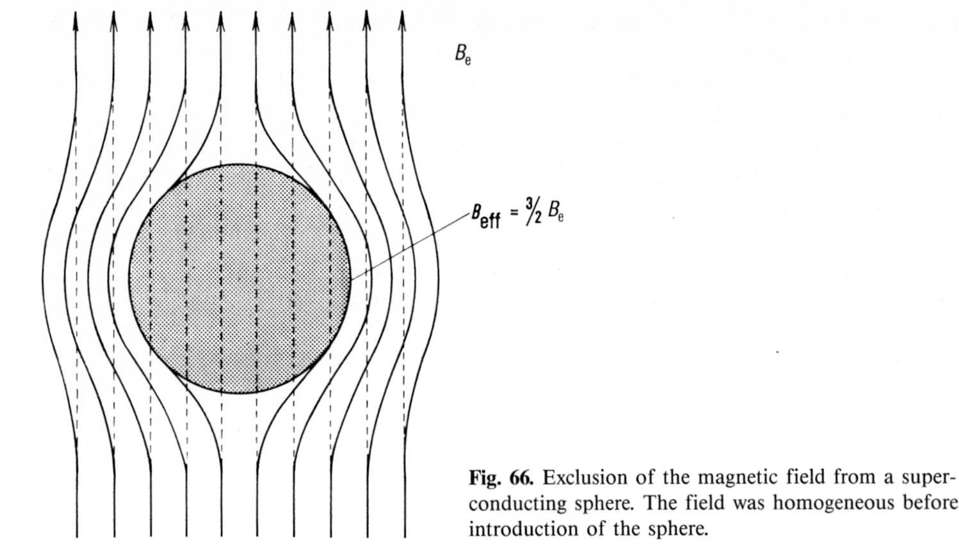

Fig. 66. Exclusion of the magnetic field from a superconducting sphere. The field was homogeneous before introduction of the sphere.

The effective magnetic field is just that field which we find at the equator immediately at the surface. So the field at the equator can be determined if the demagnetization factor n_M is known. For a sphere, for example, $n_M = 1/3$ [1]. So that for an ideal diamagnetic sphere:

$$B_{eff} = \frac{3}{2} B_e . \tag{5-27}$$

The field intensification occurring for demagnetization factors $n_M > 0$ makes it possible to find values for the external field B_e, for which the superconductor is neither completely superconducting nor completely normalconducting. Let us consider the case of the sphere more closely. We increase the external field B_e and when $B_e = (2/3)B_C$ reach the critical field B_c at the equator. If B_e is increased further the superconductivity at the equator must be destroyed. The sphere cannot become completely normalconducting, because then the field in the interior is equal to the external field and, hence, would be smaller than B_c. The superconductor goes into the intermediate state, i.e. it splits up into superconducting and normalconducting regions.

Before we discuss this separation in more detail we will first discuss the intermediate state phenomenologically. When $B_e = B_c$ the sample must be completely normalconducting. We see that exactly B_c is observed at the equator over the whole range $(2/3)B_c < B_e < B_c$. As B_e is increased the proportion of the interior of the sphere that is normalconducting increases in exactly such a manner as to ensure that the remaining field exclusion at the equator yields B_c. You might say that the demagnetization factor n_M is dependent on B_e in the intermediate state. If we determine the magnetic flux through a

1 In the case of a wire with a circular cross section in a magnetic field perpendicular to the axis $n_M = 1/2$.

coil at the equator then this flux Φ increases monotonically with B_e. Figure 67 illustrates this phenomenon.

The magnetic structure of a sphere in the intermediate state has been the subject of detailed investigation [108]. Normalconducting and superconducting regions are present simultaneously. We, therefore, have phase boundaries between the normal and the superphase. The phase boundaries are stabilized in the presence of the critical field parallel to the magnetic field. *The phase boundaries of intermediate state structures must run parallel to the magnetic field.* This statement is always true. This can lead to surprising results in the case of complicated intermediate state structures, such as superconductors subject to an external field and simultaneously carrying a current (see Section 6.1).

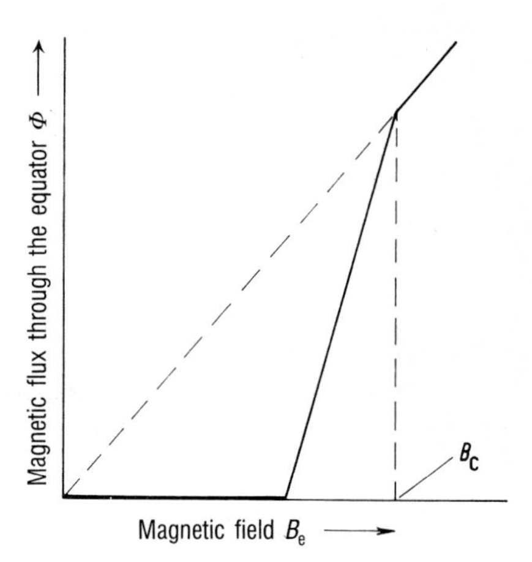

Fig. 67. Magnetic flux Φ through the equatorial plane of a sphere as a function of the external field B_e.

Since the creation of phase boundaries requires energy the partition cannot, as we have seen, be taken to any arbitrary degree of fineness. The sample must take up a state where large, normally conducting and superconducting regions are present side by side. The partition is determined by the requirement that the free enthalpy of the system is at a minimum. This requirement allows the calculation of the interfacial energy from measurements of the structure of the intermediate state.

We cannot go into the details of such calculations. We will just mention some of the more simple results. The transition of a sphere to the intermediate state should take place at $B_e = (2/3)B_c$. This transition is associated with the creation of interfaces. Because of this a finite energy contribution is necessary from the magnetic field, on account of the positive interfacial energy. This means that the transition to the intermediate state does not take place at exactly $B_e = (2/3)B_c$, but at a slightly higher value. The field is, as it were, still excluded above the critical value until the stored field energy is sufficient to create the necessary interfaces. This additional field has been observed for wires in a magnetic field perpendicular to their axis and explained in the manner described above [II].

Detailed investigations have been made of the structures that can occur in the intermediate state. Various methods were employed in these. In the case of a sphere of tin the geometric disposition of the normalconducting and superconducting states was explored by scanning with small bismuth wires [108]. For this purpose the sphere was split through

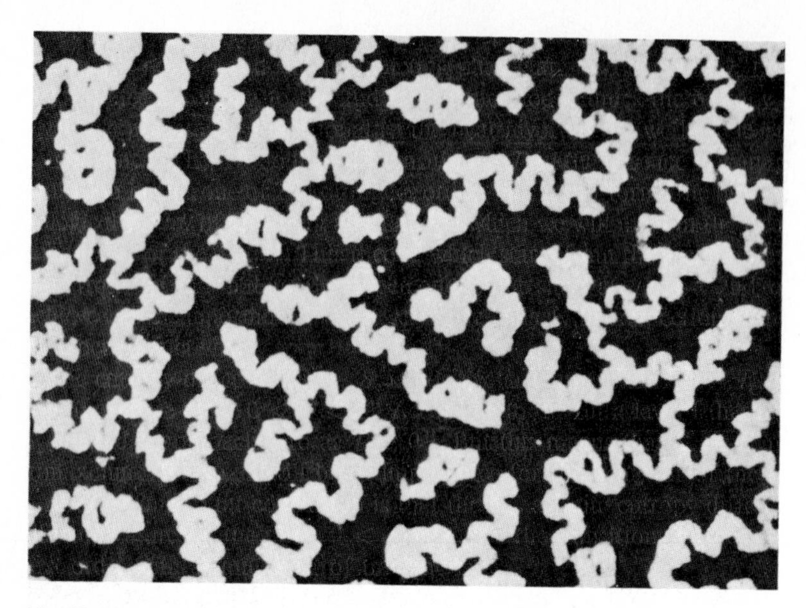

Fig. 68. Structure of the intermediate state of an indium plate. The pale structures are normalconducting regions. In purity 99.999 at ‰., thickness 11.7 mm, diameter 38 mm, $B_e/B_{cth} = 0.1$, $T = 1.98$ K; T_c of In is 3.42 K; transition N → S; enlarged 5 x; (with the kind permission of the authors from: F. Haenssler and L. Rinderer: Helv. Phys. Acta *40*, 659 (1967), Fig. 35).

a plane at its equator and the two halves were then fixed so that there was a thin gap of a few tenths of a millimeter between the two halves. The tiny bismuth wires could be inserted as field sensors into this gap. Since the electrical resistance of the bismuth is strongly dependent on the field strength it could be monitored for this purpose. Normal regions were recognized on account of the presence of a magnetic field. The superconducting regions, on the other hand, are free from field. The small gap between the two hemispheres scarcely distorts these regions whose interfaces run parallel to the magnetic field. "Maps" of the intermediate state have been produced by this technique [109].

Appreciably more direct methods, which provide an immediate overall picture of the structure, have been employed recently. These are:

1. Coating the superconducting regions with a diamagnetic powder.
2. Making the normalconducting regions visible by means of the Faraday effect.

In method 1 the sample in the intermediate state is sprinkled with a fine powder of a superconducting substance. Niobium powder is usually employed [110], because its relatively high transition temperature of 9.2 K means that it still stays fully superconducting in the magnetic fields necessary for stabilization of the intermediate states of many superconductors with lower T_c. The tiny grains of superconducting niobium are ideal diamagnets. They are thrust out of regions of high magnetic field and collect on the surfaces of superconducting regions. Figure 68 is an illustration of an intermediate state structure of a plate obtained in this way [110]. Another example is reproduced in Fig. 87 for the case of a current-stabilized intermediate state of a wire [133].

In method 2 a thin sheet of a magneto-optically active[1] material is placed on the surface of the superconductor which is prepared as a mirror. If this sheet is now irradiated with linearly polarized light and the light reflected at the mirror surface of the superconductor observed, then the plane of polarization of the reflected light is displaced somewhat from the original angle wherever there is a magnetic field, that is in the regions which are normalconducting. When suitable polarization optics are employed it is possible to make the normal and superconducting regions visible as light and dark structures. This method has the advantage over the powder-decoration method in that it allows the observation of processes of movement in the intermediate state structures, that is changes with time. Very impressive motion pictures have been made of the processes taking place as the intermediate state is passed through. Figure 69 reproduces a picture of an intermediate state structure, that was obtained by this technique. The limits of resolution of this method are about 0.5 μm at present [111]. It is not yet possible to use this method to investigate the finer structures which we will meet for type II superconductors in the next section.

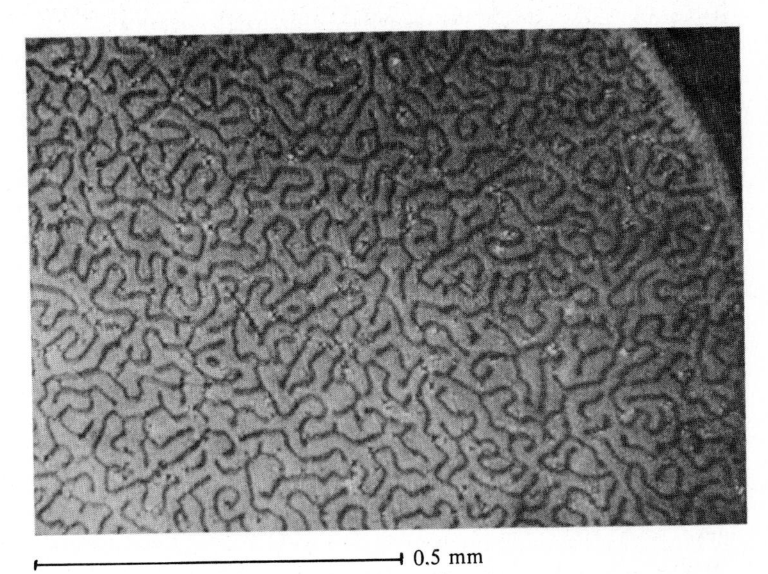

Fig. 69. Structure of the intermediate state visualized by means of the Faraday effect. Pb layer 7 μm thick, magnetooptically active layer of EuS and EuF_2 ca. 100 nm thick. Magnetic field $B = 0.77\ B_c$ perpendicular to layer. The dark portions correspond to the superconducting regions [111]. (Reproduced by kind permission of Mr. Kirchner, Siemens Research Laboratories, Munich).

1 Such materials e.g. Ce-glasses have the property of rotating the plane of polarization of light passing through the material when they are in a magnetic field. This effect is known as the Faraday effect. Michael Faraday, English physicist, 22.9.1791–25.8.1867.

5.1.5 The interfacial energy

In order to understand the behavior of type I superconductors in a magnetic field and especially in the intermediate state we have to assume that a certain finite energy per unit area is required for the formation of an interface between a superconducting and a normal region. This positive [1] interfacial energy prevents splitting up into very fine superconducting regions.

We will now look more closely at this interfacial energy. We will then see that conditions are certainly imaginable where a phase boundary does not require an energy input, but rather a state is stable where both the magnetic field and the Cooper-pair density n_C vary periodically. Superconductors that fulfil these conditions are known as type II superconductors. The consideration of the interfacial energy leads us readily to this type of superconductors whose properties will be discussed in the following sections.

First it must be emphasized once more that we assume homogeneous materials and constant temperatures. Under these conditions there must be an interface at exactly the critical field B_{cth}. In the normalconducting region it holds that $B \geqslant B_{cth}$, while in the superconducting region B decays within a thin layer with a thickness of ca. λ.

The difference between a normalconducting and a superconducting region of one and the same material lies in the fact that the density of Cooper pairs n_C in the normalconducting region is equal to zero, while the density $n_C(T)$ in the superconducting regions is finite and depends on the temperature and nature of the material. We remember that condensation to Cooper pairs results in a reduction of the free enthalpy, which makes the superconductor thermodynamically stable with respect to the normal conductor at temperatures below T_c.

Now it is decisive for our consideration that the density of Cooper pairs $n_C(T)$ cannot suddenly jump from $n_C(T)$ to zero at the phase boundary. The strong correlation between Cooper pairs (see Section 2.2) requires that a spacial variation of $n_C(T)$ can only occur over a distance that is larger than a characteristic length, which is known as the coherence length ξ_{GL}. The index GL indicates that this coherence length is a decisive quantity in the Ginsburg-Landau theory of type II superconductors.

The conditions at a phase boundary of this type are illustrated schematically in Fig. 70. In the normal region to the left ($x < 0$) the magnetic field is exactly B_{cth} or larger. So that the normal state is stabilized in this region, since the exclusion of the magnetic field would require more free enthalpy than can be supplied by transition to the superconducting state [2]. In the superconductor ($x > 0$) the density of Cooper pairs increases to the equilibrium value $n_C T$ within the coherence length. We have assumed here that the following applies to a superconductor:

$$\xi_{GL} > \lambda \,. \tag{5-28}$$

1 We refer to the energies as positive because energy must be supplied for the formation of the interface of the system, here the superconductor.

2 The boundary will only be stable when a shift to the left or right makes the magnetic field increase or decrease.

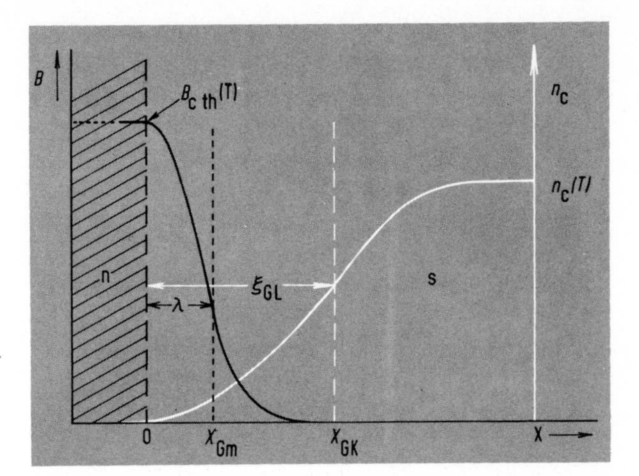

Fig. 70. Spatial variation of B and n_C in an interface between normal conductivity and superconductivity within a homogeneous material at temperature T. x_{Gm} = "magnetic boundary", x_{GK} = "condensation boundary".

We now have to consider two energy contributions to the interface, namely the energy associated with the exclusion of the magnetic field E_B and the energy released by "condensation" of the Cooper pairs E_C. In a normal conductor $E_B = E_C = 0$. The magnetic field is not excluded and there are no Cooper pairs present. When a critical field B_{cth} is applied to the interface, which was our assumption, then deep in the interior of the superconducting region we have $-E_B = E_C = (1/2\mu_0)\, B_{cth}^2 V$ (see Equation (4-15)). The full energy of "condensation" is just equal to the exclusion energy.

Neither energy reaches its full value in the boundary layer. The magnetic field is not completely screened but penetrates to a depth of λ. The exclusion energy is smaller by an amount

$$\Delta E_B = A\,\lambda\,\frac{1}{2\mu_0}\,B_{cth}^2 \tag{5-29}$$

(A = area of the interface under consideration)

than it would be on complete exclusion to the boundary (Fig. 70, $x = 0$). But the condensation energy is also reduced in the boundary layer because the Cooper-pair density is below the equilibrium value of $n_C(T)$. We can express this diminution in the energy of condensation as:

$$\Delta E_C = A\,\xi_{GL}\,\frac{1}{2\mu_0}\,B_{cth}^2\,. \tag{5-30}$$

(A = area of interface under consideration, ξ_{GL} = coherence length for Ginsburg-Landau theory).

Here we have defined λ and ξ_{GL} in such a manner that the energy contributions ΔE_B and ΔE_C take up the same value, as in the case where, on the one hand, the magnetic field penetrates completely to a depth of λ and then falls discontinuously to zero and, on the other hand, the Cooper-pair density first jumps to its full value of $n_C(T)$ at ξ_{GL}.

We have, therefore, set up, as it were, a "magnetic boundary" and a "condensation boundary". If, as here it is assumed, $\xi_{GL} > \lambda$ then:

$$\Delta E_C - \Delta E_B = (\xi_{GL} - \lambda)\, S \frac{1}{2\mu_0} B_{cth}^2 > 0\,. \tag{5-31}$$

The loss of condensation energy is then greater than the gain in exclusion energy. In order to construct such a boundary we must supply the system with an energy contribution of α_{gr} per unit of boundary, this is given by:

$$\alpha_{gr} = (\xi_{GL} - \lambda) \frac{1}{2\mu_0} B_{cth}^2\,. \tag{5-32}$$

Figure 71 illustrates the spacial variation of the energy densities ε_B and ε_C and their difference which passes through a maximum.

We see clearly that the formation of a boundary layer between the normalconducting and the superconducting regions in homogeneous matter requires energy if ξ_{GL} is greater than λ [1]. If $\xi_{GL} = \lambda$, i.e. the magnetic boundary coincides with the condensation

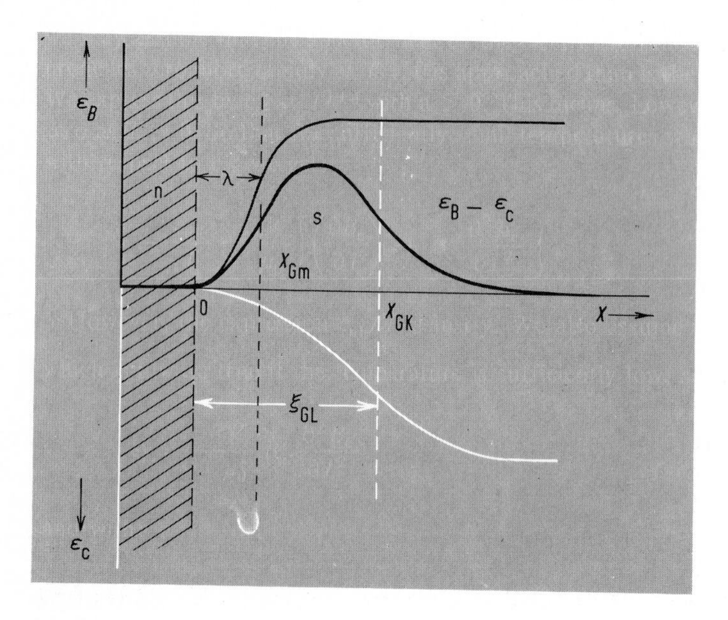

Fig. 71. Spatial variation of the exclusion energy ε_B and the condensation energy ε_C per unit volume at the interface

$$\int_0^{x \gg x_{GK}} (\epsilon_B - \epsilon_C)\, A\, dx = (\xi_{GL} - \lambda)\, A \frac{1}{2\mu_0} B_{cth}^2\,.$$

---

1 It must be mentioned that the depth of penetration of a magnetic field at a boundary layer in a metal is somewhat different than at the boundary with an insulator. The reason is the spacial variation of the density of Cooper pairs. We have not taken this difference into account until now.

Table 7. Characteristic lengths.

Characteristic length	Temperature dependence	Dependence on l^*
λ	$\lambda(T,\infty) = \lambda(0,\infty) \cdot \left\{1 - \left(\frac{T}{T_c}\right)^4\right\}^{-1/2}$	$\lambda(0,l^*) = \lambda(0,\infty) \cdot \left\{1 + \frac{\xi_{GL}(0,\infty)}{l^*}\right\}^{1/2}$
ξ_{GL}	$\xi_{GL}(T,\infty) = \xi_{GL}(0,\infty) \cdot \left\{1 - \frac{T}{T_c}\right\}^{-1/2}$ for $T \to T_c$	$\xi_{GL}(0,l^*) = \{\xi_{GL}(0,\infty) \cdot l^*\}^{1/2}$ for $l^* \ll \xi_{GL}(0,\infty)$

A third characteristic length ξ_{Co} gives the distance over which correlation to Cooper pairs is effective. This length is almost temperature-independent and in the limiting case $l^* \to 0$ becomes approximately equal to l^*.

Note: The Ginsburg-Landau coherence length ξ_{GL} is, under all conditions, greater than the mean extent of a Cooper pair ξ_{Co}. The number density of Cooper pairs can certainly not vary at distances which are smaller than the mean extent of a Cooper pair.

boundary, then $\alpha_{gr} = 0$ and if $\xi_{GL} > \lambda$ a negative boundary energy will formally be acquired. In this case a new state known as the "mixed state" is created. It must, however, be emphasized here that this treatment according to the Ginsburg-Landau theory only applies in the region of T_c.

The important new quantity in this treatment is the coherence length ξ_{GL}, which provides a measure of the "rigidity" of Cooper-pair density. We can understand this "rigidity" from our picture of the strong correlation of the Cooper pairs with each other. The historical development, however, was quite different. A coherence length was postulated by Pippard (1951) [112] to explain the results of high frequency measurements in superconductors long before the development of the BCS theory [1]. The phenomenological theory of Ginsburg and Landau [113], where the coherence length was of decisive importance, was also developed in 1950.

We have now introduced 3 characteristic lengths:

1. the depth of penetration λ
2. the mean extension of a Cooper pair ξ_{Co}
3. the coherence length ξ_{GL}.

The depth of penetration is a measure of the decay of a magnetic field in the interior of a superconductor.

The mean extension of a Cooper pair ξ_{Co} is a measure of the distance over which the correlation to Cooper pairs $\{+\vec{k}_\uparrow, -\vec{k}_\downarrow\}$ is effective.

1 With this coherence length Pippard was able to extend the London theory to the effect that the supercurrent density $j_s(x, y, z)$ at point (x, y, z) was no longer assumed to depend only on the magnetic field $B(x, y, z)$ but also on a mean field in a region with an extent of the order of magnitude of the coherence length.

The coherence length ξ_{GL} is a measure of the smallest length over which the Cooper-pair density can vary.

These three characteristic lengths depend in various ways on the temperature T and on the mean free path l^* [1]. Table 7 lists a summary of the important dependences. The dependence on temperature is immediately evident, if it is remembered that the Cooper-pair density is a function of the temperature. The influence of the mean free path is also understandable, at least in a purely qualitative sense, from the fact that the electron-phonon interaction determines both the mean free path and the correlation to Cooper pairs.

The variables listed in Table 7 can be expressed in different ways by means of other important parameters of a superconductor. There are also relationships between them, these naturally depend on the theoretical model chosen and the approximation employed. Only a few examples will be given.

For example:

a) $$\xi_0 = 0.18 \frac{\hbar v_F}{k_B T_c} \qquad \text{BCS theory} \qquad (5\text{-}33)$$

with

$$\xi_0 = \frac{2}{\pi} \xi_{Co} (T = 0, l^* \Rightarrow \infty)$$

(v_F = Fermi velocity, k_B = Boltzmann constant)

b) $$\lambda = \sqrt{\frac{m_C}{n_C 4e^2 \mu_0}} \qquad \text{London theory} \qquad (5\text{-}34)$$

(m_C and n_C are the mass and number density of Cooper-pairs)

Table 8. Experimentally determined values of the penetration depth $\lambda(T = 0)$ and the coherence length $\xi_{GL}(T = 0)$.

Material	In	Pb	Sn	Ta	Nb
λ (0, ∞) in nm	24 [a)]	32 [b)]	25– 36 * [) d)]	35 [e)]	32 [e)]
ξ_{GL} (0, ∞) in nm	360 [a)]	510 [c)]	120–230 * [) d)]	92.5 [e)]	39 [e)]

* Single crystals of Sn have considerable anisotropy. The penetration depth and the coherence length depend on the orientation of the magnetic field to the crystal axes.

References: a) P. Michael and D. S. McLachlan: J. Low Temp. Phys. *14,* 607 (1974). b) R. H. Kercher and D. M. Ginsberg: Phys. Rev. *B10,* 1916 (1974). c) J. J. Hauser: Phys. Rev. *B10,* 2792 (1974). d) P. C. L. Tai, R. M. Beasley and M. Tinkham: Phys. Rev. *B11,* 411 (1975). e) J. Auer and H. Ullmaier: Phys. Rev. *B7,* 136 (1973). Detailed information on λ and ξ_{GL} values is given in "Superconductivity Data", edited by H. Behrens and G. Ebel, Fachinformationszentrum Energie, Physik, Mathematik, Karlsruhe 1982.

Note: The experimentally determined values of $\lambda(0)$ are always (sometimes considerably) greater than those calculated from Eq. (5–34). The extent of the Cooper pairs ξ_{Co} plays a role here. This means that, in the London theory (Eq. (5–1) and (5–2)), it is necessary to reject local interrelations between $\vec{B}$ and $\vec{j}_s$.

1 The mean free path l^* is the distance an electron can travel on average between two collisions.

c) $$\xi_{GL} \simeq \xi_{Co} \frac{\lambda(T, l^*)}{\lambda(0, \infty)} . \tag{5-35}$$

Such relationships are usually only good approximations for particular regions of T and l^*. Their derivation from the microscopic theory generally requires considerable effort.

Naturally variables such as λ, ξ_{GL} and ξ_{Co} are also dependent on the characteristic parameters of solids, i.e. they are constants of matter, Equation (5-33) and (5-34) give simple examples. Relatively little is known concerning these relationships. Table 8 lists some experimentally determined values of λ and ξ_{GL}.

5.2 Type II superconductors

The consideration of a boundary layer between the normal conducting and superconducting regions (Section 5.1.5) has shown us that in the case $\lambda > \xi_{GL}$ the formation of the boundary layer can be accompanied by a gain in energy. The following conditions must be fulfilled:

$$\xi_{GL}\, A \frac{1}{2\mu_0} B_{cth}^2 - \lambda\, A \frac{1}{2\mu_0} B^2 < 0 \tag{5-31}$$

that is:

$$\xi_{GL}\, B_{cth}^2 < \lambda B^2 \tag{5-36}$$

or:

$$\frac{B_{cth}^2}{B^2} < \frac{\lambda}{\xi_{GL}} . \tag{5-37}$$

We accordingly expect that for superconductors where λ is greater than ξ_{GL}, the magnetic field will be able to penetrate the superconductor even at magnetic fields B smaller than B_{cth} and, thus, cause local variations in B and the Cooper-pair density n_C to occur, similar to those which take place in a boundary layer.

From the relationships listed in Table 7 it can be seen that we can always fulfil the condition $\xi_{GL} < \lambda$ by making the mean free path l^* sufficiently short. According to Table 7 column 3 λ increases slightly with increasing l^*; on the other hand, ξ_{GL} falls proportional to $\sqrt{l^*}$

A reduction of the mean free path is easily achieved by adding a small quantity of impurity to the superconductor under observation. The foreign atoms will scatter the electrons and, thus, reduce their mean free paths. Alloys are in fact – as we will see – generally type II superconductors.

The unusual behavior of superconducting alloys was recognized experimentally in the 1930s. DeHaas and Voogd [114] found that lead-bismuth alloys still remained superconducting in magnetic fields of up to ca. 2 T, that is in fields more than a factor of 20 greater than the critical field for pure lead. Attempts were made to explain these high critical fields by means of the sponge model of Mendelssohn [115]. According to this a network of fine precipitates, say at the grain boundaries, were the cause of the high critical field. If these pricipitates were small with respect to the penetration depth in at least one direction then the high critical field could be understood qualitatively on geometrical grounds alone (see Section 5.1.3). We know today that this sponge model certainly has its justification in many alloys, but that homogeneous superconductors also exist, namely the type II superconductors, which stay superconducting up to very high magnetic fields.

It is very easy to understand qualitatively why a superconductor can tolerate high magnetic fields if it allows the magnetic field to penetrate it at least to some extent. The exclusion energy, which increases the free energy of the superconducting state in a magnetic field, will be reduced on penetration of the magnetic field. So that larger external fields will be required to raise the free enthalpy of the superconducting state until it becomes the same as that of the normal state. We were able to understand the increase in the effective critical field for thin superconductors, into which the magnetic field is able to penetrate, by using just such arguments.

Naturally we have to expect (see Section 5.1.5) that when the field penetration occurs there will also be a reduction in the condensation energy. However, the gain on account of the reduction of the exlusion energy can, if $\lambda \gg \xi_{GL}$, be appreciably greater than the loss of condensation energy.

Type II superconductors are quantitatively described by means of what is known as the GLAG theory. With the phenomenological theory of Ginsburg and Landau [113], which is an extension of the London theory [73] – the salient points of this theory are sketched out in Appendix B – as his starting point Abrikosov [116] found concrete solutions for certain conditions. Finally Gorkov [117] set up the connection with the BCS theory. The resulting theory, the GLAG theory, owes its name to the four scientists who developed it.

This phenomenological theory has acquired great importance not least because it covers the technologically interesting superconductors. The large superconducting magnets that have already been built (see Section 9.1) would certainly not have been feasible if it had not been possible to acquire a fundamental understanding of type II superconductors. We will treat the basic properties of type II superconductors in the sections that follow.

5.2.1 The magnetization curve of type II superconductors

The differences between type I and type II superconductors are particularly apparent in the shape of the magnetization curve. The magnetization curve of a type II superconductor is schematically illustrated in Fig. 72. We are again considering a rod-shaped sample, whose demagnetization factor is practically zero. At a certain field B_{c1} the gain in expulsion energy on penetration of the field becomes greater than the loss in condensation energy because of the spacial variation this causes in the Cooper-pair density n_C. The magnetic field then penetrates the superconductor. We will discuss the appearance of the field distribution later in this section. Here we will just consider the magnetization curve.

The penetration of the magnetic field causes the magnetization of the superconductor to decrease monotonically with increasing field. The magnetization becomes zero at field

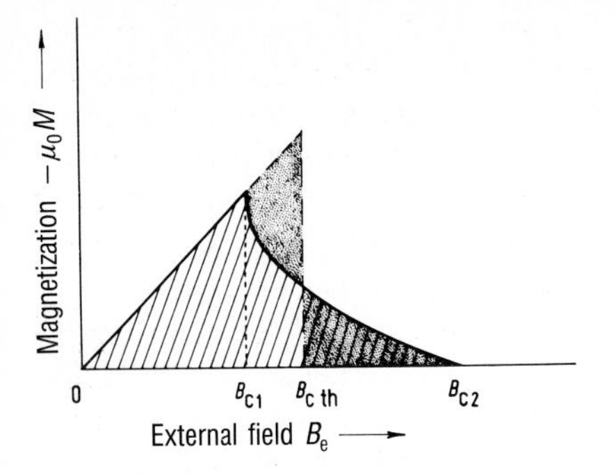

Fig. 72. Magnetization curve of a type II superconductor. Rod-shaped sample with $n_M \approx 0$. The definition of B_{cth} requires that the dotted regions be equally large.

B_{c2}, the superconductivity has been abolished by the external field[1]. The magnetic fields B_{c1} and B_{c2} are known as the lower and upper critical fields.

From general thermodynamic considerations (see Sections 4.1 and 5.1.1) the difference between the free energies at constant temperature T and constant pressure p is given by:

$$G_n - G_s = -\int_0^{B_{c2}} MV\,dB\,. \tag{4-13}$$

If we ignore (as in Section 4.1) the very small changes in volume V we can extract V from the integral as a constant. The integral is proportional to the area under the magnetization curve:

$$\int_0^{B_{c2}} M\,dB = \mu_0 A_M \tag{5-38}$$

A_M is the area under the magnetization curve[2].

Let us take for comparison a type I superconductor, that possesses the same difference in free energies, its magnetization curve is then given by the broken line. The areas under the two magnetization curves must be identical. We can now see clearly the difference between a type I and a type II superconductor. While the type I superconductor completely excludes the magnetic field up to the thermodynamic critical value B_{cth}, that is remains in the Meissner phase up to this field, the type II superconductor goes at the lower critical field of B_{c1} into a state with field penetration, into the Shubnikov phase. In this Shubnikov phase, also known as the mixed phase, the magnetization then decreases monotonically with increasing field and only disappears completely at the upper critical field B_{c2} which can be considerably higher than the thermodynamic field of the corresponding type I superconductor.

---

1 Under certain conditions superconductivity can be retained in a thin surface layer for fields $B_{c2} < B_e \leqslant 1.7\,B_{c2}$. We will ignore this surface superconductivity here.

2 In Fig. 72 (just as in Fig. 59) we have plotted the quantity $\mu_0 M$ instead of the magnetization M. In the case of ideal exclusion $M = -B/\mu_0$ or $-\mu_0 M = B$. So for the conditions chosen here with the same units for abscissa and ordinate the plot is a straight line at 45°.

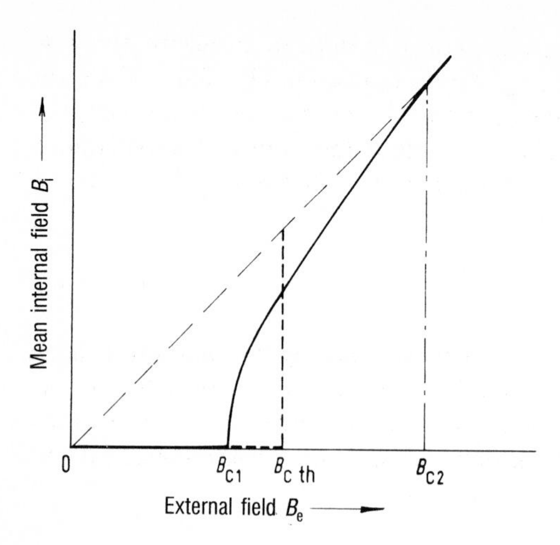

Fig. 73. Mean magnetic field in the interior of a type II superconductor as a function of the external field.

This important result is illustrated once more in Fig. 73. Here the mean magnetic field in the interior of the sample is plotted against the external field. Figure 73 is completely analogous to Fig. 60 (Section 5.1.1). At B_{c1} the field begins to penetrate the superconductor, but it is only when B_{c2} is reached that the mean internal field equals the external field or, expressed another way, the magnetization equals zero. The behavior of a corresponding type I superconductor is illustrated here too with a broken line. As a "corresponding" type I superconductor we mean again a type I superconductor with the same free enthalpy difference $G_n - G_s$ as the type II superconductor.

As we have already stated we would expect to be able to convert a type I superconductor into a type II superconductor by shortening the mean free paths of the electrons sufficiently. This prediction of the Ginsburg-Landau theory has been confirmed in a very convincing manner. All type I superconductors can be converted to type II superconductors by alloying into them foreign atoms which reduce the mean free paths of the electrons. Figure 74 gives one example from many. Here the magnetization curves of pure lead and of a lead-indium alloy containing 13.9 atom% In are reproduced [118]. The pure lead is a type I superconductor. The alloy exhibits the typical behavior of a type II superconductor. The superconductivity of the alloy is only completely abolished by an upper critical field of 0.24 T (2400 G), while at the temperature of 4.2 K chosen for this experiment the thermodynamic field of pure lead is only 0.055 T (550 G).

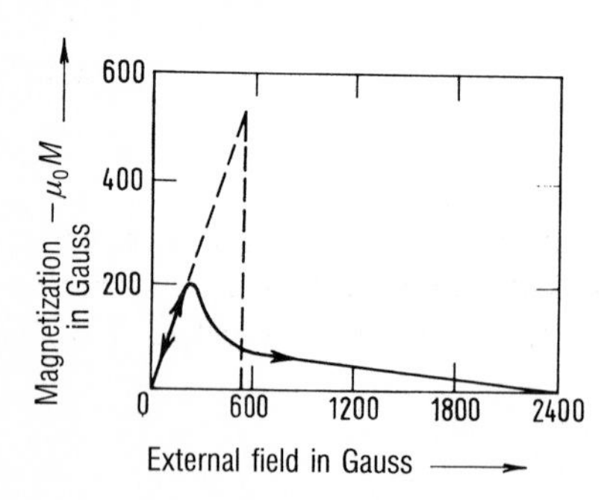

Fig. 74. Magnetization curve of lead containing 13.9 atom% indium. Rod-shaped sample with small demagnetization factor (after [118]) (1 G = 10^{-4} T).

After this rather qualitative approach we must now introduce some quantitative relationships between the fields B_{c1}, B_{c2} and B_{cth}. All these relationships follow from the GLAG theory. It should always be remembered in this context that this theory only applies in the neighborhood of T_c. The decisive variable is the ratio of the penetration depth λ to the coherence length ξ_{GL}. This ratio is known as the Ginsburg-Landau parameter κ.

$$\kappa = \frac{\lambda}{\xi_{GL}} . \tag{5-39}$$

It should be pointed out once more that the extension made by the Ginsburg-Landau theory over the London theory lies in the assumption of a new characteristic length ξ_{GL}. The London theory only postulated the penetration depth λ, which was coupled with a spacially constant density of superconducting electrons n_s. With the introduction of the Cooper pair the following expression is obtained:

$$\lambda = \sqrt{\frac{m_C}{n_C 4e^2 \mu_0}} . \tag{5-34}$$

In the Ginsburg-Landau theory the quantity n_s is allowed to vary locally – today we call it the Cooper-pair density n_C. With brilliant insight the proposers of this theory took account of the strong correlation of superconducting electrons by means of a characteristic length ξ_{GL} for the spacial variation of n_C.

We can formulate all the quantitative relationships quite simply with the aid of the Ginsburg-Landau parameter. The upper critical field B_{c2}, for example, is given by the simple relationship:

$$B_{c2} = \sqrt{2}\ \kappa B_{cth} . \tag{5-40}$$

Where B_{cth} is defined for every type II superconductor in terms of the differences of the free enthalpies (*cf.* Equation (4-15)):

$$G_n - G_s = \frac{1}{2\mu_0} V B_{cth}^2 . \tag{4-15}$$

The calculation of B_{c1} is appreciably more complex. For the limiting case $\kappa \gg 1/\sqrt{2}$ he following is obtained according to Abrikosov [116]:

$$B_{C1} = \frac{1}{2\kappa} (\ln \kappa + 0.08) B_{cth} . \tag{5-41}$$

We see that as κ increases, B_{c1} becomes smaller and B_{c2} larger.

Since in an alloy the mean free path l^* decreases monotonically with increasing content of foreign atoms, we will be able to give a certain "critical" percentage content for every system where the host metal becomes a type II superconductor. The transition is laid down by the requirement:

$$B_{c2} \geqslant B_{cth} .$$

According to Equation (5–40) this requirement is equivalent to:

$$\kappa \geqslant 1/\sqrt{2}\ .$$

So type I and type II superconductors are distinguished simply by the size of κ.

$$\begin{aligned}&\text{type I superconductors:} \quad \kappa < 1/\sqrt{2}\\&\text{type II superconductors:} \quad \kappa > 1/\sqrt{2}\end{aligned} \qquad (5\text{–}42)$$

Strictly this distinction is only applicable in the neighborhood of T_c. When $T < T_c$ then there is, at values of κ that are slightly greater than $1/\sqrt{2}$, a transition into a state where Meissner and Shubnikov phases are present side by side [130].

Some values for κ in mixed In-Pb crystals are listed in Table 9. The transition to the type II superconductivity takes place at ca. 1.5 atom % Bi. Similar critical concentrations are found for other systems. It can, therefore, be seen how easy it is to produce a type II superconductor. The κ_1 values in line 2 were obtained from determinations of the upper critical field B_{c2}. The values in line 3, on the other hand, were calculated from a formula derived by Gorkov and Goodman. This represents approximately the relationship between κ and the mean free path of the electrons. The formula is [119]:

$$\kappa = \kappa_0 + 7.5 \times 10^3 \, \frac{\text{cm}^2\,\text{Grad}^2}{\text{Ohm} \cdot \text{erg}} \, \varrho \, \gamma^{1/2}\ . \qquad (5\text{–}43)$$

Where κ_0 is the Ginsburg-Landau parameter of the pure superconductor, i.e for the limiting case where $l^* \rightarrow \infty$, γ is the Sommerfeld coefficient for the specific heat of the electron system in erg/cm^3 per degree^2 and ϱ is the specific resistance in the normalconducting state in ohm cm. The mean free path is expressed in this formula by the quantities ϱ and γ.

A comparison of lines 2 and 3 of Table 9 shows that the values of κ, determined in quite different ways, are in good agreement with each other.

In order to be able to apply Equation (5–43) it is necessary to know the value of κ_0. The value of κ_0 can be obtained by extrapolation of the κ values of alloys back to zero foreign atom concentration (Table 10).

Table 9. Values for κ for In-Bi mixed crystals.

Atom-% Bi	1.55	1.70	1.80	2.0	2.5	4.0
κ_1 at T_C	0.76	0.88	0.91	1.10	1.25	1.46
κ_2 at T_C	0.74	0.85	0.88	1.15	1.29	1.53

$$\kappa_1 = \frac{B_{c2}}{\sqrt{2}\, B_{cth}}\ ; \qquad \kappa_2 = \kappa_0 + 7.5 \times 10^3 \varrho \gamma^{1/2}$$

(after T. Kinsel, E.A. Lynton and B. Serin: Rev. Mod. Phys. *36*, 105 (1964)).

Table 10. Values of κ_0 for superconducting elements.

Element	Al	In	Pb	Sn	Ta	Tl	Nb	V
κ_0 at T_C	0.03	0.06	0.4	0.1	0.35	0.3	0.8	0.85

See also: "Superconductivity Data", edited by H. Behrens and G. Ebel, Fachinformationszentrum Energie, Physik, Mathematik, Karlsruhe 1982.

The question arises as to whether the constant κ_0 has any meaning for a type I superconductor. In fact the GLAG theory does endow it with a meaning. Equation (5–40) defines for this case a field B_{c2}, that is smaller than B_{cth}. This field provides us with an absolute limit for so- called "supercooling experiments".

The transition from the normal state to the superconducting state in a magnetic field is a 1st order transition (see Section 4.1). Supercooling and superheating effects can occur in such transitions. For example, it is possible, by cautious cooling, to bring water several degrees below its freezing point without ice forming.

Translated to superconductivity this means, for example, that it is possible to lower a magnetic field below the critical level without the superconducting state immediately appearing. The size of this effect depends fortuitously on the conditions of the experiment. The normal state is indeed unstable at $B < B_{cth}$, but the new phase cannot form. A seed must first appear for this phase so that it can grow. The prediction is that in such experiments the field cannot fall below $B_{c2} = \sqrt{2} \times \kappa_0 B_{cth}$. So if we have a superconductor, whose κ_0 is very close to $1/\sqrt{2}$, then it is practically impossible to observe supercooling effects. In the case of superconductors such as pure Al or pure In, however, the range of magnetic supercooling that can take place is very large. This prediction has been confirmed experimentally.

Finally yet another means of determining κ should be mentioned. The gradient of the magnetization curves at B_{c2} is determined by κ. For:

$$(\mathrm{d}M/\mathrm{d}B)_{B_{c2}} = -\frac{1}{1.16\,(2\kappa^2 - 1)}\,. \tag{5–44}$$

We now have discussed four possibilities for the determination of κ. All 4 determination methods yield close agreement for the value of κ in the neighborhood of T_c. On the other hand, the different methods give different values of κ when $T \ll T_c$. This is a result of the temperature dependence of the various quantities[1]; this means that the Ginsburg-Landau parameter is also temperature-dependent.

The two critical fields B_{c1} and B_{c2} are also naturally functions of temperature on account of their dependence on B_{cth}. Here, however, it must also be remembered that, from what has been said above, κ is also a function of temperature. The 3 critical fields of a type II superconductor are represented schematically in Fig. 75 neglecting all com-

1 The GLAG theory (Appendix B) was specifically developed for use in regions close to T_c. It is, therefore, not very surprising that extrapolation to lower temperatures leads to very complicated relationships.

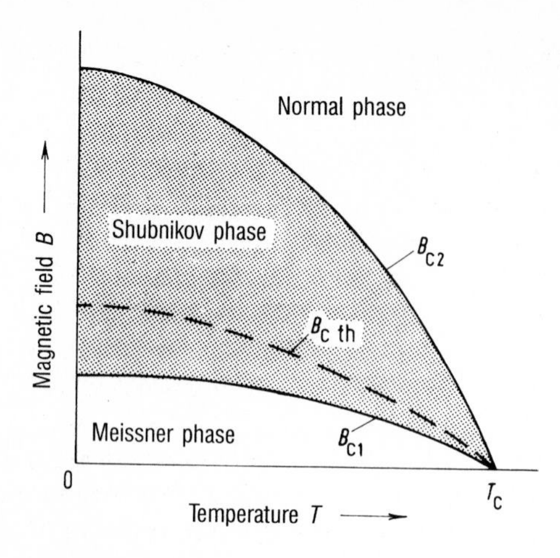

Fig. 75. Phase diagram of a type II superconductor (schematic representation).

plicating details. Here we can clearly recognize the regions of stability of the various phases. The Meissner phase, the phase with complete exclusion, is stable below B_{c1}. The mixed state, the Shubnikov phase, is stable between B_{c1} and B_{c2} and finally we have the normal phase above B_{c2}. Figure 76 reproduces this phase diagram for the indium-bismuth alloy (In + 4 at% Bi) [120].

To conclude this discussion of the magnetization curves of type II superconductors we cannot emphasize too strongly that all of the discussion in this section has been concerned with magnetization curves which can be passed through reversibly. That is, the same curve must be passed through with increasing magnetic field as is passed through with diminishing magnetic field. This is only the case when the thermodynamic equilibri-

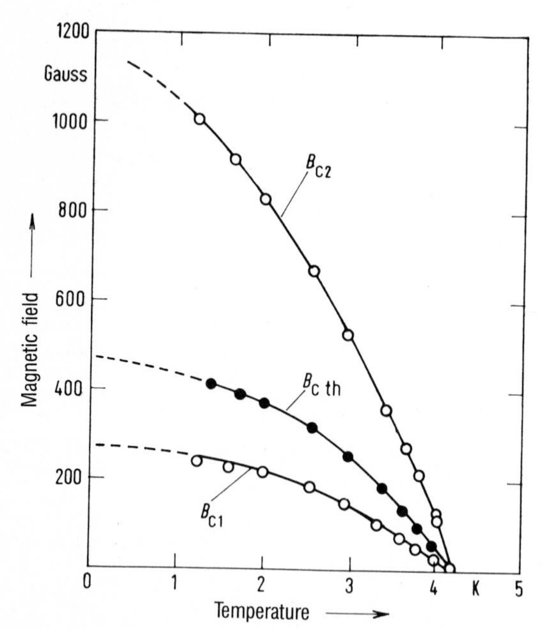

Fig. 76. Critical fields (1 G = 10^{-4} T) in an indium-bismuth alloy In + 4 atom % Bi (after [120]).

um state can be set up for each value of B_e. As we will see in the next chapter when discussing the hard superconductors this is not generally the case, but rather that inhomogeneities of all types can prevent the equilibrium distribution of the field. In such cases the shape of the magnetization curve is dependent on the previous treatment of the sample. Only the value of B_{c2}, where the superconductivity is completely suppressed, is still unequivocal.

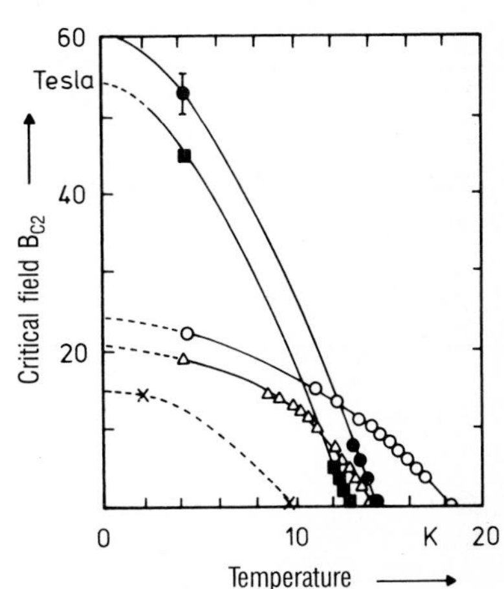

Fig. 77. Upper critical fields of some high field superconductors.
-○-○-○- Nb_3Sn, wire diameter 0.5 mm [121]
-△-△-△- V_3Ga, sintered sample [121]
-×-×-×- $Nb_{50}Ti_{50}$ [122]
-■-■-■- $PbMo_{6.35}S_8$ [124]
-●-●-●- $PbGd_{0.3}Mo_6S_8$ [124]
(see also Ø. Fischer: Proceedings LT 14, Otaniemi 1975, Volume 5, North-Holland Publishing Co. 1975).

This upper critical field can take on considerable values for some substances; these are known as high-field superconductors. In the case of the Chevrel phases (ternary molybdenum sulfides) materials have been discovered that can be brought into extreme fields of ca. 60 T (600 kG), without losing their superconductivity [124, 125, 126]. The B_{c2} values for some high-field superconductors are reported as a function of temperature in Fig. 77. The critical fields for such substances can be several hundred times greater then those for type I superconductors (Fig. 46). Here lies their technological importance. The ternary molybdenum sulfides have opened up a new dimension.

We will deal with the critical magnetic fields and the critical current densities of the new superconducting oxides separately in Section 7.4.

5.2.2 The Shubnikov phase

The magnetic flux penetrates into the superconductor in the case of the Shubnikov phase which, for a type II superconductor, is stable in the region between B_{c1} and B_{c2}. Since in a superconductor a magnetic field differing from zero is always associated with a supercurrent [1)] currents must flow in the interior of the superconductor in the Shubnikov phase. These currents must naturally flow in closed circuits, since only then can they form stationary states. The supercurrents, for their part, alter the magnetic field.

1 The relation between magnetic field and supercurrent density is given by the London equation:

$$\mathrm{curl}\,(\Lambda \vec{j}_s) = -\vec{B} \qquad (5\text{-}2)$$

So in the Shubnikov phase we expect a spacial variation of both the magnetic field strength and the supercurrent density. Only very specific field and current configurations are possible on account of the strict phase coherence of the Cooper pairs.

Abrikosov [116] produced such a configuration as a solution of the Ginsburg-Landau equations. Here the magnetic flux Φ, which penetrates the superconductor, is divided up into individual elementary magnetic flux quanta Φ_0 which are, in addition, arranged in a regular lattice. The lowest enthalpy is obtained for an arrangement of flux quanta in the corners of equilateral triangles.

Figure 78 illustrates this structure of the Shubnikov phase schematically. The superconductor is penetrated by tiny tubes of magnetic flux, each containing an elementary quantum of flux and sited at the corners of equilateral triangles. This is the magnetic structure that would be set up in an ideal homogeneous type II superconductor for fields $B_{c1} < B_e < B_{c2}$. Each flux quantum is made up of a system of ring currents which are indicated for two flux tubes in Fig. 78. These currents together with the external field create the magnetic field through the tube and exclude the magnetic field somewhat from the space between the tubes. They are, therefore, referred to as flux vortices. As the external field B_e is increased the distance between the flux tubes becomes smaller.

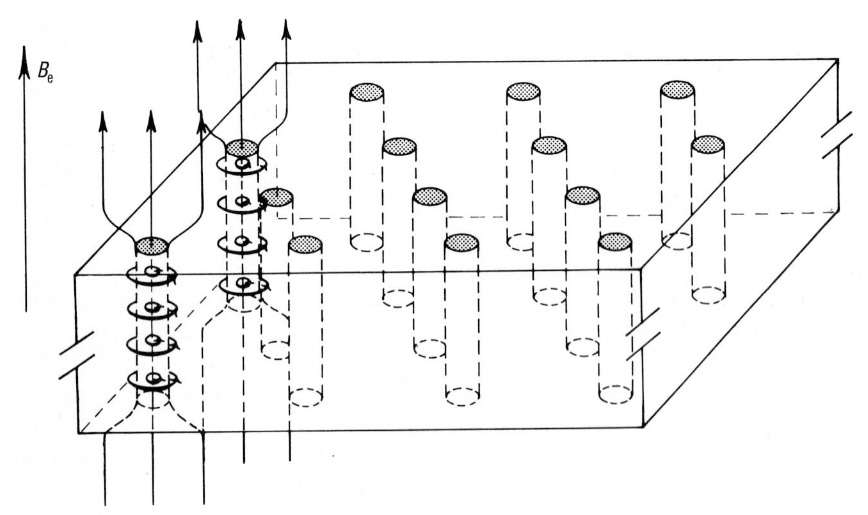

Fig. 78. Schematic representation of the Shubnikov phase. Magnetic field and supercurrent are only illustrated for two flux tubes.

The first demonstration of a periodic structure of the magnetic field in the Shubnikov phase was made in 1964 by means of neutron diffraction by a group at the Nuclear Research Center Saclay [127]. It was only possible then to demonstrate the basic periodicity of the structure. Very fine neutron diffraction experiments were made with this magnetic structure at the Nuclear Research Center at Jülich [128]. True "pictures" of the Shubnikov phase were obtained by Eβmann and Träuble [129] with a beautifully developed decoration method. Figure 79 illustrates an example for a lead-indium alloy. These wonderful illustrations of magnetic structure are obtained in the following manner. Iron is evaporated from a glowing wire above the superconducting sample. The iron atoms dif-

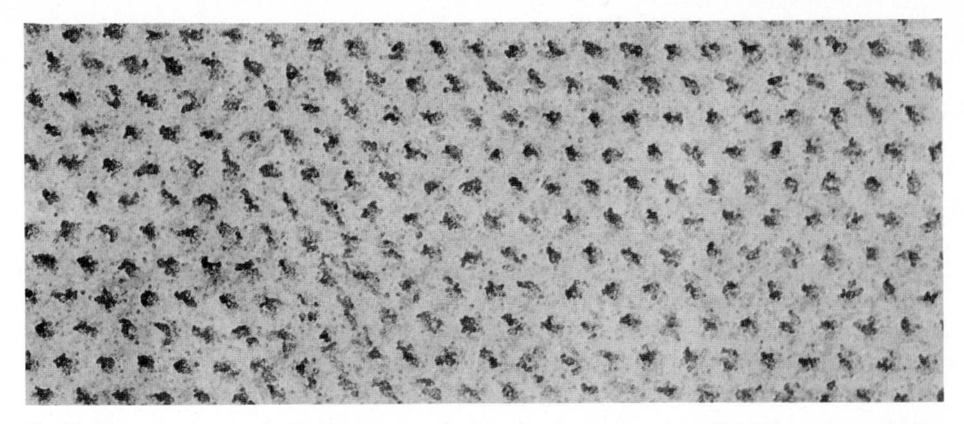

Fig. 79. Electron micrograph of a flux quantum lattice after decoration with colloidal iron. Frozen-in flux with field zero. Material: Pb + 6.3 atom % In, temperature: 1.2 K, sample shape: cylinder 60 mm long, diameter 4 mm, magnetic field B parallel to axis, enlargement: 8300-fold (reproduced by kind permission of Dr. Eßmann).

fuse through the helium gas in the cryostat to form colloidal iron particles. These colloids having a diameter of less than 50 nm sediment slowly through the helium onto the surface of the superconductor. The flux tubes of the Shubnikov phase, each containing one flux quantum Φ_0, emerge from this surface (indicated in Fig. 78 for two flux tubes). The ferromagnetic colloidal iron now deposits at those places where the flux tubes emerge from the surface since the strongest magnetic fields are to be found here. It is, thus, possible to decorate the flux tubes. It is then possible to use the usual print-taking methods of electron microscopy to make this structure visible. The pictures in Fig. 79 were obtained in this manner. It was an excellent experimental achievement to confirm the presence of these theoretically predicted structures in such a convincing way.

The decoration of magnetic structures with colloidal iron is exactly the same, in principle, as the method involving the sprinkling of niobium powder, which was employed to make intermediate state structures visible (Fig. 68). But, in contrast to niobium powder, colloidal iron is preferentially attracted to areas of high field strength and because of its smaller size it allows the resolution of finer structures.

The question that is still to be answered is whether the decorated portions of the surface are really the ends of flux tubes made up of only *one* flux quantum. To decide this question it is necessary to count the flux tubes and, at the same time, determine the total flux, say by means of an induction experiment. Then the size of the magnetic flux through a flux tube is obtained by dividing the total flux Φ_{tot} by the number of flux tubes. Such evaluations have, in fact, unequivocally demonstrated that, in the case of very homogeneous type II superconductors, each flux tube only contains one elementary flux quantum $\Phi_0 = 2 \times 10^{-7}$ G cm^2.

We ought to say a little more precisely that each flux tube contains just one fluxoid (see Section 3.2). When we pass round a flux tube and, hence, form the ring integral over the vector potential $\vec{A}$, we get the magnetic flux that goes through the closed path. If supercurrents flow along the ring path then we must also (Section 3.2) form the integral over the current density $\oint \Lambda \vec{j}_s \, d\vec{l}$ and it is then the sum of the magnetic flux and the current integral that provides the quantum conditions for the fluxoid.

The Cooper-pair density, the field distribution and the supercurrent density of a flux tube are represented schematically in Fig. 80. The Cooper-pair density equals zero at the axis of the tube and at about a distance of ξ_{GL} it equals the equilibrium value $n_C(T)$. The magnetic field is maximal at the axis and decreases outwards, whereby the diminution of B is controlled by the penetration depth. The core of the flux tube is encompassed by superconducting ring currents, which are what determines the variation of the magnetic field.

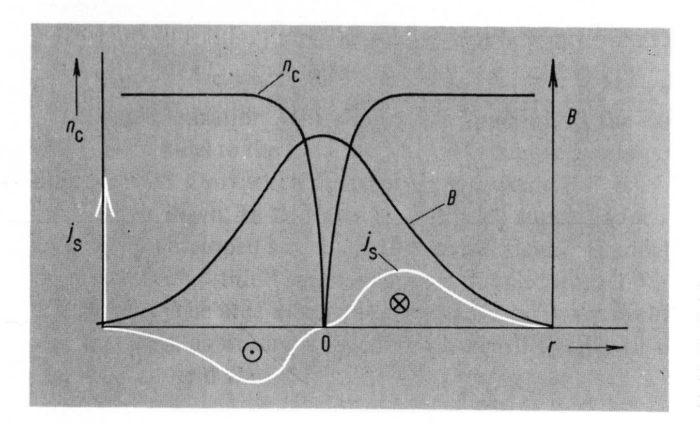

Fig. 80. Spatial variation of Cooper-pair density, magnetic field and supercurrent density for a planar section through a flux vortex (schematic).

A long calculation is required to demonstrate that it is just the state with an triangular lattice and just one flux quantum per flux tube that possesses the lowest enthalpy and, hence, is the stable state. We can, however, readily perceive that only a structure, with flux tubes containing exactly integer multiples of the elementary flux quanta, can occur. In Section 3.2 we derived the necessity for flux quanta from the requirement that the wave function of the Cooper pairs should superimpose on itself for one passage round a superconducting ring. This requirement naturally also holds for the flux tubes in the Shubnikov phase. From this it follows that each flux tube can only contain an exact whole multiple of a flux quantum. The result, that to obtain the state with the lowest enthalpy for an ideal homogeneous type II superconductor only one flux quantum must be taken per flux tube, can only be obtained from a quantitative consideration. Figure 81 illustrates the spacial variations of Cooper-pair density and magnetic field in one direction. As the external field B_e is increased the distance between the flux tubes becomes smaller, but, at the same time, there is a decrease in the mean Cooper-pair density $n_C(T)$. In the case of external fields, which lie a little below B_{c2}, the distance separating the flux tubes has been reduced to ca. 2 ξ_{GL}. As a result of the large degree of overlap of the current systems it is no longer possible to speak of individual flux vortices. Simultaneously the Cooper-pair density approaches zero as B_{c2} is approached[1].

1 The transition to the normal state at B_{c2} is a 2nd order transition where the order parameter, here the Cooper-pair density, falls continuously to zero (Appendix B).

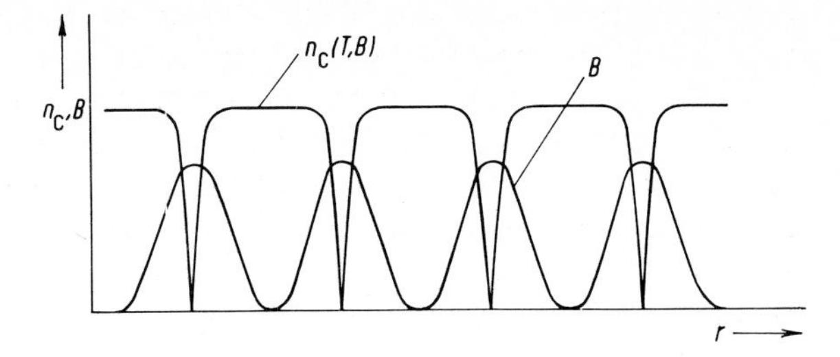

Fig. 81. Spatial variation of Cooper-pair density and magnetic field in the Shubnikov phase (schematic). The Cooper-pair density takes up the equilibrium value associated with T and B between the flux vortices. Since $\lambda > \xi_{GL}$ the magnetic field between the flux tubes is not completely excluded with increasing field.

All the deliberations in this section have assumed that the sample had a demagnetization factor equal to zero. Unlike in the intermediate state experiments (see Section 5.1.4) the penetration of the magnetic flux was *not* dependent on the geometry of the sample [1]. The question now arises as to whether an intermediate state of some sort can exist for a type II superconductor and what phases can coexist in this intermediate state. So long as a type II superconductor is in the Meissner phase it excludes the field just like a type I superconductor. If now the field B_{c1} is reached at the surface of the sample then, also in the case of a type II superconductor, the flux must penetrate it. A state is then reached where macroscopic regions of Meissner phase and Shubnikov phase exist side by side. Figure 82 illustrates an example of this new state. Instead of the normal phase which is present in type I superconductors we have here, for type II superconductors the Shubnikov phase.

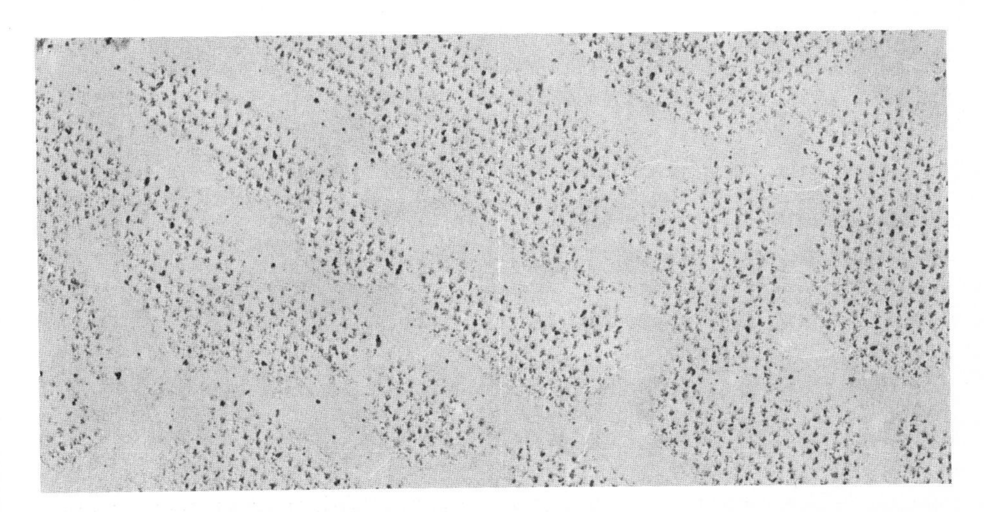

Fig. 82. Coexistence of Meissner phase and Shubnikov phases in a type II superconductor with κ close to $1/\sqrt{2}$. Material: Pb + 1.89 atom% Tl, $\kappa = 0.73$, $T = 1.2$ K; sample shape: disc 2 mm diameter, 1 mm thick; external field: $B_e = 365$ G, magnification 4800-fold. This state can be effected for samples with finite demagnetization factors. (Reproduction by kind permission of Herr Dr. Eßmann.)

1 Figure 78 only illustrates a section of the structure, whereby the sample can be extended as desired in the z direction.

Such structures are only found for κ values which are slightly greater than $1/\sqrt{2}$. Neumann and Tewordt [130] have extended the GLAG theory to temperatures below T_c. They found that, in principle, for all temperatures $T < T_c$ there exists a κ region above and very close to $\kappa = 1/\sqrt{2}$ in which the transition from the Meissner phase with complete exclusion of the flux does not go directly into the Shubnikow phase, but rather passes through a state in which Meissner and Shubnikov phases exist side by side. It is only at values of κ, which are sufficiently higher than $1/\sqrt{2}$, that there is a direct transition from the Meissner to the Shubnikov phase.

6. Critical currents in type I and type II superconductors

We recognized in Section 2.2 that the strong correlation of Cooper pairs leads to the existence of a critical velocity and, hence, of a critical current density. The system of Cooper pairs is unable to interact with the lattice for current densities below this critical value j_c. If this critical value is exceeded Cooper pairs are broken up. We will now discuss some of the effects which the existence of a critical current density has on the current that can be carried by a superconductor. Here we will limit ourselves to simple geometrical conditions.

These deliberations concerning the critical current are of the utmost importance with respect to technical applications of superconductivity. We have, to be sure, in type II superconductors materials which remain superconducting in the presence of technically interesting magnetic fields. But as far as applications are concerned it is just as important that these superconductors are able to support sufficiently high currents without any resistance in high fields. Here, as we will see, further problems exist and these are only solved when we come to the hard superconductors.

6.1 Type I superconductors

We will consider a wire of circular cross section with a current I flowing through it as the simplest case geometrically. The wire will be in the Meissner phase if the current is sufficiently small. In this phase there can be no magnetic field in the interior of the superconductor. This also means that no current can flow in the interior otherwise the magnetic field created by the current would also be present. It follows then that the current through a superconductor is limited to the thin surface layer in which the magnetic field is able to penetrate the Meissner phase. We name currents flowing through a superconductor transport currents to distinguish them from the screening currents which flow in the superconductor as ring currents.

Figure 83 illustrates schematically the spacial distribution of the transport currrent in a circular wire with current density plotted as a function of the radius. The total current is obtained by integrating the current density over the whole cross section:

$$I = \int_A \vec{j}_s \, d\vec{a} \, . \tag{6-1}$$

The magnetic field associated with this current is also illustrated in Fig. 83. As early as 1916 F.B. Silsbee [131] put forward the hypothesis that in the case of "thick" superconductors, i.e. superconductors with a completely developed screening layer, the critical current is just reached when the field created at the surface, as a result of the current, reaches the value B_{cth}. This hypothesis has been amply confirmed. Put slightly differently it says: Magnetic field and current density are strictly correlated in the case of a surface with a completely developed screening layer (Equation (5-17)). The critical value of the current density is associated with a particular critical field, namely B_{cth}, whereby it is completely irrelevant whether the current density results from a transport current or from a screening current.

The validity of the Silsbee hypothesis now makes it very simple to calculate from the critical fields, e.g. Fig. 46, the associated critical current strength for wires of circular cross section. The magnetic field at the surface of such a wire, through which a current I is flowing, is given by:

$$B_0 = \mu_0 \frac{I}{2\pi R} \tag{6-2}$$

where B_0 is the field at the surface, I the total current, R the radius of the wire, $\mu_0 = 4\pi \times 10^{-7}$ Vs/Am.

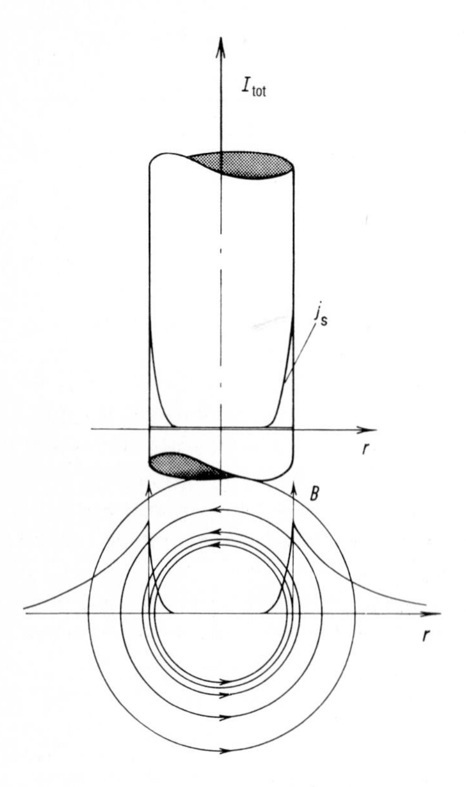

Fig. 83. Current density and magnetic field distribution in a superconducting wire with transport current. The surface layer has only the thickness of the penetration depth $\lambda \simeq 30$ nm.

Only the cylindrical symmetry of the current distribution is required. The radial dependence of the current density is not relevant.

It follows immediately from Equation (6–2) that the critical current possesses the same temperature dependence as the critical magnetic field. Figure 84 reproduces as an example the temperature dependence of the critical current for a tin wire with a diameter of 1 mm. The critical field of ca. 300 G at 0 K corresponds, according to Equation (6–2), to a critical current strength of $I_{co} = 75$ A. Since the whole current flows in the thin

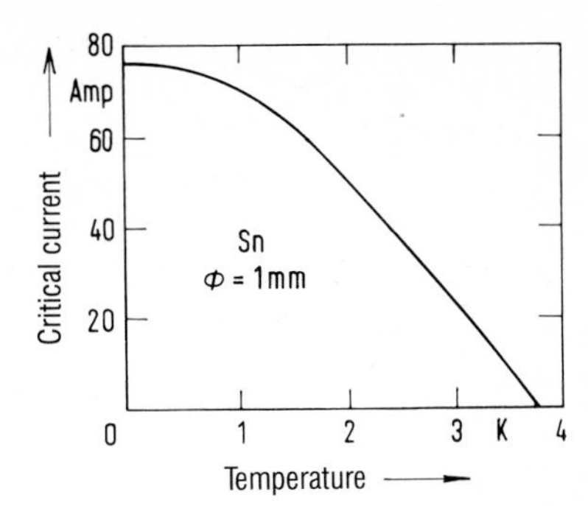

Fig. 84. Critical current density for a tin wire with a diameter of 1 mm.

screening layer this critical current strength only increases proportionately with the radius of the wire.

We can also calculate the critical current *density* at the surface. Here we replace the expontentially diminishing current density (Fig. 83) by the distribution, whereby the full current density on the surface remains constant to a depth λ, the depth of penetration, and then suddenly falls to zero. The depth of penetration λ is just so defined that we still get the same total current [1]. According to this method of consideration we find, for example, that the critical current density for the tin wire at 0 K is:

$$j_{co} = \frac{I_{co}}{2\pi R \lambda_0} = 7.9 \times 10^7 \text{ A/cm}^2 \qquad (6\text{–}3)$$

($R = 0.5$ mm, $\lambda_0 = 3 \times 10^{-6}$ cm (according to Table 8), $I_{co} = 75$ A).

The critical current densities are very high and would allow considerable transport currents to flow if it were possible to circumvent the screening which also leads to the limitation of the current to a thin layer. Such substances have been developed in the form of the hard superconductors.

We can calculate the critical current for a superconductor in an external magnetic field by means of the Silsbee hypothesis. We merely have to carry out a vector addition of the external field and the field produced by the transport current on the surface. The critical current density is reached when this resultant field has the critical value. In the case of a wire with radius R in the external field B_e, perpendicular to the axis of the wire [2], it follows that:

$$B_{tot} = 2B_e + \mu_0 I/2\pi R\,. \qquad (6\text{–}4)$$

The magnetic field distribution and the critical current strength as a function of the external field at a fixed temperature, i.e. for a fixed B_{cth}, are presented in Fig. 85a and b.

1 Since the depth of penetration is only a few times 10^{-6} cm it is always true for macroscopic wires that $R \gg \lambda$. This means that, as far as the questions of interest here are concerned, we can regard the surface of such wires as being planar.

2 The demagnetization factor n_M is $1/2$ for a wire perpendicular to the field. For this reason we have $B_{eff} = 2\,B_e$ at the generating lines exposed to the highest field strength.

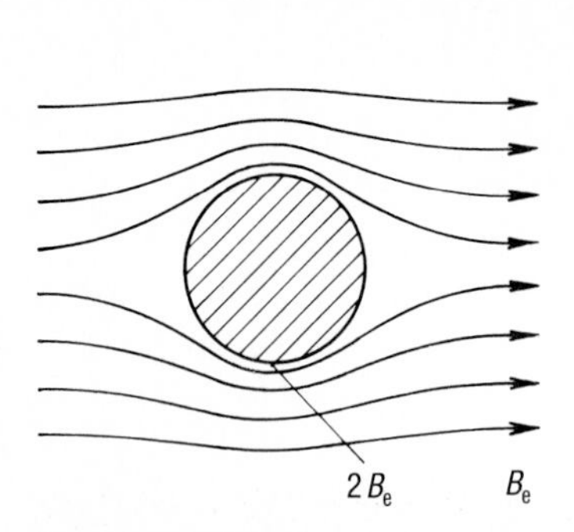

Fig. 85a. Field distribution around a superconducting wire in the Meissner phase without current load.

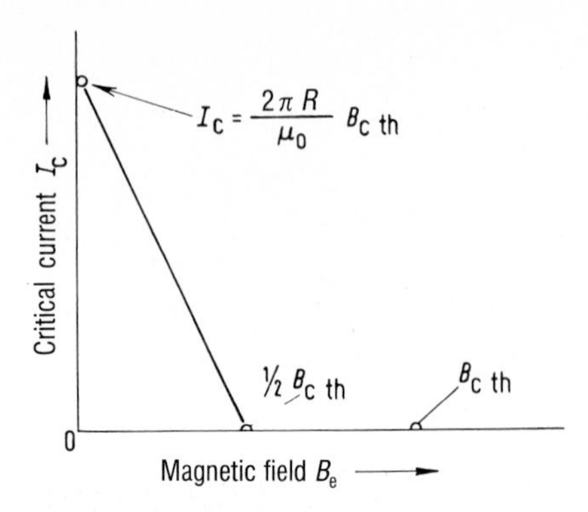

Fig. 85b. Critical current for a wire of circular cross section in external field B_e perpendicular to the axis of the wire (after (6–4)).

We will now address the question of how the superconductor goes into the normal state when the critical current density is reached. To do this we will again consider a wire of circular cross section. When the critical current strength is exceeded the Meissner phase with complete exclusion becomes unstable. We would now expect the superconductor to become a normal conductor. But then the current flowing through it would be distributed over the whole of its cross section. The field strength at the surface would not be affected by this redistribution. However, we would have throughout the superconductor a current density that would be lower than the critical one. Since we view the critical current density as the decisive quantity for the stability of the superconducting state we would expect that the transition could not occur in such a way that the current density everywhere became subcritical.

Experiment confirms this presumption. When the critical current strength is exceeded the superconductor goes into an intermediate state, i.e. normal regions appear. Several models have been developed for this intermediate state. For these the arrangements of the normal and superconducting regions are sought so that the critical field strength B_{cth} occurs over as much of the interfaces as possible. In the case of the macroscopic structure of the intermediate state this critical field strength also determines the critical current density as a consequence of the completely developed screening effect. Figure 86 reproduces a model of this type [132]. Since the magnetic field of the load current is made up of circular field lines the phase boundaries have to adjust itself to these field lines. The requirement that the field strength at each radius must equal B_{cth} means that

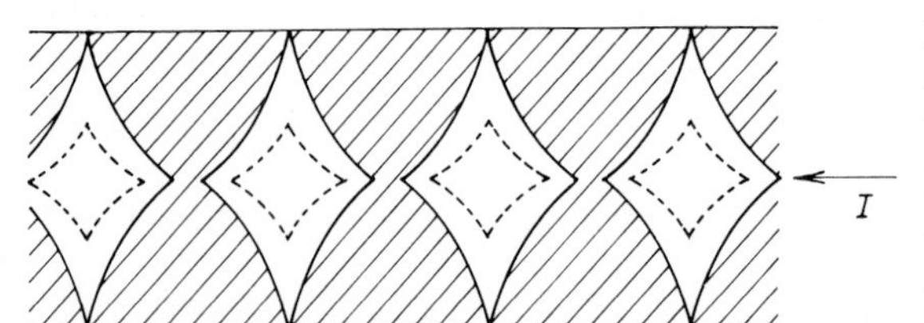

Fig. 86. Intermediate state structure of a wire of circular cross section at the critical current. Hatched regions are normalconducting. The structure is rotationally symmetrical about the axis of the cylinder. At a current load of $I > I_c$ the superconducting areas shrink (broken lines after [132]).

the current density must increase towards the axis of the wire. This is achieved by the thickness of the superconducting lamellae also increasing as they approach the axis of the wire.

The individual details of the shapes can only be found as a result of calculations which involve making certain additional assumptions. The various models for this current-stabilized intermediate state differ in these additional assumptions.

The lamellar structure of this intermediate state has been demonstrated very beautifully by powder decoration (see Section 5.1.4). Figure 87 illustrates an example of this structure [133].

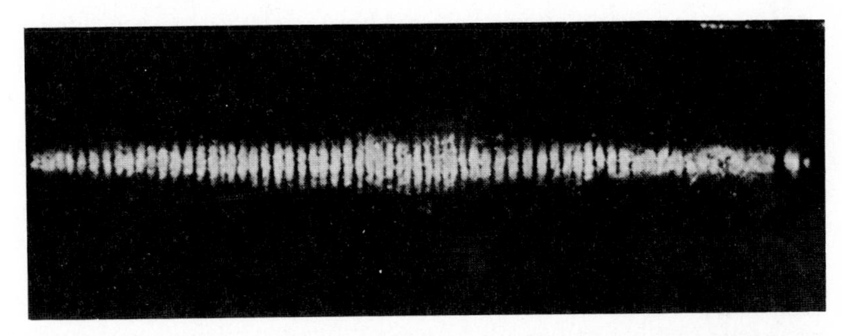

Fig. 87. Intermediate state structure of an In cylinder carrying current. The light zones correspond to normalconducting regions. Length 38 mm, diameter 6 mm, current load 30 A, external field B_e perpendicular to cylinder axis 0.01 T, $T = 2.1$ K (T_c of In is 3.42 K), transition N → S (reproduced by kind permission of the authors of [133]).

When the critical current density is exceeded the superconducting wire jumps into a state where the superconducting lamellae still reach as far as the surface. As the current is increased further a mantle of normalconducting phase is produced surrounding a core made up of the intermediate state; the thickness of this mantle increases as the current strength is increased. Figure 88 contains a plot of the electrical resistance of the wire as a function of the current load [132]. A finite resistance appears suddenly at I_c, but the resistance does not immediately jump to the full normal resistance, this only appears after a further increase in the current I. Such determinations are very difficult because, in the current-stabilized intermediate state with a finite resistance, Joule heat can readily lead to a temperature increase and, hence, to instabilities [1)]. As far as the stabilizability of the intermediate state is concerned a "thick" wire carrying a current load corresponds to a sample with a demagnetization factor differing from zero, which is subjected to an external field (see Section 5.1.4). In both cases the superconductor is able, with transition into the intermediate state, to respond to the "coercion" of the external variables, current or magnetic field by splitting up into normal and superconducting regions. In this manner it is possible to maintain the critical quantities constant over a range of values of the external variables.

1 If the resistance of the external circuit is low enough the characteristic can be run through in a stable manner but the Joule heat produced results in an increase in the average temperature.

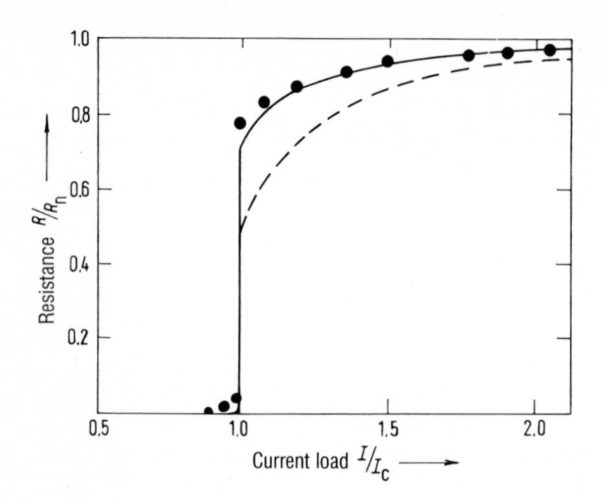

Fig. 88. Electrical resistance as a function of current load. Solid curve: model calculation after [132], broken line: model of F. London [II]; measured points: B.R. Scott: J. Res. Nat. Bur. Stand. *41*, 581, 1948 (see also L. Rinderer: Helv. Phys. Acta *29*, 339 (1956)).

A "thin wire" ($R < \lambda$) with a current load would correspond to the case of a long cylinder in a magnetic field parallel to its axis (demagnetization factor = 0). The current density in it would be practically constant over the cross section. Complete normal conductivity ought to begin when the critical current is reached, since the thin superconductor has no possibility of creating a new distribution of the current over its cross section on maintaining the critical current density, while allowing passage of a greater current. Here too it is possible to stabilize an intermediate state by means of a suitable external circuit [134]. Such comparative considerations of the stability of the intermediate state can contribute a great deal to an understanding of the phenomenological behavior of a superconductor [1]).

We have employed a static model for the intermediate state until now. The superconducting and normal regions have been assumed to be fixed in the sample apart from the thermal variations. Gorter [135] has proposed a dynamic model (see Section 6.2, last paragraph).

Very complicated intermediate states are obtained in the case of wires in a longitudinal magnetic field carrying large transport current. Here the longitudinal external field is superimposed on the circular field of the current so that the phase boundaries, which always have to run parallel to the field, take on a helical shape. This then causes the load current to flow in a screw-thread path. This leads to unexpected effects, e.g. an intensification of the field in the interior of the sample, but these can be understood in terms of the structure of the intermediate state and, thus, demonstrate very vividly the general rule (see Section 5.1.4) that the phase boundaries must always run parallel to the magnetic field [136].

1 In the case of a long cylinder in a magnetic field parallel to its axis it is necessary to use an experimental arrangement for the stabilization of the intermediate state such that the appearance of normal regions can react on the external magnetic field, e.g. by induction, if this field is produced by a persistent current through a superconducting coil.

6.2 Type II superconductors

After our consideration of type I superconductors we must now turn our attention to type II superconductors, since they differ fundamentally from the type I superconductors in one important aspect. In small magnetic fields and – from what has been said above – when carrying small supercurrents type II superconductors are to be found in the Meissner phase too. In this phase they behave just like type I superconductors, i.e. they exclude the magnetic field and the current from all except a thin layer at their surfaces.

A difference from type I superconductors will first occur if the magnetic field on the surface exceeds the value B_{c1}. The type II superconductor must then go into the Shubnikov phase, i.e. flux tubes must penetrate into the superconductor (see Section 5.2.2).

It has been found that the ideal type II superconductor has a finite electrical resistance in the Shubnikov phase even at very low current loads [137]. In order to understand the reason for this resistance let us first consider an arrangement with a somewhat different geometry, namely a rectangular plate through which a current flows parallel to the plane of the plate and which is held in a magnetic field $B_e > B_{c1}$ perpendicular to the plane of the plate in the Shubnikov phase (Fig. 89).

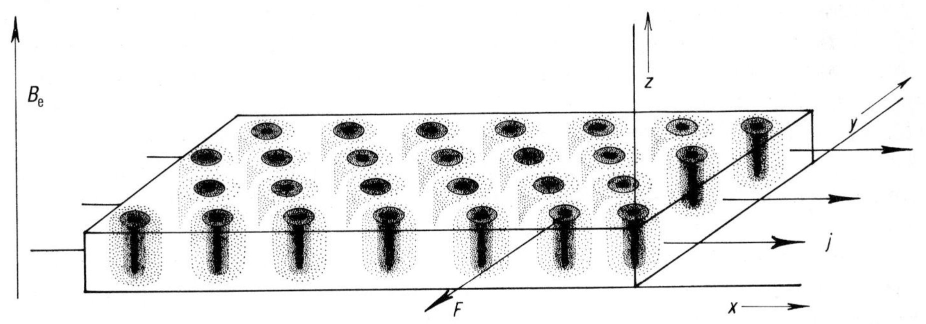

Fig. 89. Shubnikov phase with transport current density *j*. A force *F* acts on the flux tubes, and moves them here in the $-Y$ direction. The magnetic field distribution about the flux tubes is indicated by shading.

The first important result discovered in such an experiment is that under these conditions the transport current *I* is distributed over the entire cross section of the plate, i.e. it is no longer wholly limited to a thin surface layer. With the penetration of the magnetic flux into the superconducting sample a current can also flow in the interior of the superconductor.

A very important interaction between the transport current and the flux tubes appears here. The current, assumed say to flow in the *x* direction, also flows through the flux tubes, i.e. the regions in which a magnetic field is present [1)]. A force known as the

---

1 When representing the Shubnikov phase cylindrical flux tubes are drawn for the sake of simplicity (Fig. 78 and 89). This could lead to the false assumption that the transport current could simply flow round the flux tubes and, thus, not enter at all the regions with a magnetic field. It must, however, be remembered that the magnetic field varies slowly in comparison with the density of Cooper pairs (Fig. 81) and that, therefore, transport current also flows in regions where there is a magnetic field.

Lorentz force acts between every current and every magnetic field. For a current I along a wire of length L perpendicular to a magnetic field B the modulus of force will be equal to:

$$F = ILB\,. \tag{6-5}$$

It is directed (Figure 90) perpendicular to B and to the current (determined here by the axis of the wire). It is this Lorentz force that drives our electric motors.

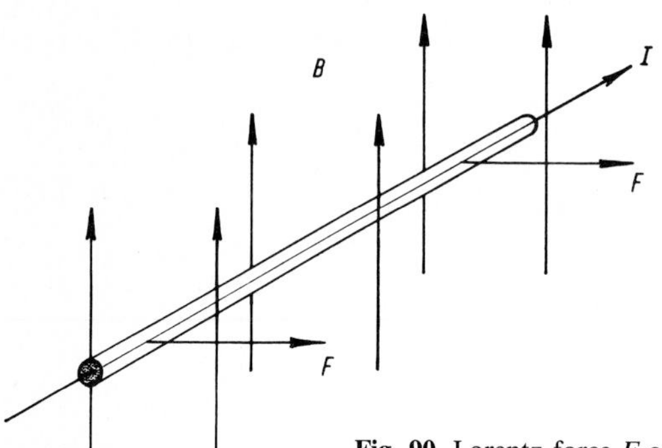

Fig. 90. Lorentz force F on a wire carrying a current in a magnetic field.

This force acts between the flux tubes and the current in a Shubnikov phase with a transport current. Since the transport current is held immovable by the limits of the plate, the flux tubes must migrate perpendicular to the direction of the current and perpendicular to the magnetic field, that is to their own axes, under the influence of the Lorentz force [138]. This migration ought to take place with the smallest of forces, that is for the smallest of currents, in the case of ideal[1)] type II superconductors, for which after all completely free movement of the flux vortices is possible.

The migration of the flux tubes through the superconductor causes the occurrence of dissipation, i.e. electrical energy is converted into heat. This energy can only be obtained from the transport current. This means that the sample exhibits an electrical resistance.

The conversion of electrical energy to heat as a result of the migration of a flux tube can basically take place as a result of two different processes. The first mechanism for losses depends on the occurrence of local electrical fields. Let us consider a certain point P in a superconductor, over which a flux tube migrates with a velocity $\vec{v}$. The migration of the flux vortex through point P causes the magnetic field to change with time. As the flux tube approaches P the magnetic field there increases, as it migrates away again the magnetic field falls. However, a change in the magnetic field with time at point P results in an electrical field there. This field then accelerates the unpaired electrons. They can

1 An ideal type II superconductor was defined in Section 5.2.1 as one whose magnetization curve was run through reversibly on magnetization and demagnetization. For this it must be possible to make any arbitrarily small change to the equilibrium concentration of the flux tubes, i.e. an arbitrarily small displacement. Put in another manner: The ideal type II superconductor should be completely homogeneous with respect to the positions of the flux tubes.

pass their energy, taken from the electrical field, on to the lattice and, hence, produce heat.

Alongside this very transparent process of energy dissipation, which is connected with the spacial variation of the magnetic field in a flux tube, we also have a second possibility, which depends on the spacial variation of the Cooper-pair density in the flux tube. When a flux tube migrates over point P there is also a change with time of the Cooper-pair density n_C at this point, because n_C increases outwards from zero in the core of the flux tube. Now it must be expected that the re-establishment of the equilibrium value of n_C after such a deviation from this equilibrium will require a finite time τ – such time which is required to restore an equilibrium is known as a relaxation time[1] – (see also Section 8.2). If n_C changes very slowly, i.e. in times that are large with respect to τ, then the system passes continually through equilibrium states. Then the energy consumed for the break-up of Cooper pairs at the front of the flux tube will be released by the recombination of Cooper pairs at the back, so that, in practice, no heat is produced. However, if the flux tubes migrate so rapidly that the Cooper-pair density cannot follow through equilibrium states then energy will be dissipated as a result of the change in n_C with time, i.e. heat will be produced.

We can clarify this in the following way. The large magnetic field of the core of the flux tube migrates so rapidly that it is not possible for the superconductor to adjust to the equilibrium concentration of Cooper pairs associated with each value of the field. The Cooper pairs break up in too large a magnetic field at the front. Conversely the Cooper pairs are reformed at the rear in a magnetic field that is too small for the concentration involved. Now since the heating required for the break-up of a Cooper pair decreases with increasing magnetic field[2] the break-up will consume less heat than will be released by the reformation. In this manner heat is produced when the variation of n_C with time is so rapid that deviations appear from equilibrium between all parameters.

What has been described here, with Cooper-pair density as an example, is nothing more or less than the mechanism of any relaxation process after the change of external parameters during times comparable with the relaxation time. Other well-known examples include the losses on polarization of a dielectric in an alternating electrical field or the magnetization of a ferromagnetic material in a magnetic field.

So we have established that as soon as the flux tubes in the Shubnikov phase begin to migrate under the influence of the transport current a dissipation mechanism goes into action and an electrical resistance is created. Since in the case of an ideal type II superconductor even the smallest transport current can lead to migration of the flux tubes the critical current for an ideal type II superconductor is equal to zero in the Shubnikov phase [137]. Thus, such superconductors are unsuitable for technical applications in spite of their high critical fields B_{c2}. The critical currents of a Shubnikov phase can only be finite if the flux tubes are fixed in their positions in some manner. This fixing (pinning)

1 It is assumed that the system recovers from the displacement, so that Δn_c decays exponentially.

$$\Delta n_C = \Delta n_{Co} \exp\left(-\frac{t}{\tau}\right).$$

2 As the magnetic field increases density of Cooper pairs in the Shubnikov phase and, hence, their binding energy are reduced.

of flux tubes is actually possible. Type II superconductors with pinning centers are known as hard superconductors. They are substances of current technological interest (Chapter 7).

We have explained the occurrence of resistance in the Shubnikov phase on the basis of the motion of the flux tubes and the associated dissipation effects. The significance of the motion of flux tubes for the creation of an electrical potential has been the subject of animated discussion for some time. A range of elegant experiments have been proposed and carried out in this context. We will only deal with one example here.

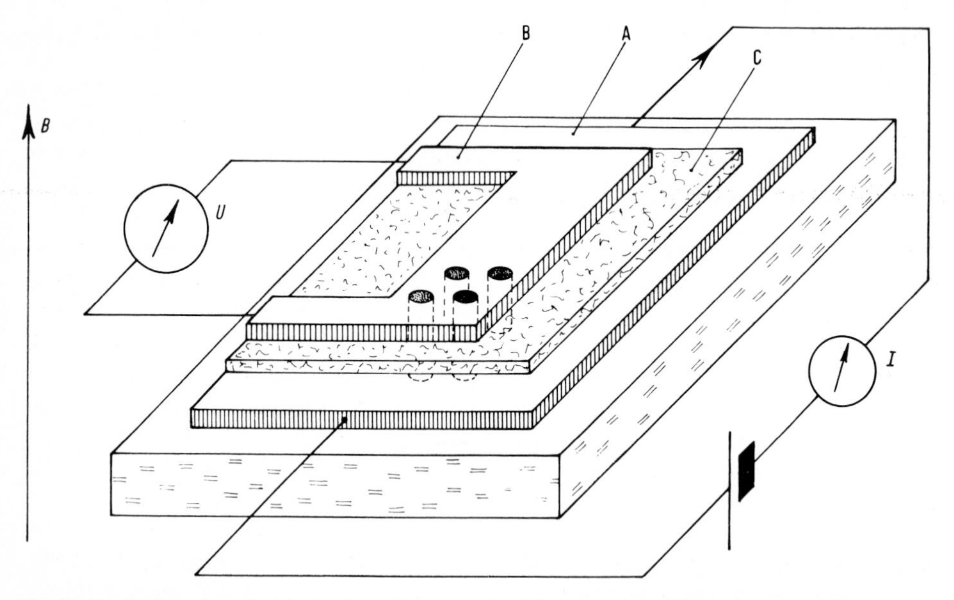

Fig. 91. Production of an electrical voltage U as a result of the migration of flux tubes. A and B are superconductors. C is an insulating layer. All layer thicknesses are greatly exaggerated (after [139]).

An experiment was carried out in 1966 by I. Giaever [139], whose fundamental idea will be explained on the basis of the device illustrated in Fig. 91. A thin film of a superconductor A is brought into the Shubnikov phase by means of an external magnetic field and a load current is passed through it. The migration of flux tubes should now begin. In order to demonstrate this migration Giaever superimposed on film A a second thin film of superconductor B, which was, however, completely separated from film A electrically by as thin an insulating layer as possible. However, the magnetic flux tubes in the two films are coupled [1)]. If the flux tubes in superconductor A migrate under the influence of the transport current then the coupling means that the flux tubes in layer B will be taken along too. If – postulated Giaever – the migration of the flux tubes causes an electrical voltage across the ends of the superconductor then this voltage ought to be

1 The magnetic field distribution of the flux tube system in superconductor A penetrates the thin insulating layer and causes a similar distribution of flux tubes in superconductor B.

measurable in film B, which is completely insulated from the primary circuit. Giaever carried out the experiment without an external magnetic field. In this modification too the field of the transport current, which was passing through film A, penetrated both films in the form of flux tubes. These flux tubes migrate from both sides of film A to the middle where they mutually cancel each other out. The migration of the flux tubes yielded the expected voltage across the completely separate film B. If the films were normalconducting no electrical voltage could be observed across film B under otherwise comparable conditions. This experiment makes evident the importance of flux tube migration for the creation of electrical voltages in the Shubnikov phase [1].

We will discuss some of the properties of this state with flux tubes in motion, which is referred to as the "resistive" state, in the next chapter.

In considering the critical current in type II superconductors we have until now employed an external magnetic field to stabilize the Shubnikov phase. We will now return to the question with which we started concerning the critical current *in the absence* of an external magnetic field. Let us take a wire of circular cross section through which a current I is flowing. When the value $I_c = B_{c1} \times 2\pi R/\mu_0$, which just serves to create the field B_{c1} on the surface, the superconductor goes into the Shubnikov phase. Since with this geometry the magnetic field of the transport current passes in circles around the axis of the wire closed circular flux tubes are also created. They migrate under the influence of the Lorentz forces to the axis of the tube, becoming ever smaller as they do so and vanishing when they arrive there. So we expect for an ideal type II superconductor [2] of this geometry a critical current I_c which is determined by B_{c1} just as it is determined by B_{cth} for type I superconductors (see Section 6.1). Since B_{c1} is smaller than B_{cth} the critical current for type II superconductors is always smaller than for a corresponding type I superconductor [3].

To conclude this chapter it must be mentioned once more that when type I superconductors in the intermediate state (see Section 5.1.4) are put under a current load a migration of the regions occurs that leads to resistance. It was Gorter [135] who pointed out this possibility of a dynamic model of the intermediate state. Here the phase boundaries lie in this model parallel to the direction of flow of the transport current. In the static model (see Section 6.1) they lie perpendicular to the direction of current flow. Decoration experiments (see Section 5.1.4) with Nb powder have indicated that when the current load is large enough the regions actually do migrate through the superconductor at right angles to the current [140].

---

1 Since we represent the magnetic field in the flux tube by lines of force (e.g. in Fig. 78) the question is often asked in connection with the migration of the flux tubes through the superconductor as to what happens to the many lines of force that are forced through the superconductor during such an experiment. The answer is very simple. A flux tube is made up of a system of ring currents (Figs. 78 and 80) which create exactly the additional field of the flux tube. These ring currents are set up on one side of the superconductor when a flux tube originates and disappear at the other side when they have passed through the superconductor.

2 Since we designate type II superconductors with pinning centers as hard superconductors we can omit the adjective "ideal". We often then simply refer to hard superconductors when pinning centers are deliberately introduced into a type II superconductor.

3 "Corresponding" means in this context that the superconductor is to have the same geometry and the same value for B_{cth}.

7. Hard superconductors

In Section 5.2.1 we met the magnetization curves of type II superconductors (Fig. 72). Here we assumed ideally homogeneous substances, i.e. substances in which the flux tubes of the Shubnikov phase could be freely displaced, that is do not have any energetically preferred positions. This assumption represents a limiting case, which can only be more or less well approximated by real samples.

In this section we will consider just those superconductors where the flux tubes in the Shubnikov phase are bound very strongly to energetically preferred positions. These superconductors, which we call hard superconductors, are the technically applicable substances.

As is often the case in solid state physics, substances which are ideal from the theoretical point of view are not the best for real applications [1]). Rather it is actually the deviation from ideal behavior that give them their technical importance.

7.1 The magnetization of hard superconductors

If the flux tubes of the Shubnikov phase are bound to particular sites in the material then they cannot adjust to the magnetization by the external field necessary for thermodynamic equilibrium. For a freely displacable flux tube is necessary for this to happen. We would, therefore, expect to observe quite different magnetization curves.

Figure 92 reproduces as an example the behavior of Nb-Ta alloy [141]. Niobium and tantalum are miscible in all proportions. Careful tempering yields very homogeneous mixed crystals. An almost reversible magnetization curve can be observed for them (Curve *a*).

If such a mixed crystal is deformed, as occurs in the drawing process during the manufacture of a wire, then many dislocations are created in the lattice structure, which can act as pinning centers for flux tubes. A magnetization curve with a quite different shape is then obtained (Curve *b*). The first thing that is apparent is that the degree of magnetization is substantially greater. This means that the deformed sample can have virtually ideal diamagnetic screening even in external fields greater than B_{c1}. In addition every trace of reversibility has completely disappeared. As the field strength is reduced the magnetic flux remains "imprisoned" in the sample even when $B_e = 0$ [2]).

Only the upper critical field B_{c2} remains unchanged. This is understandable when it is remembered that mixed crystal of the alloy $Nb_{55}Ta_{45}$ already possesses a very small mean free path l^*. The additional dislocations are not able to cause appreciable changes in $\varkappa$ and B_{c2} (see Section 5.2.1).

---

1 Thus, the fluorescence and phosphorescence phenomena, such as we employ, for instance, in video screens, are dependent on particular impurity centers, that is on deviations from ideal behavior. The excellent mechanical properties of structural materials too, e.g. of duralumin, are dependent on particular precipitates, that is inhomogeneities.

2 This imprisoned flux is directed parallel to the external field. In Fig. 92 we have followed the convention of plotting the negative magnetization.

If we assume that the flux tubes in the disordered material are more or less fixed in their positions then it is easy to understand this magnetization curve at a qualitative level. If we follow the path from zero field there is no difference until B_{c1} is reached; the sample is in the Meissner phase, which is scarcely affected by dislocations [1].

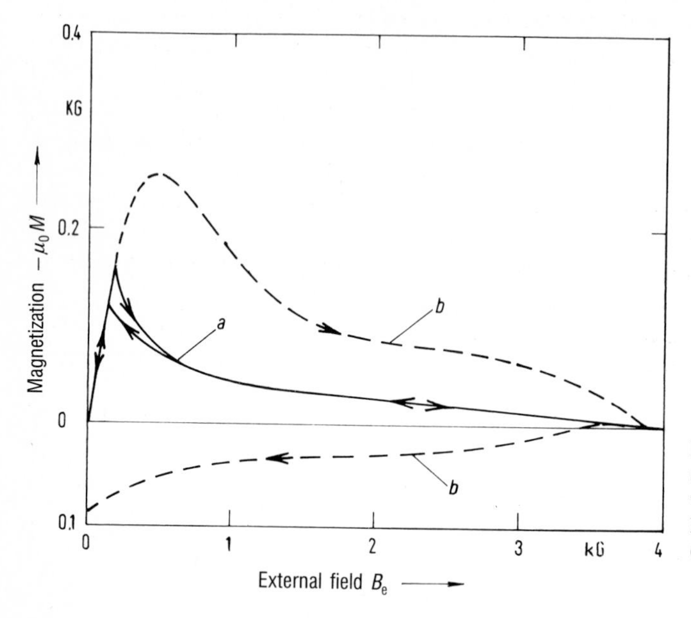

Fig. 92. Magnetization curve of a $Nb_{55}Ta_{45}$ alloy. *a)* very well annealed, *b)* with many lattice defects (after [141]) (1 kG = 0.1 T).

When B_{c1} is exceeded the flux tubes penetrate into the sample from the surface. These flux tubes are now fixed to their places at first immediately below the surface and cannot distribute themselves evenly over the whole volume as in the case of homogeneous material. But there are flux tubes present in the surface layer and screening currents can flow. This means that the total screening current, which determines the diamagnetic behavior, can be greater than is the case for the Meissner phase where supercurrents can only flow in a layer the thickness of the penetration depth. The situation can be expressed rather differently: The penetration of the flux tubes into the surface of the sample increases the effective thickness of the screening layer and, thus, the total screening current.

At B_{c2} the Cooper-pair density falls to zero and superconductivity is abolished [2]. The magnetic field penetrates the superconductor homogeneously. When the external field is reduced below B_{c2} the sample again goes over to the Shubnikov phase. The magnetic flux is again distributed in multiples of the flux quantum Φ_0 and the flux tubes are more or less tightly bound to the lattice disorder. So their emergence from the material as the magnetic field is reduced is inhibited. Even at $B_e = 0$ there is a magnetic flux fro-

1 The slight influence of the dislocations on the penetration depth λ can be ignored here.

2 Here we ignore the surface superconductivity which can occur immediately under the surface for parallel fields up to $B_e = 1.7\ B_{c2}$.

zen in the previous direction of the field. The size of this frozen-in paramagnetic field[1] depends, as does the shape of the magnetization curve, on the individual details of the disorder present, whose quantitative description is still not possible at present [XIV].

Normalconducting precipitates in a superconducting matrix form particularly effective "pinning centers". Figure 93 illustrates an example which was observed for a lead-bismuth alloy containing 53 atom % Bi [142]. On cooling from the melt two limiting phases precipitate out from this system, namely practically pure Bi and what is known as the ε-phase, an alloy with its own lattice structure and a stability range between ca. 18 and 33 atom % Bi at room temperature. The ε limiting phase has a transition temperature of ca. 8.6 K. The values of B_{c1} and B_{c2} are 250 and ca. 15 000 G. The reversible magnetization curve, such as would be expected for the ε-phase, is illustrated in Fig. 93.

Magnetization curves *a*, *b* and *c* for the mixed phases (ε and pure Bi) are associated with various degrees of precipitation. More and more Bi precipitates out as the alloy is stored at room temperature. This evidently results in the creation of ever more effective pinning centers. After a precipitation period of 19 days the irreversibility has become very great (Curve *c*). Considerable magnetic flux remains frozen in when the field is returned to zero after taking it up to B_{c2}.

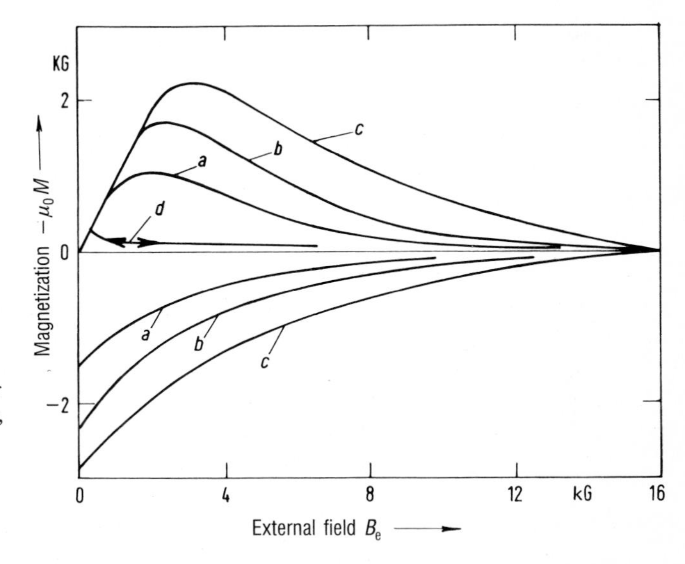

Fig. 93. Magnetization curves of a Pb-Bi alloy in various states of precipitation. a) after 1 day at room temperature, b) after 5 days at room temperature, c) after 19 days at room temperature, d) reversible magnetization of the pure ε phase (after [142]) (1 kG = 0.1 T).

If the direction of the field is now reversed we can pass through a whole hysteresis loop. Figure 94 illustrates such a hysteresis curve for the same Pb-Bi alloy (53 atom % Bi) [142]. The discontinuities in the magnetization are attributable to what are known as flux jumps. Here whole bundles of flux quanta or whole regions of a flux quantum lattice evidently loose themselves from their pinning center and allow sudden jumps in the magnetization in the direction of thermodynamic equilibrium. Such flux jumps are very dangerous in the case of superconducting magnet coils because the heat that is generated can

1 The flux is referred to as being paramagnetic because it is directed parallel to the external field.

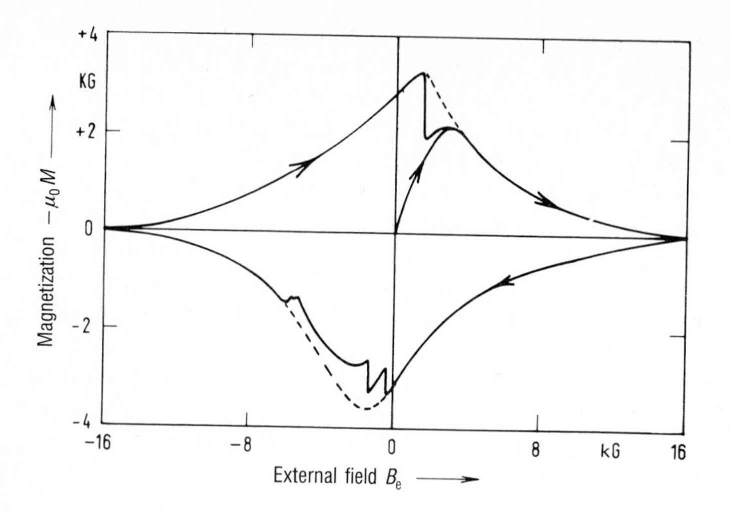

Fig. 94. Complete magnetization cycle of a Pb-Bi alloy (53 atom % Bi). The broken curve is the one that should be followed if no flux jumps occur (after [142]) (1 kG = 0.1 T).

lead to the superconductor becoming normalconducting across its whole cross section (see Section 9.1.2) with the result that the magnetic field breaks down.

We have explained the irreversibility of the magnetization curve and in particular the frozen-in flux at $B_e = 0$ on the basis of the existence of pinning centers for flux tubes. This explanation is certainly justified. There is no doubt that pinning centers exert the influence that has been described upon the magnetization curve. But it is difficult to separate this influence from purely geometrically determined irreversibility which always occurs when superconducting rings or loops – in general superconductors with multiple connection [1)] – are brought into a magnetic field.

We saw in Section 1.1 (Fig. 5) that a persistent current can be induced in a superconducting ring and, hence, a magnetic flux frozen into it. Let us now observe the behavior of such a ring in the magnetic field more closely with the aid of Fig. 95. We choose a ring with a circular cross section in a magnetic field perpendicular to the plane of the ring. We plot the magnetic moment m_M of the ring as a measure of its magnetic state, since it is not possible to speak of the magnetization of a single ring through which a persistent current is flowing. The concept of magnetization only becomes intelligible when very many such rings contribute to the magnetic behavior of a sample.

Starting from a situation without any supercurrent and a field zero (point 0) we begin to increase the external field. Currents begin to flow in the superconducting ring which exclude the magnetic flux from the ring (from the whole area encircled by the ring), or put another way maintain the flux at its initial value of zero (Lenz's rule). Here the magnetic moment of the field directed against the external magnetic field increases more rapidly the greater the surface of the ring (Path OA in Fig. 95). This flux exclusion results in the magnetic field at the external edge of the ring being increased. If this field exceeds

---

1 A multiple connection exists when closed curves can be described in a body which cannot be drawn down to a point within the body. In a ring, for instance, all closed curves which run round the opening possess this property. However, if *all* imaginable closed curves *within a body* can be drawn down to a point then the body is referred to as being single connected.

the critical value B_c [1], then the superconductor must, at least temporarily, enter the intermediate state, resistance appears and the supercurrent is reduced. This causes magnetic flux to penetrate the ring which leads to a reduction in the field at the edge of the ring. In this way the superconducting ring current decreases with increasing field, so that there is always just the critical field at the edge of the ring (Curve branch AB). The interior of the ring becomes thereby more and more filled with magnetic flux. The macroscopic ring current has fallen to zero by $B_e = (1/2)B_c$. The critical field is now generated at the directrix of the ring by the screening of the ring material itself [2]. If the field is further increased the ring goes into a stationary intermediate state [3]. The superconductivity disappears completely when $B_e = B_c$. If the field is now reduced the ring first passes through the intermediate state and then becomes completely superconducting. The Curve branch BC is a reversible branch. If B_e is lowered further a supercurrent is set up in the ring which maintains the flux present in the ring, i.e. the ring current amplifies the field in the interior. We then have the maximum field at the inner edge of the ring and it is maintained at B_c. When the external current is zero a persistent current flows in the ring (see Section 1.1, Fig. 5).

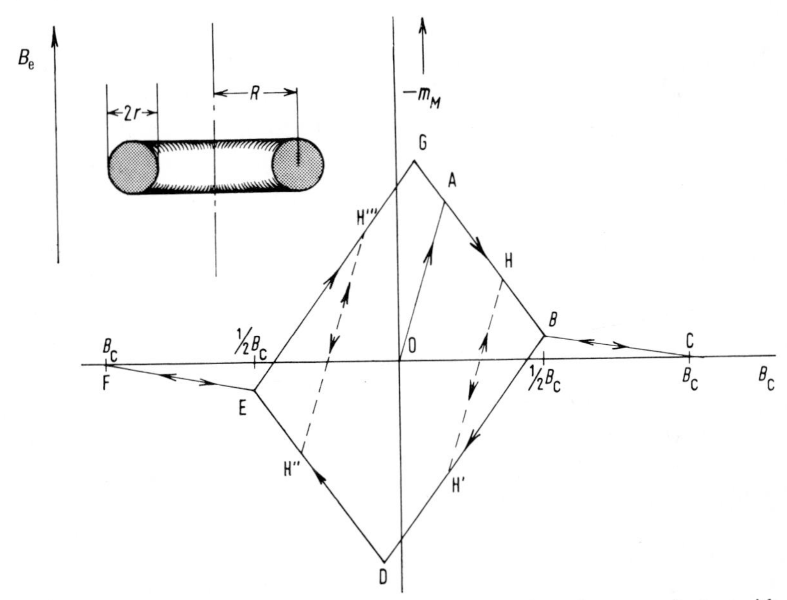

Fig. 95. Superconducting ring in a magnetic field – to the magnetic behavior of superconductors with multiple connection (schematic representation). m_M is the magnetic moment of the ring.

Now let us increase the external field in the reverse direction. The fact, that the magnetic moment carries on increasing until point D, depends on the somewhat different situations at the outer and inner edges of the ring and does not require further consider-

---

1 For simplicity's sake we will consider a type-I superconductor.

2 A wire of circular cross section in a perpendicular field has a maximum field strength of $B_{max} = 2B_e$ on account of the demagnetization factor $n_M = 1/2$ (see Section 5.1.4).

3 The ring was fully superconducting up to point B($B_e < (1/2)B_c$).

ation [1]. As the field increases further the magnetic moment decreases to the extent that the ring current is reduced. The situation along DE is absolutely analogous to that along GB. The maximum field that is maintained at the critical value is now at the outer edge of the ring [2].

The ring is in the intermediate state along Path EF. Finally Path EG is completely analogous to Path BD. If we reverse the direction of the field at a point where the ring is completely superconducting (points H, H', H'', H''') then the moment changes in the same manner as along OA. The ring initially reacts fully, i.e. unhampered by the critical field at the surface, to the field change.

The hysteresis curve of such a superconducting ring has many similarities with the magnetization curve illustrated in Fig. 94. It must certainly be assumed, in alloys with topological multiple connections, caused for instance by precipitates of higher T_c than the matrix (Pb-Bi alloys), that these play a considerable part in the irreversibility of the magnetization curve. The sponge model developed by Mendelssohn [115] took such effects into account.

7.2 Critical currents in hard superconductors

As we saw in Section 6.2, an ideal type II superconductor cannot carry a current without dissipation in the Shubnikov phase perpendicular to the direction of the magnetic field since the flux tubes come into motion under the influence of the Lorentz force and dissipative processes occur. Now the flux tubes in a real superconductor can *never* be freely displaced. A force F_P, even though it may be a small one, is always necessary to remove the flux tube from preferred centers. As long as the Lorentz force F_L is less than this holding force F_P the flux tubes are unable to migrate. So we will be able to observe resistance-free current (supercurrent) in the Shubnikov phase of real type-II superconductors. If the current exceeds the critical value where $F_L = F_P$ then the flux tubes will begin to migrate i.e. resistance then appears. The critical current is, thus, a measure of the force F_P which "anchors" the flux tubes to energetically preferred sites.

Figure 96 reproduces the current-voltage characteristics of two samples of a Nb_{50}-Ta_{50} alloy having different degrees of internal disorder [143]. Both samples were investigated in the Shubnikov phase. In the case of the more disordered sample 2 no electrical voltage was observed up to a current load I_{C2} of 1.2 A [3], while a voltage and, hence, a resistance

---

1 We have also represented the situations at points B and E in a somewhat simplified manner (see D. Shoenberg: Superconductivity, Cambridge, The University Press (1952)).

2 Whenever the field of the ring current in the interior is antiparallel to the external field we have the maximum field at the outer edge of the ring. If we neglect the details occurring at D and G this statement amounts to saying that with increasing field, independent of its direction (increasing contribution of the external field), the maximum field is to be found at the outer edge of the ring.

3 The statement "no voltage" is naturally a great simplification. Since even at currents $I < I_c$ dissipation can normally occur even if it is usually small and we will also observe small voltages here. The characteristics in Fig. 96 indicate this in that they gradually enter the linear portion. It has not yet been completely explained what effects are responsible for this behavior.

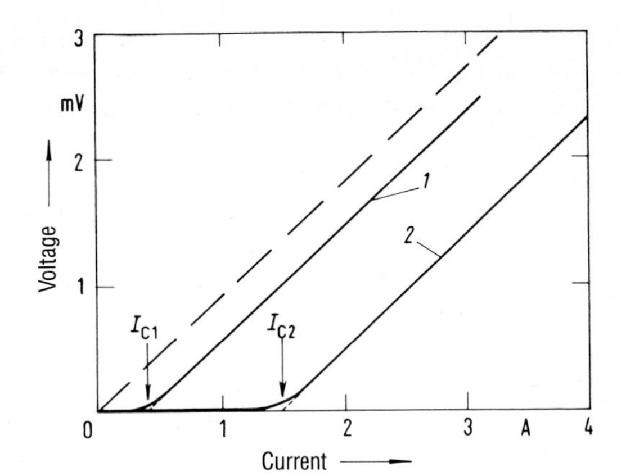

Fig. 96. Current-voltage characteristics of a $Nb_{50}Ta_{50}$ alloy in the Shubnikov phase. T = 3.0 K, external field B_e = 0.2 T, T_c in field zero is 6.25 K (after [143]).

appeared at only ca. 0.2 A in the case of the less disordered sample 1. We call the currents I_{c1} and I_{c2}, the critical currents for both samples. An "ideal" sample of the same material, that is a completely homogeneous sample, would, under the same conditions, have the current-voltage characteristic indicated by the broken line. The point in the case of hard superconductors is that the pinning force F_P is to be made as great as possible in order to achieve as large a resistance-free transport current as possible. Before we treat some examples of hard superconductors we will briefly consider the linear part of the characteristics in Fig. 96.

In this part of the characteristic the voltage U across the superconductor is created by the movement of the flux tubes. The differential resistance $\mathrm{d}U/\mathrm{d}I = R_{fl}$, the "flow" resistance so produced is evidently the same for both samples, hence, it does not depend on the pinning forces of the flux tubes.

This finding can be understood as follows. When the flux tubes have been released from the pinning centers they move through the material under the influence of the force difference $\vec{F}^* = \vec{F}_L - \vec{F}_P$. As a result of the dissipation processes which set up a type of "friction" for the flux tubes in the superconductor, the flux tubes move at a velocity v that is proportional to F^*. So that:

$$v \propto F^* \propto I - I_c \,. \tag{7-1}$$

The electrical voltage U, on the other hand, is proportional to v, so that:

$$U \propto v \propto I - I_c \,. \tag{7-2}$$

Hence:

$$\mathrm{d}U/\mathrm{d}I = \text{const.} \tag{7-3}$$

The equality of $\mathrm{d}U/\mathrm{d}I$ for samples with different degrees of disorder (Fig. 98) shows that the pinning centers evidently have no influence on the relationship between F^* and v. The same F^* is associated with the same velocity v and, hence, the same voltage U

independently of the nature of the pinning centers. The characteristic is merely displaced along the I axis.

According to this finding there are 2 contributions to the energy dissipation UI, namely UI_c, on the one hand, and, on the other hand, $U(I - I_c)$. The UI_c part is attributable to a dissipation process at the pinning centers, which will be gone into more detail in Section 7.3. However, the portion $U(I - I_c)$ is evidently determined by the velocity of migration.

The major part of the energy losses caused by the migration of the flux tubes results from the occurrence of local electrical fields which interact with the unpaired electrons. This makes it clear why the flow resistance is dependent on the normal resistance of the material. It proves that R_{fl} is proportional to R_n [1)].

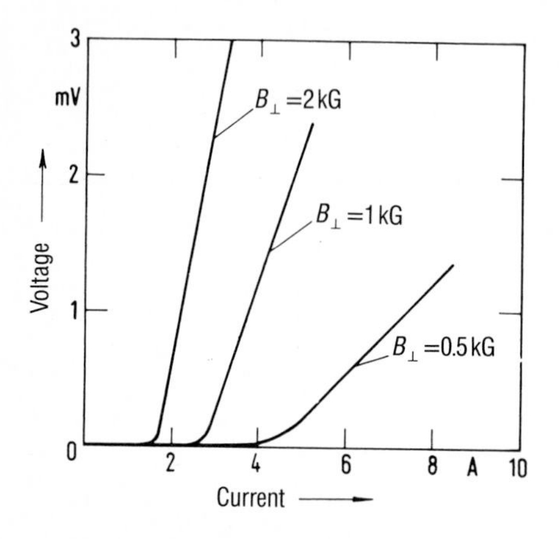

Fig. 97. Current-voltage characteristics of a Pb-In alloy in the Shubnikov phase. Material: Pb + 17 atom % In, T = 2.0 K, transition temperature in field zero is 7.1 K (after [143]).

In addition the same dissipative effect occurs at each flux tube. For this reason the flow resistance is also proportional to the density (number per unit area) of flux tubes. This density increases with increasing external field B_e. The voltage-current characteristics become steeper as the magnetic field is increased. Figure 97 illustrates the dependence of the flow resistance on the external field for a lead-indium alloy (Pb + 17 atom % In) [143].

Here the critical current I_c is reduced as the field is increased. This reduction can have several causes. On the one hand, all flux vortices cannot be pinned to the same degree

---

1 As R_{fl} increases under otherwise equivalent conditions the velocity v of the flux tubes, under the influence of the force exerted on them $F^* = F_L - F_P$ (i.e. under the given current $I = I_c + I'$), is greater. This can be demonstrated by means of a simple but instructive consideration of the power consumption. If a flux tube is moved by force F^* at velocity v through the material then the force F^* develops the power $P_{F^*} = F^*v$. This power has to be converted into heat by the local electrical field E. In the normal state the electrical power P_{el} is proportional to E^2/ϱ_n when the specific resistance is ϱ_n. We have, on the other hand, already seen that E is proportional to v. So that the requirement $P_{F^*} = P_{el}$ results in:

$$E^2/\varrho_n \propto F^* v \propto v^2/\varrho_n \qquad \text{that is}$$

$$F^* \propto v/\varrho_n \,.$$

When F^* is fixed v increases with increasing ϱ_n.

as the density of flux tubes increases. This means that the mean pinning force decreases. On the other hand, the pinning force at one and the same center decreases as the external field is increased and approaches zero as $B_e \to B_{c2}$. We can see the behavior of the critical current when the external field increases particularly clearly in a plot of the dependence of I_c against field (Fig. 98 and 99).

We already concluded in the discussion of Fig. 96 that the critical current I_c depends on the degree of internal disorder of the sample. One of the ways that this internal disorder can be produced is by plastic deformation. For instance, if a metal wire (e.g. Cu) is pulled through an orifice at room temperature, thereby reducing its cross section, then this process produces very many internal faults, i.e. areas where the periodic structure of the metal lattice is greatly disturbed, in the wire. Such areas are, for example, the grain boundaries, that is the transition regions between the individual crystalline grains. If the wire is heated then these improperly ordered regions gradually "order", i.e. are converted more and more into ordered regions.

If regions of disorder act as pinning centers for the flux tubes then a plastically deformed type II superconductor should have a particularly high critical current immediately after preparation. After heating, the critical current should fall as a result of the gradual loss of the disorder.

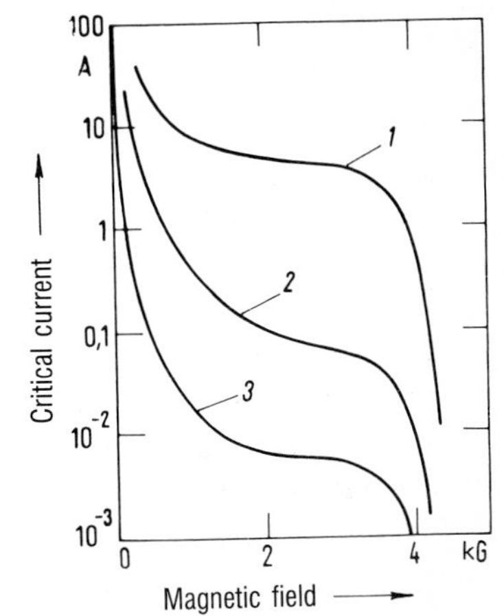

Fig. 98. Critical current of a $Nb_{55}Ta_{45}$ alloy in an external magnetic field (1 kG = 0.1 T) perpendicular to the current. Wire diameter: 0.38 mm, measurement temperature: 4.2 K. 1: immediately after cold forming, 2: after 24 h at 1800 K, 3: after 48 hours at 1800 K (after [141]).

Figure 98 illustrates this effect very clearly [141]. Here the critical current of a $Nb_{55}Ta_{45}$ alloy is plotted as a function of the external magnetic field perpendicular to the current for different degrees of disorder. Immediately after the mechanical drawing process, that is in a state with very high disorder, the critical current in the Shubnikov phase is large[1)]. As the disorder is removed by annealing the pinning centers for the flux tubes disappear more and more. As expected the critical current is sharply reduced. The corresponding magnetization curves are reproduced for the same alloy in Fig. 92. Here the pinning centers lead to great irreversibility of the magnetization curve.

---

1 The steep drop in the critical current in very small magnetic fields takes place in the Meissner phase, in which the superconductor is to be found for fields $B_e < B_{cl}$ (*cf.* Fig. 85b).

Some critical current values for technically important high-field superconductors are reproduced in Fig. 99 [121, 144]. Since, on account of the dependence on disorder already described, the critical current is very much affected by the history of the material the values reported in Fig. 99 can only be taken as a guide. The examples have been chosen arbitrarily from a wealth of data. However, they show that, with suitably prepared superconductors, it is possible to obtain dissipation-free supercurrents of from more than 10 A up to 100 A through a wire only 0.5 mm in diameter in a magnetic field of 10 T. Multicore leads made up of many filaments can carry even higher supercurrents. Such superconductors are naturally of great importance for the construction of superconducting magnets (*cf.* Section 9.1).

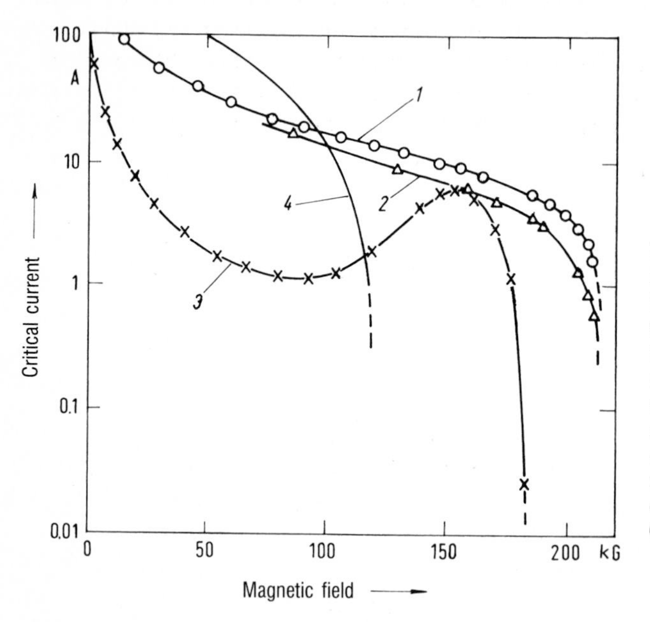

Fig. 99. Critical currents of high field superconductors (1 kG = 0.1 T). Measurement temperature: 4.2 K; 1: V_3Si; 2: Nb_3Sn; 3: V_3Ga; wire thickness of these samples uniformly 0.5 mm. The compound was only present in a surface layer produced by the diffusion of the second component into the basic material Nb or V. 4: Nb-Ti, wire thickness 0.15 mm (after [121, 144]).

Figure 99 also includes a curve (V_3Ga) where there is a maximum, what is known as a peak effect, in the critical current in the neighborhood of B_{c2}. Evidently with increasing external field conditions can be set up which lead to a higher efficiency of the pinning centers. Models have been set up for such peak effects which will be discussed briefly here because they can deepen our understanding of the Shubnikov phase.

We cannot expect the flux tubes of the Shubnikov phase of a hard superconductor to be arranged in such a regular lattice as they are in an ideal type II superconductor. Rather the flux tubes will seek to occupy the energetically most favorable sites insofar as this is possible in view of the mutually repulsive forces. In this situation not all flux tubes will be optimally pinned. When the flux tube density now increases with increasing field it can happen that regions of field occur where the flux tube lattice is particularly well arranged with respect to the arrangement of pinning centers [145]. The critical current will be particularly high in these regions. Since the peak effect, which – it must be emphasized – does not occur in by any means all samples, always occurs in the region of

B_{c2} it could also be caused by a weakening (reduction in repulsive forces) of the flux tube lattice.

Another significantly more trivial explanation is that, in very inhomogeneous superconductors, normalconducting regions appear in the neighborhood of B_{c2} and that these function as additional pinning centers (*cf.* Section 7.3) and, thus, allow an increase in the critical current.

To complete this consideration of the critical current in hard superconductors we will now discuss another model, that of C.P. Bean [146], proposed for an approximately quantitative description of the hard superconductors in a magnetic field and bearing a current load. This model also forms the basis for the quantitative determination of critical currents from magnetization curves. Since, in the case of substances such as Nb_3Sn, there are certain difficulties involved in the direct application of currents of 100 A it is a good thing to have a method of determining the critical current without the application of electrodes. This method is of decisive importance in the study of sintered material that is usually not available in the form of wires or sheets and, hence, cannot be investigated by the usual methods.

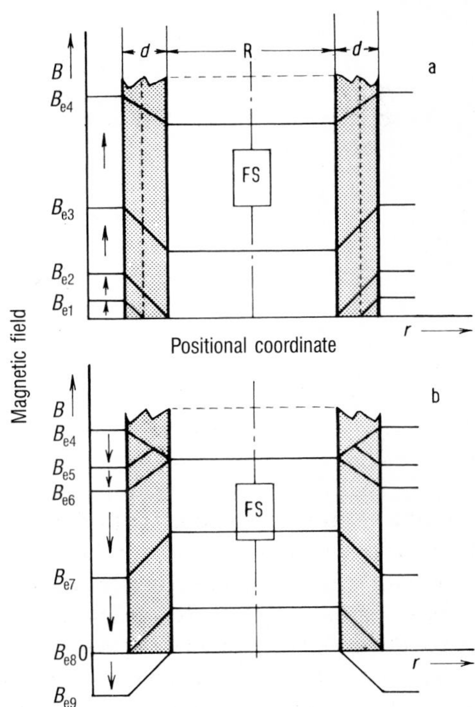

Fig. 100. Variation of magnetic field in a hollow cylinder for a hard superconductor. For a discussion of the Bean model [146]. As B increases j_c decreases; in the case of B_{e4} this decrease is indicated by a flattening of the field variation [147].
FS = field sensor.

We will consider a very simple geometry namely a long hollow cylinder in a magnetic field B_e parallel to its axis (Fig. 100). When field B_e is applied a ring current will be set up around the outside of the superconducting cylinder shielding the interior of the cylinder. As long as the hard superconductor remains in the Meissner phase the screening current only flows within the penetration depth under the external surface of the sample cylinder. If the value of B_e exceeds B_{c1}, which we can assume here to be very small, then the superconductor has to allow the penetration of magnetic flux in the from of flux tubes. The pinning centers of the hard superconductor now become very important for its whole behavior. For the flux tubes do not distribute themselves over the whole volume, they are rather first immobilized by the pinning centers below the surface. Here the Bean model makes the simple assumption that the region of the superconductor into which the flux tubes have penetrated carries a critical current with the homogeneous critical current density j_c. In the simplest approximation the current density is, therefore, assumed to be constant, i.e. independent of the magnetic field in the superconductor. This state with critical current density is known as the "critical state".

A further increase of B_e above the value B_{c1} will lead to the formation of an ever thicker layer filled with flux tubes below the surface of the cylinder. The total screening current increases in proportion to the thickness of this layer, which we have assumed is loaded to a constant current density j_c. The critical state grows into the interior of the superconductor. The screening current also makes the magnetic field in the superconductor decrease towards its axis. If the radius R is large compared with the thickness of the cylinder wall d then this can be treated to a good approximation as a planar layer. The magnetic field then diminishes linearly towards the axis [1]. This field behavior is reproduced for several external fields in Fig. 100a. At B_{e1} only a portion of the mantle is filled with flux tubes (broken line, diagram a). At B_{e2} the screening current has acquired its maximum value, because the whole cylinder wall is now subject to a homogeneous flow of current density j_c. But the field B_i inside the cylinder is still zero. It is only when B_e is increased further that a magnetic field appears in the interior. In our simple model this field B_i should equal $B_e - B_{e2}$ (for $B_e > B_{e2}$).

Our assumption that j_c = const. can naturally not apply at higher fields, for j_c must approach zero as B_e approaches B_{c2}. So we must improve our model by asssuming reduction of j_c as B_e increases. But then the difference $B_e - B_i$ will gradually fall to zero as field B_e increases. This diminution of I_c is visible in Figs. 98 and 99.

Figure 101 reproduces the field B_i in the interior of the cylinder as a function of the external field B_e for the case of the high-field superconductor $V_3(Ga_{0.54}Al_{0.46})$. At B_{e2} the value of B_i is still zero. Then when j_c = const $B_i = B_e - B_{e2}$ (broken line). However, if j_c decreases as the field increases the Curve $B_i(B_e)$ approaches the linear $B_i = B_e$ more and more closely. It is, therefore, possible to determine the value of the critical current density of a material from the course of $B_i(B_e)$ on the basis of the model described, that is by assuming the existence of a critical state. This elegant method is frequently employed in the study of new high-field superconductors.

We will now consider here the case where from about B_{e4} we reduce the external field. Then with what are now negative field changes ($\Delta B < 0$) an inductive process occurs which abolishes the screening current on the outside of the cylinder and sets up a critical current in the opposite direction. Figure 100b reproduces the variation of the magnetic field for some consecutive values of the magnetic field B_{e5} to B_{e7}. We again have the whole cylinder in a critical state with this time the density of flux tubes increasing with

1 On account of the flux tube structure there is a local variation in the magnetic field in the superconductor. In the discussion of this model the field in the superconductor stands for the mean magnetic field. When this mean field is decreased then this means that the density of flux tubes decreases.

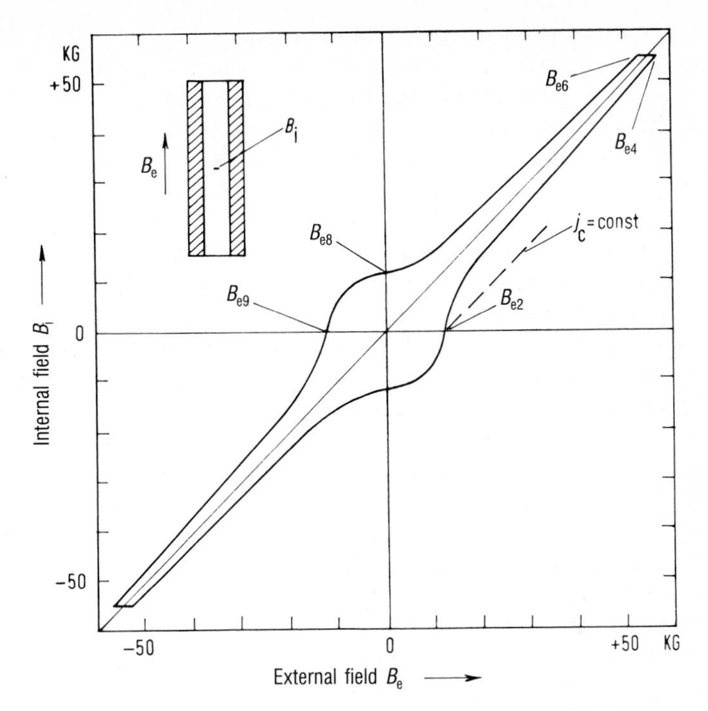

Fig. 101. Magnetic shielding of a hollow cylinder of $V_3(Ga_{0.54}Al_{0.46})$. Record of the internal field B_i as a function of the external field B_e (1 kG = 0.1 T), measurement temperature: 4.2 K, transition temperature: 12.2 K (reproduced by kind permission of Dr. H. Voigt, Siemens Research Laboratories, Erlangen).

the distance from the outer edge. We could say, somewhat picturesquely, that the flux tubes are running out of the cylinder.

The value of B_i remains constant until B_{e6} is reached. After this B_i falls. At $B_e = 0$ we have frozen in a certain magnetic flux and, hence, a field B_i. The points concerned are marked in Fig. 101.

Let us now make the external field increase in the opposite direction, we must reach exactly the field $B_i = 0$ at $-B_{e9} = B_{e2}$. This is quite evident from Fig. 100b. Figure 101 reproduces the associated variation of B_i together with the whole hysteresis loop that can be travelled round in such an experiment. In the case of many hard superconductors this dependence corresponds very well to the assumption [147]:

$$j_c = \frac{\alpha_c}{B + B_0} \tag{7-4}$$

where the constants α_c and B_0 are characteristic for the material concerned.

A systematic study of hard superconductors has made it possible to develop some very useful materials empirically. At the moment we do not have sufficient insight into the quantitative aspects for a proper basic understanding of the phenomena that occur here.

7.3 Pinning centers for flux tubes

We dealt in Section 7.1 and 7.2 with the influence of the pinning centers on the magnetization curves and the critical currents of hard superconductors. In doing this it was not necessary to make any statements concerning the physical nature of these pinning centers. We will now take the opportunity in this section of discussing the occurrence of pinning centers, in somewhat greater detail.

We saw with the example of a lead-bismuth alloy (Fig. 93), that normalconducting precipitates can evidently serve as pinning centers. Faults in the crystal structure can also lead to the flux tubes being pinned to preferred sites (Fig. 92 and 98). How can we understand the effects of these pinning centers?

It is simplest to come to a qualitative understanding from a consideration of the energy. The formation of a flux tube requires a certain amount of energy. This energy manifests itself in the ring current (Fig. 78 and 80), that must flow round the core of each flux tube. It is immediately clear here that under fixed conditions a flux tube is associated with a certain amount of energy per unit length, i.e. the longer the flux tube the greater the energy required to create it.

An estimate of this energy per unit length, which we will call ε^*, can be obtained from the lower critical field B_{c1}. The magnetic flux begins to penetrate type II superconductors at this field. The exclusion energy that is thereby gained is sufficient to create flux tubes in the interior. For simplicity's sake let us once more consider a "long" cylinder with a field parallel to its axis, i.e. a geometry where the demagnetization factor equals zero. The penetration of the magnetic flux at B_{c1} means that we create n flux tubes per unit area. Each flux tube carries just one flux quantum Φ_0. So the energy required is:

$$\Delta E_F = n \in^* L\, A \tag{7-5}$$

where n = the number of flux tubes per unit area, ε^* = the energy of a flux tube per unit length, L = the length of the sample, A is the area of cross section of the sample.

The gain in magnetic expulsion energy is

$$\Delta E_M = B_{c1}\, \Delta M\, V \tag{7-6}$$

where ΔM = change in the magnetization of the sample, V = volume of sample, $V = L\,A$

We can express ΔM in terms of the penetrating flux quanta. When

$$\Delta M = n\, \Phi_0/\mu_0 \qquad \mu_0 = 4\pi \times 10^{-7}\,\mathrm{Vs/Am} \tag{7-7}$$

we, therefore, get

$$\Delta E_M = \frac{1}{\mu_0} B_{c1}\, n\, \Phi_0\, L\, A \tag{7-8}$$

for the gain in exclusion energy.

If we now make both energy differences equal, in accordance with the definition of B_{c1}, then when $\Delta E_F = \Delta E_M$ we get:

$$n \epsilon^* L S = \frac{1}{\mu_0} B_{c1} n \Phi_0 L A \tag{7-9}$$

and, hence, [1)]

$$\epsilon^* = \frac{1}{\mu_0} B_{c1} \Phi_0. \tag{7-10}$$

The pinning effect of normalconducting precipitates is now easily understood knowing the flux-tube energy ε^*. If a flux tube can pass through a normal region it can reduce its length in the superconducting phase and, hence, its energy. Figure 102 illustrates this schematically. The hatched region indicates a normalconducting region. A flux tube in position a has an energy lower than one in position b by an amount $\varepsilon^* l$. This means that this amount of energy $\varepsilon^* l$ must be provided to move a flux tube from a to b. A force must be employed to effect a change of position from a to b.

If many such pinning centers are present the flux tubes will try to take up the energetically most favorable positions. This will result, as illustrated in Fig. 103, in the production of bends such that the total energy has the most favorable value. The increase in length resulting from the bending must be more than compensated by the shortening which is brought about by passing through normalconducting regions. In the case of a lattice of flux tubes, such as is present in the Shubnikov phase, the total balance also has to take account of the repulsion occurring between the flux tubes.

It is possible, in principle, to understand other pinning centers, such as lattice defects, in the same manner. Since the density of Cooper pairs in the interior of a flux tube is reduced in comparison with that in the pure superconducting phase and a state, therefore, exists which is closer to normal conductivity, every inhomogeneity of the material which is less favorable for superconductivity – normal conductivity is a limiting case – will act as a pinning center. If, for example, a precipitate is still superconducting, but its transition temperature is lower then it will, in general, act as a pinning center.

If the superconductivity of a particular material is associated with an expansion of its lattice (see Section 4.3) then regions with a contracted lattice will be less favorable for superconductivity and will act as pinning centers. This can be the case at the grain

---

1 It is also possible to calculate ϵ^* by integrating over the ring current (see, for example, P.G. DeGennes, Chapter 3, page 57 ff.) The following expression is then obtained:

$$\epsilon^* = \frac{1}{\mu_0} (\Phi_0/\lambda)^2 \ln \lambda/\xi$$

where λ = depth of penetration (see Section 5.1.2) and ξ = coherence length (see Section 5.1.5).
It is evident here that ε^* increases as the square of the magnetic flux in a tube. Flux tubes with more than one flux quantum are, therefore, energetically unfavorable and can only occur in type II superconductors or hard superconductors if they are favored by other conditions, such as inhomogeneities in the material.

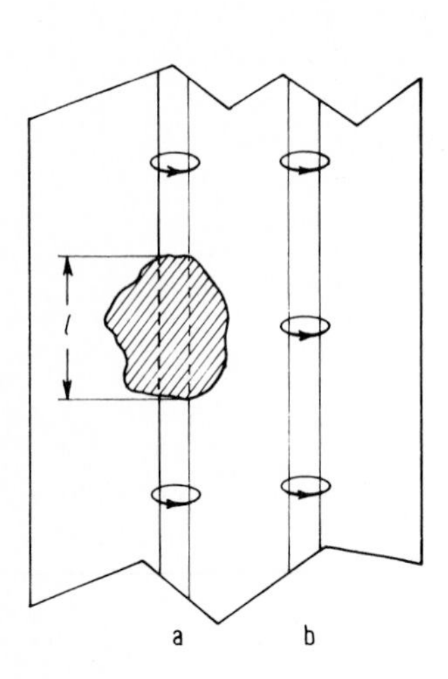

Fig. 102. The pinning effect of normalconducting precipitates. In position a the flux tube has reduced its effective length in comparison to that in position b, since no ring currents are present in the normalconducting region.

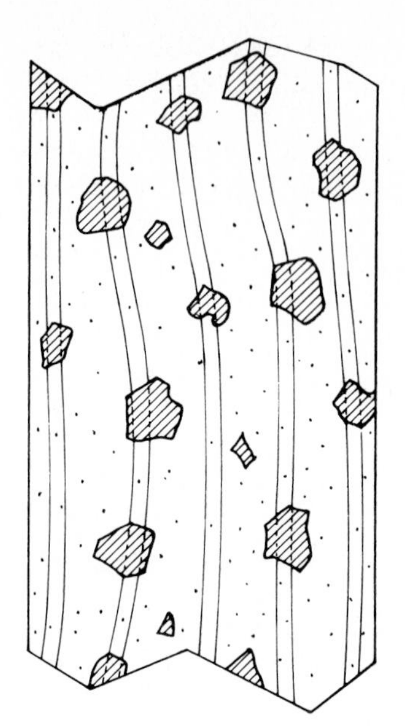

Fig. 103. Arrangement of the flux tubes in a hard superconductor. The hatched regions are pinning centers. The dots represent atomic dislocations. They do not have any pinning effect since the flux tube extends over many point defects along its length.

boundaries and agglomerations of dislocations, such as can be produced by plastic deformation.

Atomic faults in the lattice constitute a particular problem with respect to pinning forces. Since the flux tubes have a diameter of the order of magnitude of the coherence length ξ_{GL} one might expect that atomic faults would not be effective pinning centers. However, it has proved possible to demonstrate that the scattering of electrons at the sites of atomic faults leads to a considerable pinning effect [148]. Since, however, each flux tube normally encompasses many atomic faults there is no effective pinning force when the atomic faults are distributed homogeneously. No positions in the material are then favored over others. This is only the case when there are spacial variations in the density of atomic faults.

The extremely small coherence lengths of the new superconductors (see Section 7.4.1) mean that the situation is completly different here. Here centers taking up very little space can also act as pinning centers in many different ways. Very little is known concerning the nature of the pinning centers in the superconducting oxides.

To conclude this brief discussion of pinning centers we will now deal with a problem that occurs in conjunction with the voltage-current characteristics (Fig. 96). It has been found experimentally that the incorporation of pinning centers displaces the characteristic parallel to a larger current by an amount I_c. This means, however, that in a hard superconductor with increasing I_c, the electrical power $P_{el} = UI$ transformed into heat also increases, or put another way, that there must be a dissipation effect present that increases with increasing pinning force.

We will use a somewhat more abstract model of the pinning effect to explain this finding. A pinning center corresponds to a potential well. The flux tube lies in its most favorable position, just like a ball, when it lies at the deepest point in the dish. A force is needed to overcome the increase in potential energy required to displace the ball from this position. To remove the ball completely from its favored position it is necessary to provide enough energy to remove it from the well. Figure 104a illustrates this.

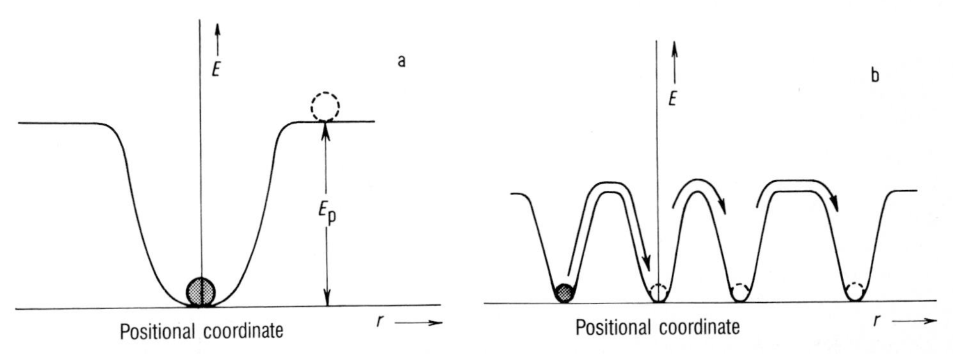

Fig. 104. Representation of the potential at a pinning center. E_p = depth of potential well (binding energy). The arrows in Fig. 104b are intended to indicate the movement of a flux tube through the potential well.

There are many potential wells in the case of a flux tube; we will, for the sake of simplicity, assume that they all possess the same pinning energy, E_p. This is represented in Fig. 104b by a series of equally deep potential wells. If the flux tube is now moved through the material under the influence of the Lorentz force, set up by the transport current, it means in our model that the flux tube must pass through the potential wells. The pinning force[1] has then to be overcome when the flux tube is raised out of the well. It then falls into the next well. It could thereby, in principle, regain the same amount of energy, that had been required to remove it. If this were the case then passing from well to well would not involve an energy dissipation. In particular, the energy dissipation should be independent of the depths of the wells.

The experimental results force us to assume that a process of dissipation is occurring that is associated with the pinning centers and particularly with their depth. Such a process is readily comprehensible and has already been reported [149]. When a flux tube falls back into a potential well it must take place in such a manner that part of the energy is converted into heat. One can imagine that, say, the flux tube vibrates in the well and the energy of motion is converted into heat. Even entering and leaving the well at different velocities leads to the dissipation of energy since the dissipative effect will depend on the velocity. The dependance on the depth of the potential well will also be comprehensible. The deeper the well the larger can be the variation in the velocity of passing

1 The pinning force has a maximum value of dE/dr.

through it. It is only when such processes are considered that it is possible to understand the voltage-current characteristics of flow resistance.

The representation of pinning centers as potential wells for flux tubes offers us the possibility of describing the phenomenon of "flux creeping" in a very graphic manner. Flux creeping is a thermally activated migration of the flux tubes. When transport current I is flowing which is smaller than I_c then it is not possible for the flux tubes to be removed from the pinning centers by the Lorentz force alone. The Lorentz force is not sufficient to provide the binding energy E_p (Fig. 104). However, when the superconductor is at a finite temperature T there is a probability $w = e^{-\Delta E/k_B T}$ that random thermal variations could make up the energy difference ΔE[1]. This means that, in principle, flux tube migration in a hard superconductor can always take place at finite temperature even when the Lorentz force is smaller than the pinning force [150].

Until now we have only considered the pinning force for one flux tube. Since, however, flux tubes repel each other (see Footnote 1 p. 175) the pinning force per unit volume does not just depend on the individual pinning force. Rather the pinning force per unit volume is dependent in different ways on the individual pinning force and on the concentration of pinning centers, depending on how strongly the individual pinning force and the interaction of the flux tubes with each other contribute to the pinning force per unit volume [XIV]. Fundamental questions still remain to be answered here.

A pinning force also occurs at the boundary between superconducting and normal regions. This "magnetic pinning force" requires regions that are larger than the penetration depth λ. Magnetic pinning is without importance in the case of more or less optimized hard superconductors. A very good description of hard superconductors is given in [XIV].

7.4 Critical fields and critical currents in the new superconducting oxides

When the sensationally high transition temperatures of $YBa_2Cu_3O_7$ were reported it was quite evident that the technical application of superconductivity might receive an enormous boost. It is not just that cooling with liquid nitrogen is a factor of 50 to 100 times more economical than it is with liquid helium; liquid nitrogen is also very familiar in industry. It is much easier to handle than liquid helium.

For this reason great efforts were made from the beginning to determine the quantities of importance for technological applications such as magnet construction. The complex structure (see Section 8.7.4) and the high specific resistance already indicated that they ought to be type II superconductors. The marked layer structure of these perowskite-like materials[2] also pointed to a strong anisotropy of the characteristic parameters.

1 Here the thermodynamic variations can either reduce the depth of the potential well or provide the flux tube with sufficient energy.

2 The mineral perowskite is $CaTiO_3$. TiO_6 octahedra are fundamental structural units. In the case of the new superconducting oxides the structural units are CuO_6 octahedra, but they are connected somewhat differently.

The decisive variables as far as application is concerned are the critical field and the critical current density. We will now learn some experimental facts. For since there are always difficulties, particularly in the preparation of well-defined samples, the quantitative results are still somewhat unreliable.

However, it is possible to report on the orders of magnitude and on the problems which remain to be solved. The evidence we have is basically for $YBa_2Cu_3O_7$ which has come to be a model substance for the new superconductors.

7.4.1 The critical fields for the new oxides

Even the first investigations revealed that the new materials are, in fact, superconductors with marked type II properties. The depth of penetration λ is between 100 and a few 100 nm. The coherence length is surprisingly short. It is between 2 nm and 3 nm in the plane of the layers (see Section 8.7.4) and shorter than 1 nm at right angles to the layers. This gives, from Equation (5–39), values of $\varkappa$ of 100 and more (see, for example, [151]).

The bulk samples of the new materials have mainly been prepared by mixing the oxides, carbonates or nitrates of the metals and then sintering a compacted sample[1]. The samples that are produced – they are often referred to as a ceramic – are made up of tiny crystals, which are in contact with thin regions of more or less disordered material of usually unknown composition.

Such samples allow the penetration of the magnetic flux even with external fields of only a few gauss. However, these fields are not identical with B_{c1}, rather they are more associated with the material between the grains. The upper critical field of the grains of such materials can be determined from the magnetization curves (see, for example, Fig. 94) [152]. It is not possible to determine the lower critical field of such samples since the first penetration of the magnetic field is also dependent on the shapes of the grains due to the influence of the demagnetization factor. However, it was possible to determine both B_{c1} and the anisotropy of the critical fields after small single crystals of the new materials had been prepared.

Here we will describe, as an example, the determination of B_{c2} for a very good single crystal sample of $YBa_2Cu_3O_7$ [153]. Figure 105 reproduces the upper critical fields perpendicular to the plane of the layers (see Section 8.7.4) $B_{c2\perp}$ and parallel to the plane of the layers $B_{c2\parallel}$ as a functions of temperature. The strong anisotropy can be understood when it is remembered that the screening current can flow far better in the layers of oxygen octahedra with the central Cu ion than it can perpendicular to the layers[2]. If the external field is perpendicular to the layers then screening currents which increase G_s can be set up in the layers. These screening currents are much smaller when the external field is parallel to the layers; this means that the field has to be raised to substantially higher values in order to make $G_s > G_n$ (see Section 4.1).

---

1 The process of pulverization, mixing and sintering is usually carried out several times.

2 The conductivity in the normal state is also very anisotropic. It is much greater along the layers than across them [84].

As can be seen from Fig. 105 the critical fields are extremely large. Extrapolations to temperature $T = 0$ K yield values for $B_{c2\parallel}$ of much greater than 100 T. It is also difficult to measure the values of B_{c1} in single crystals on account of the influence of pinning centers[1)] and of the demagnetization. At 11 K values of 0.012 T were found for $B_{c1\parallel}$ and 0.069 T for $B_{c1\perp}$ [151]. Other authors [153] have reported considerably higher values for the B_{c1} fields.

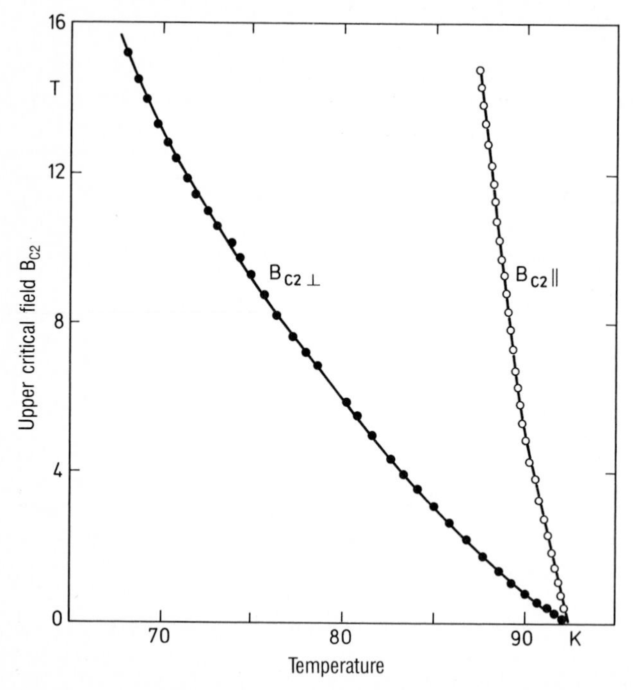

Fig. 105. The critical field B_{c2} (in Tesla) of a $YBa_2Cu_3O_7$ single crystal as a function of the temperature (in Kelvin) (after [153]).

7.4.2 The critical currents of the new oxides

The first determinations of the critical current on sintered samples yielded disappointingly low values for the critical current densities. However, it very soon became clear that this was dependent on the contact between the grains. Under special preparation conditions it has been possible to achieve j_c values of a few 10^3 A/cm^2 at 77 K (boiling point of liquid nitrogen) [154]. But these values of the critical current density are too low for employment of the material say in magnet construction.

A substantial increase in the critical current density has been observed in films which have been grown on a crystalline substrate, mainly with the c axis perpendicular to the

---

1 For the determination of B_{c1} it is necessary to use samples with reversible magnetization curves. Pinning centers always cause irreversibility.

substrate. Critical current densities of more than 10^5 A/cm^2 were determined at 77 K and of more than 10^6 A/cm^2 at 4.5 K [155]. If the components are condensed under a moderate pressure of oxygen (ca. 0.1 mbar) onto a single crystal of $SrTiO_3$ as substrate the single crystalline films of $YBa_2Cu_3O_7$ are immediately formed, having their copper oxide layers parallel to the substrate. The current densities j_c are considerable even at 77 K and in external magnetic fields. Figure 106 reproduces the results [156] which have also been obtained by other groups [157]. They confirm that it is possible to reach technically exploitable current densities, in principle. The task is now to develop conductor configurations which possess the evidently necessary alignment of the crystals relative to the magnetic field produced and where the flexibility of the conductor necessary for the construction of magnets is guaranteed (see also Section 9.1).

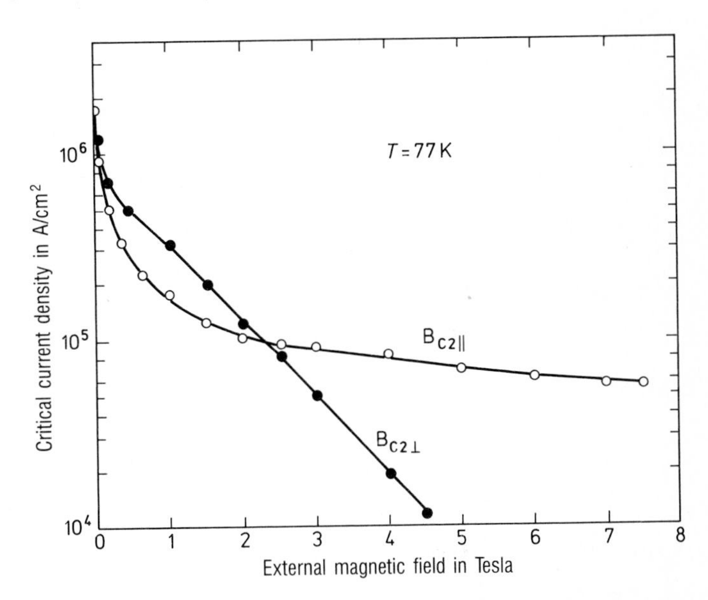

Fig. 106. The critical current density of a $YBa_2Cu_3O_7$ single crystal at 77 K as a function of the external magnetic field (after [156]). Recent results have shown that the j_c-values in high magnetic fields can be improved remarkably (K. Watanabe et al., Appl. Phys. Lett. 54, 575 (1989)).

Since it is difficult to form good contacts with the ceramic superconductors the critical current densities of the compact samples, e.g. single crystals, were determined from the magnetization curves (see, for instance, Fig. 94). The Bean model [146] (Fig. 100) can then be employed to calculate the supercurrent density from the magnetic moment of the sample. However, this method of determination is somewhat questionable, since the requirements of the Bean model, e.g. homogeneous current density over the cross section of the sample, are only approximately fulfilled.

These problems are not so serious in the case of thin films. It is possible to construct a small link between two broader regions and study the current-determined transition to normal conductivity in it. When this geometry is employed the load at the contact can remain very small.

8. Further properties of superconductors

8.1 Fluctuation phenomena

We learned in Chapter 2 that the decisive characteristic of the state of superconductivity is the existence of Cooper pairs. There is a finite concentration of such Cooper pairs below the transition temperature T_c. This equilibrium concentration approaches zero as T_c is approached. There are no Cooper pairs in thermodynamic equilibrium at temperatures above T_c.

Now we know from numerous observations that at finite temperatures physical systems are subject to fluctuations that can bring the system out of equilibrium. However, the state of disequilibrium, which is produced by such a statistical fluctuation, cannot remain as it is. Processes are set in train that lead the system back to equilibrium. Continuous fluctuations are, therefore, set up about the equilibrium state.

A simple example will serve to illustrate this general assertion. Let us consider a gas which is enclosed in a container. We know that a particular quantity of gas M^* at a given volume V and a fixed temperature T[1] is associated with a particular equilibrium pressure P. In the case of an ideal gas this pressure is given by the equation of state $P = M^*R\,T/V$ (R = gas constant = 8.3 Nm/K mol). Now if we monitor the pressure sufficiently sensitively at a small part of the surface of the vessel wall we will discover that it fluctuates statistically about the equilibrium pressure P. This fluctuation is directly comprehensible in terms of the idea that gas atoms undergo completely random motion. As a result of this purely statistical motion it is quite possible that during time t_1 somewhat more atoms arrive at the surface element under observation and with a somewhat greater velocity than is the case during observation period t_2. The pressure fluctuates. The equation of state only gives the mean value about which these fluctuations take place[2].

The occurrence of fluctuations is particularly easy to understand in this example. If, however, we remember that every system at finite temperature is fundamentally subject to a random statistical motion of its components the general character of such fluctuations will be immediately apparent.

Now for a superconductor the existence of thermodynamic fluctuations means the following: Deviations from equilibrium can occur above T_c, that is in the normal state, which can result in the existence of the superconducting state in certain regions, i.e. the creation of Cooper pairs. Such deviations from equilibrium are naturally not stable. They disappear more or less quickly (Section 8.2).

We must, therefore, suppose that because of fluctuations swarms of Cooper pairs will also appear and disappear here and there at temperatures above T_c. This will occur the

---

1 Here the requirement of a fixed total energy is of equal importance to that of a fixed temperature. It is of decisive importance to know which variables are to be kept constant when describing thermodynamic processes.

2 This mean can, in principle, be measured as accurately as desired by suitably extending the period of measurement.

more rarely the higher the temperature, since as the temperature rises the normal state becomes ever more stable with respect to the superconducting state and ever greater deviations from equilibrium are required to cause the formation of the superconducting state.

When we also consider that the swarms of Cooper pairs can move through the superconductor without dissipative processes it will immediately become clear that an excess conductivity can be created in the normal state above T_c as a result of the statistical occurrence of Cooper pairs and that this must increase sharply as T_c is approached.

We expect, therefore, from our understanding of the superconducting state and taking into account the existence of thermodynamic fluctuations that even above T_c the superconductivity will be detectable in terms of an additional diminution in resistance (increase in conductivity).

This influence of fluctuation phenomena has been detected unequivocally for a range of superconductors. Figure 107a reproduces the transition curve of a bismuth film (see Section 8.5.4) in the region of the transition temperature T_c [158]. It is quite evident that completely normal resistance is only reached at temperatures appreciably above T_c. In Fig. 107b it is the conductivity that is plotted instead of the resistance. The excess conductivity σ' resulting from statistically produced and re-evaporating Cooper-pair swarms is particularly clearly visible here [1]. Note that here and in the following treatment we do

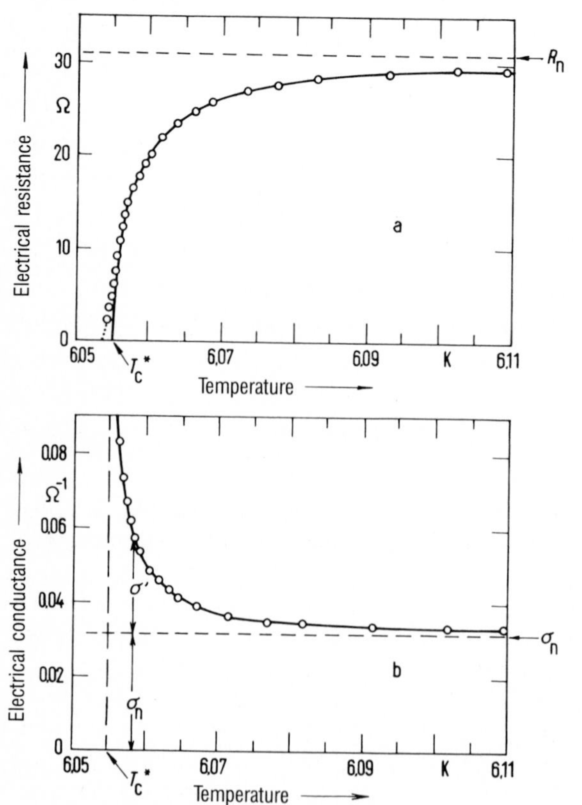

Fig. 107. Transition curve of an amorphous bismuth film. a) resistance, b) conductivity, thickness: 47 nm. The solid curves correspond to Equations (8–1) and (8–2) (after [158]).

1 In Figs. 107 to 109 the resistance and the conductivity have been normalized to a square film geometry (length l = breadth b).

not consider the "specific" conductivity σ^*, but the conductivity $\sigma = \sigma^* d$ (d = thickness) in the dimensions A/V $= \Omega^{-1}$.

Sensitive resistance measurements indicate that this effect of fluctuations continues to be detectable up to rather high temperatures. Figure 108 reveals this for the Bi film [158] whose behavior in the region of T_c is illustrated in Figs. 107a and b. The conductivity scale has been expanded by a factor of 100 here. At this magnification it can be seen that even at $T > 2T_c$ the conductivity σ is still somewhat higher than in the normal state without fluctuating Cooper pairs. It is not always easy to determine this residual conductivity σ_n, which corresponds to the residual resistance R_n, since the normal temperature dependence of the conductivity is already becoming detectable at $T > 2T_c$. In favorable cases there is a temperature range where the conductivity is practically independent of T. The conductivity in this range then corresponds to the residual conductivity σ_n of the normal conductor.

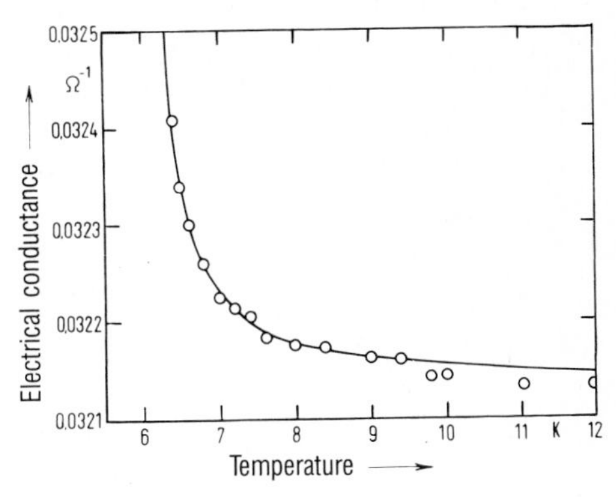

Fig. 108. Electrical conductance of an amorphous Bi film between 6 K and 12 K. Thickness: 47 nm, transition temperature $T_c^* = 6.055$ K (after [158]).

The excess conductivity which is caused by Cooper pairs can be calculated by combining the theory of fluctuations with the present theory of superconductivity. In the case of a thin film this yields [159]:

$$\sigma'(T) = \frac{e^2}{16\hbar} \frac{T_c^*}{T - T_c^*} \frac{b}{l} \tag{8-1}$$

where e = the elementary charge, $\hbar$ = Planck's constant/2π, T_c^* = transition temperature which leads to optimum fit of Equation (8-1) to the measurements points, b = breadth, l = length of the film.

A film is regarded as being "thin" in this context if its thickness is less than the coherence length ξ_{GL} (see Section 5.1.5). Such samples are referred to as being "two-dimensional". Later we will return briefly to the influence of the dimensionality. The excess conductivity σ' must, hence, be proportional to $1/(T - T_c^*)$. This prediction of the theory has been well confirmed experimentally. The curves illustrated in Figs. 107a, b and 108 correspond to this dependence. In addition the excess conductivity σ' should, according to Equation (8-1), be independent of all material properties. This theoretical prediction has also been confirmed experimentally. Figure 109 reproduces the results obtained with 7 very different films [158]. Here the excess conductivity has been normalized to a square

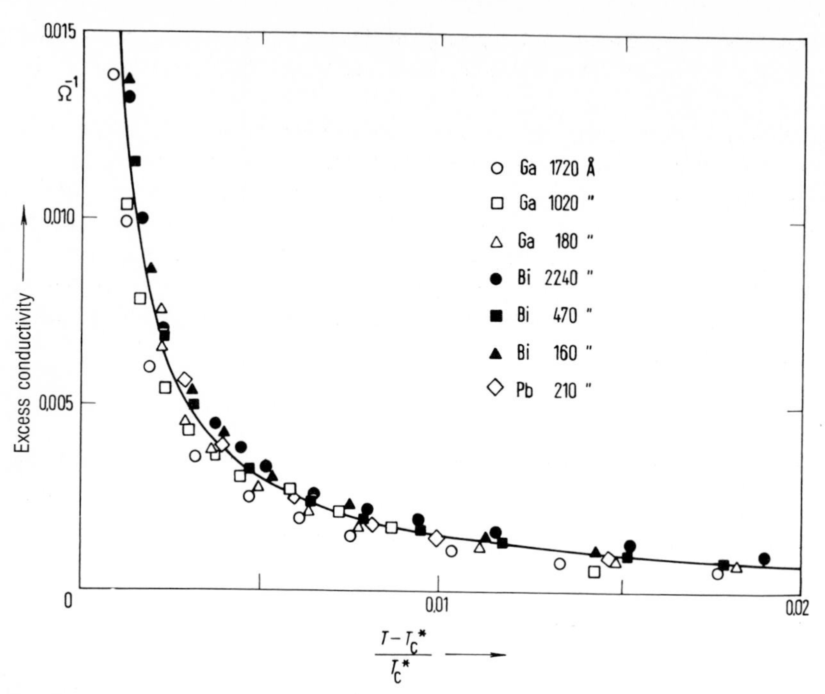

Fig. 109. Excess conductivity of seven different films as a function of the reduced temperature (after [158]).

film geometry (breadth b = length l). All films follow the general law which is represented by the solid line[1]. The agreement goes as far as to include the absolute size of the excess conductivity. In the case of square film geometry Equation (8–1) becomes:

$$\sigma' = \frac{e^2}{16\hbar} \frac{T_c^*}{T - T_c^*} . \tag{8-2}$$

The constant $e^2/16\hbar$ has the value $1.52 \times 10^{-5}\ \Omega^{-1}$. The experiments yielded a result of $1.51 \times 10^{-5}\ \Omega^{-1}$. This is a proof that the experimentally observed excess conductivity in the examples given here actually is a result of fluctuation phenomena.

This very critical discussion of such experiments is, therefore, necessary because inhomogeneities in the sample, which lead to regions with differing transition temperatures, can also yield similar transition curves. When, as in the case of tin for example (see Section 8.5.3), the transition temperatures can be displaced by several degrees by the incorporation of defects, then it is also possible to explain an extension of the transition curve to higher temperatures by the existence of such distorted regions. In such a case the resis-

---

1 The relatively thick films of Bi (2240 Å) and Ga (1720 and 1020 Å) also fulfil the condition $d < \xi_{GL}$ in the temperature range under consideration near T_c since ξ_{GL} approaches infinity on approaching T_c (Table 7, p. 133).

tance gradually falls, because more and more regions become superconducting one after the other as a function of the transition temperatures determined by their degree of disorder.

The amorphous films employed by R.E. Glover for the detection of fluctuation phenomena have the decisive advantage that their extreme disorder makes them very homogeneous with respect to their transition temperatures[1]. Crystalline films can exhibit very large deviations from the relationships applying for fluctions as a result of their inhomogeneity, e.g. as a result of differing stresses in the individual crystallites.

Equation (8–1) gives us the excess conductivity of 2-dimensional samples. The following relationships hold for the dependence of the excess conductivity of 3- and 1-dimensional samples on temperature:
3-dimensional, l, b and d all large compared with ξ_{GL},

$$\sigma' = \frac{e^2}{16\hbar} \frac{1}{2\xi_{Co}} \left(\frac{T_c^*}{T - T_c^*}\right)^{1/2} \frac{db}{l} \tag{8-3}$$

1-dimensional l large compared with ξ_{GL}, b and d small compared with ξ_{GL},

$$\sigma' = \frac{e^2}{16\hbar} \pi \xi_{Co} \left(\frac{T_c^*}{T - T_c^*}\right)^{3/2} \frac{1}{l} \tag{8-4}$$

where e = the elementary charge, $\hbar$ = Planck's constant/2π, ξ_{Co} = the mean extension of a Cooper pair (Table 7, p. 133), d = thickness, b = breadth and l = length of the film.

It is easy to see qualitatively that the dimensions of a sample must have an effect on the size of the fluctuations in it. We discussed in Section 5.1.5 the fact that the Cooper-pair density can only vary over lengths of the order of magnitude of ξ_{GL}. Steeper local variations require relatively high energies and do not, therefore, occur in practice. In the case of a sample which is large in all three dimensions it is possible for the Cooper-pair density to vary spacially in all three dimensions. All these possible configurations have to be taken into account in the calculation of the excess conductivity. For a 2-dimensional sample the Cooper-pair density is always constant along the direction of smallest extension. The determination of the variation of the Cooper-pair density in this direction is superfluous. For a 1-dimensional sample the determination is superfluous in both directions in which the sample is thin with respect to ξ_{GL}. The statistics are, therefore, restricted by the sample geometry which is expressed by the different formulae for the excess conductivity.

Experience has shown that the transition curves are very sharp in the case of 3-dimensional samples such as wires whose thickness is very great compared with ξ_{GL}, i.e. the effects under discussion cannot be observed. The reason for this is not the absence of fluctuation phenomena but rather the relatively high residual conductivities of 3-dimensional samples. A pure tin wire with a diameter of 1 mm has, for instance, a residual con-

---

1 A further advantage of these amorphous films is their extremely low residual conductivity. Since the excess conductivity is independent of the material it can be determined all the more readily the lower the residual conductivity.

ductivity at least 8 powers of ten higher than an amorphous bismuth film of the same length having a thickness of 100 nm and a breadth of 1 mm[1]. So in order to measure the excess conductivity that is present according to Equation (8–3) it is necessary that the factor $T_c^*/(T - T_c)$ have a magnitude of ca. 10^{15}. But this means nothing more or less than that such samples can have extremely sharp transition curves and the influence of the fluctuations of the excess conductivity remains unobservable.

Until now we have only considered how the fluctuations affect the electrical conductivity. When Cooper-pair swarms appear in a statistical manner above T_c then this must also express itself in other phenomena. We know that a superconductor excludes small magnetic fields from its interior below T_c, that it becomes an ideal diamagnet. It is to be expected that, as with the conductivity, something of this property ought to be observable above T_c on account of the fluctuations that occur.

The Cooper-pair swarms should lead to the diamagnetic behavior of the superconductor above T_c being temperature-dependent in a characteristic manner. The additional diamagnetism already becomes very small even a few hundredths of a degree above T_c and only corresponds to the displacement of a few flux quanta. Nevertheless, the size of this effect has been unequivocally determined [160], by exploiting a superconducting magnetic field sensor (see Section 9.5.4).

Finally the fluctuations should also affect the specific heat (see Section 4.2) by producing an increase in c even above T_c. This effect has also been demonstrated [161]. Fluctuation phenomena have also been observed in the new superconductors [162].

The results obtained concerning the influence of the fluctuations confirm in a wonderful manner the internal consistency of our general concepts concerning the superconducting state.

8.2 States outside thermodynamic equilibrium

The fluctuations treated in Section 8.1 were the result of the superconductor spontaneously departing from thermodynamic equilibrium. The superconductor then returns from these states to thermodynamic equilibrium by means of "relaxation processes".

However, it is also possible by means of external influences to bring a superconductor into a state that departs more or less from the thermodynamic equilibrium state. If the influence of such a parameter is then switched out "instantaneously"[2], then the superconductor returns to the equilibrium state via relaxation processes. We saw an example of such a process in Section 6.2 associated with the migration of flux tubes.

Experience has taught that very many such equilibration processes take place with time according to an exponential law. If the equilibrium value of a variable x is designated by x_0 then the exponential decay of the perturbation $(x(t) - x_0)$ take place according to[3]:

---

1 The sample geometry provides a factor of approximately 10^4. In addition the specific conductivity of the amorphous material is about 10^4 less than the pure metal in the region of the residual resistance.

2 By "instantaneously" we mean in this context switch-off times that are small with respect to the time within which equilibrium can be set up.

3 Such an exponential decay is always obtained when the rate of change of a variable $a(t)$ is proportional to the instantaneous value of the variable that is when $\mathrm{d}a(t)/\mathrm{d}t \propto a(t)$.

$$(x(t) - x_0) = (x(0) - x_0)\, e^{-\frac{t}{\tau}}. \qquad (8\text{–}5)$$

Such relaxation processes are described by a constant, the decay time τ. In order to describe and understand the process it is necessary to determine this decay time τ.

There is a whole range of such relaxation systems in the case of our superconductors. One can bring the electron system, i.e. the electrons which remain unpaired at $T_c > T > 0$ into a nonequilibrium state (see Fig. 20), e.g. by irradiating with microwaves – in particle language with photons of suitable energy. We will look more closely at this possibility and the phenomena that result from it.

We can also alter the density of the Cooper pairs by irradiation with photons whose energies are greater than the energy gap (i.e. $h\nu > 2\Delta$ must apply to their frequency ν). In Section 6.2 we mentioned a process where the Cooper-pair system could have its equilibrium perturbed (by migration of the flux tubes) by means of a magnetic field that was changing rapidly enough.

The phase of the Cooper-pair system, which is, for example, of decisive importance for the Josephson effects (Section 3.4), can be perturbed from its equilibrium and returned to equilibrium by a relaxation process after switching off the external parameter.

A characteristic time is associated with all these processes, e.g. τ_s for the equilibration of the one-electron system, or τ_r for the recombination of single electrons to Cooper pairs. These times are of importance in the exploitation of superconducting tunnel diodes as phonon sources (Section 9.5.3).

Finally the system of quantized lattice vibrations, the phonons, can be perturbed from their equilibrium by irradiating at a sufficient intensity with phonons of a particular frequency, or by creating such phonons within the superconductor by the recombination of single electrons to Cooper pairs (Section 9.5.3). The phonon system can also return to the equilibrium state if the "perturbation" is switched off. Here in accordance to the process i, which is particularly important here, a distinction is also made between the characteristic times τ_i. The excess phonons can, for instance, simply escape from the superconductor: the characteristic time for this is known as τ_{Ph}^{esc}. The characteristic time τ_{Ph}^{s} is decisive for equilibration by means of internal scattering processes within the phonon system. Nowadays it is possible, in principle, to determine all these times.

In recent years with the increasing theoretical understanding of superconductivity there has also been a greatly increased interest in the quantitative treatment of nonequilibrium phenomena. Here we can only look at two simple examples from the wealth of results that are available (see [163]). First we will describe the determination of the recombination time of single electrons to Cooper pairs. Secondly we will consider a very astonishing result, namely the "improvement of superconductivity" in a state away from the equilibrium state.

In order to be able to determine relaxation times it is first of all necessary to be able to set up a well-defined perturbation of the variable under consideration from its equilibrium value and then to determine this perturbation. Here a very high time resolution is necessary on account of the very short relaxation times to be expected for electron systems. Both can be achieved with a tunnel device (Section 3.3.3). Experiments of this type were carried out rather soon after the development of the BCS theory [164]. We will only discuss one possibility here [165]. Two tunnel diodes 1 and 2 of the type Al-Al_2O_3-Al are set up so that the central Al layer is a component of each diode (see Fig. 44). A voltage $U_1 > 2\Delta_{Al}/e$ is now applied to one diode so that single electrons can tunnel into the central Al layer. This makes the concentration of single electrons in this layer higher then the equilibrium concentration. The new concentration that is set up naturally depends on the size of the tunneling current which transports the electrons to the central layer. But it also depends on the lifetime of single electrons in this layer. The shorter the lifetime, i.e. the more rapidly the excess electrons recombine to form Cooper pairs, the smaller will be the concentration that results. If it is possible to measure the change of the concentration on switching on the tunneling current by means of diode 1 then it is possible from this to calculate the lifetime of a single electron, or put in another way the recombination time in the central layer.

The changes in concentration of single electrons in the central Al layer can now be detected with the aid of the diode 2. To do this a voltage $U_2 < 2\Delta_{Al}/e$ is applied to this diode. It is now only the single electrons present in the central Al layer that can contribute to the tunneling current through diode 2. Only very few individual electrons are available at sufficiently low temperatures in equilibrium. The current I_2 through diode 2 is accordingly small. If tunneling current I_1 is switched on the increase it causes in the concentration of single electrons in the central layer will result in an increase in I_2. This makes it possible to follow the concentration changes quantitatively.

A pulse technique is also employed in addition to this stationary method. Here it suffices to investigate a single tunnel diode [165]. A brief voltage pulse with $U > 2\Delta_{Al}/e$ is used to produce a nonequilibrium concentration of single electrons, whose decay after the pulse can be observed directly with the aid of the tunneling current at a voltage of $U < 2\Delta_{Al}/e$.

Both methods yield the same results. Figure 110 illustrates the results of Gray, Long and Adkins [165]. The lifetime of the single electrons, i.e. the recombination time τ_r to Cooper pairs, increases proportionally to $\exp(\Delta(T)/k_B T)$ as the temperature falls. Thereby τ_r varies from ca. 10^{-7} s at temperatures somewhat below T_c to ca. 5×10^{-5} s at 0.2 K. These results can only be approximate since the determination of the recombination time is influenced by several parameters that can vary from experiment to experiment.

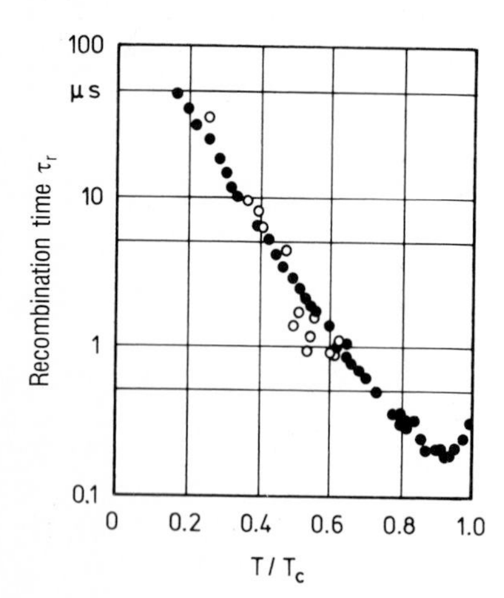

Fig. 110. Recombination time in Al films as function of the reduced temperature. Transition temperature T_c = 1.27 K, Δ_0 = 0.195 meV; thicknesses: 102 nm, 76 nm and 64 nm. ••• values measured in steady state. ○○○ values measured by pulse technique (after [165]).

Thus, it is, for example, of decisive importance whether the phonons produced on recombination (Section 9.5.3) "disappear" rapidly, i.e. travel out of the superconductor, or can decay. For these phonons are themselves able to break up Cooper pairs and, hence, create single electrons. Under some circumstances this could lengthen the average life of single electrons very greatly. It is the thickness of the layer investigated and the substrate (the latter on account of its acoustic adaptation) that is of importance for the disappearance of the phonons. The effective recombination rates can be changed by an order of magnitude and more as a result of this effect.

In contrast to such complex influences it is very easy to understand the exponential increase in τ_r as the temperature falls. The concentration of the single electrons decreases exponentially with decreasing temperature (Section 9.5.2). Hence, the probability of a single electron finding a suitable partner to form a Cooper pair also decreases.

Finally it should be mentioned once again here that the change in the electron concentration can also be effected by irradiation with electromagnetic waves (photons) of suitable frequency and that the changed concentration can also be detected with electromagnetic waves $h\nu < 2\Delta$ by determination of the reflectance or the surface resistance [166]. We met this possibility in Section 3.3.1.

In the second example, that of "improving" the superconductivity in a nonequilibrium state, we consider a case where the destruction of the equilibrium is carried out by irradiation with suitable photons. The at first astonishing result is that the irradiation can lead to an increase (!) in the transition temperature and to a broadening of the energy gap.

The experiments began with the observation that the current through a region where there is a weak linkage between two superconductors – a Josephson junction is just such a region – can be influenced by irradiation with photons. Giaever [62] exploited this ef-

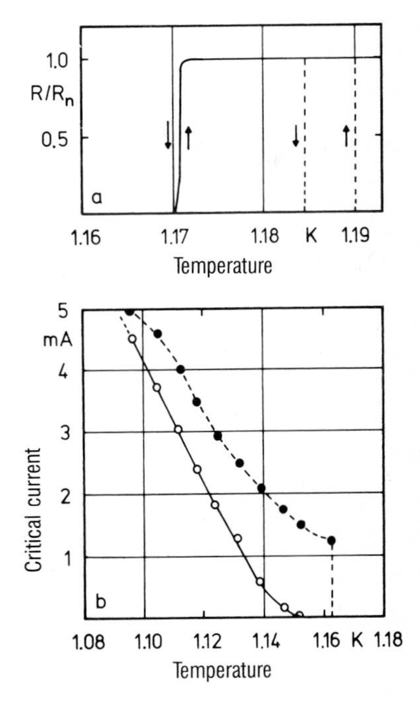

Fig. 111. Transition curves a) and critical currents b) of Al films without and with high frequency radiation ($\nu = 3 \times 10^9\ s^{-1}$).
–○–○– without high frequency
–•–•– with high frequency
Size of films

Length l	Width b	Depth d
a: 2.90 mm	3.8 µm	0.4 µm
b: 2.92 mm	3.5 µm	1.0 µm

(after [168]).

fect to demonstrate the Josephson alternating current. It is possible, under favorable conditions, to increase the critical current considerably [167]. Klapwijk and Mooij [168] finally succeeded in demonstrating in a beautiful experiment that this phenomenon is a general property of the superconducting state and not specifically restricted to regions of weak linkage. We will describe this very instructive experiment briefly.

The investigations were carried out on thin condensed Al strips (length ca. 3 mm, width ca. 3–5 μm and thickness 0.2–1 μm). The critical current I_c and the transition temperature T_c were determined with and without high frequency irradiation (frequency ν between 10^7 s^{-1} and 10^{10} s^{-1}). The characteristic results obtained for the frequency 3×10^9 s^{-1} are presented in Fig. 111.

Figure 111 illustrates the transition to the superconducting state as followed by simple resistance determination (part a). The transition temperature was appreciably increased by high frequency irradiation. The critical current was also bigger under the influence of high frequency radiation (part b).

This surprising result – one would expect that the supply of energy would be unfavorable for the superconductiviy – can be very readily understood, at least qualitatively, and had already been conjectured by Parmenter [169] in 1961. A quantitative description has been provided by Eliashberg and others [170]. The basic idea is as follows. The Cooper pairs $\{\vec{k}_\uparrow - \vec{k}_\downarrow\}$ are the more firmly bound the more states there are available into which they can be scattered. If we now have single electrons in the region of the Fermi energy, then on account of the Pauli exclusion principle these "block" the scattering states for Cooper pairs [1] and, thus, reduce the binding energy of the pairs. Expressed in another manner: The energy gap will be smaller as the number of single electrons increases.

Now if it is possible to remove single electrons from the states in the upper edge of the energy gap (Fig. 20), where they disturb the Cooper pairs most sensitively, then this should lead to an increase in the size of the energy gap, just as reducing the temperature also does, since this also brings about a reduction in the concentration of single electrons. High frequency irradiation has just this effect. The single electrons are excited, i.e. increased in energy, and, hence, displaced from the upper edge of the energy gap. As a result additional scatter states become available for Cooper pairs, thus, leading to an increase in the energy gap. The increase in the critical current can be immediately understood (see Section 2.2). The increase in the transition temperature can also be explained [170]. However, a quantitative treatment is required for this since the irradiating photons can also break up Cooper pairs in the region of T_c.

Further nonequilibrium processes occur when the critical current is reached in a thin superconductor (e.g. in a whisker [2]). The transition to normal conductivity takes place in discrete steps [171], which are explained with the occurrence of "phase slip centers" [172]. A phase slip center separates two superconducting regions whereby the phase between the two Cooper-pair systems increases monotonically. This phase change is associated with an electrical voltage between the two regions (see Section 3.4.2, Equation (3-52)).

---

1 It is quite evident here that the building blocks of Cooper pairs are fermions.

2 Whiskers are very thin single crystals (diameter a few μm).

Nonequilibrium processes also occur when an electrical current flows between a normal conductor and a superconductor, say between tin and copper. The normal current has to be converted to supercurrent at the interface. The manner in which this occurs has been investigated by means of beautiful experiments [173]. A summary of the last nonequilibrium processes dealt with here is to be found in [174].

We have examined these nonequilibrium processes in some detail because, on the one hand, they have acquired increasing importance and, on the other hand, because their treatment can yield new insight into the state of superconductivity.

8.3 The high frequency behavior

Persistent current experiments (see Section 1.1) have shown that a sufficiently small *direct* current ($I < I_c$) can pass through a superconductor without any resistance at temperatures below T_c. A quite different situation exists for high frequency alternating currents. In 1940 London [175] was able to show that there is a distinct measurable resistance present at temperatures below T_c for frequencies $\nu >$ ca. $10^9\ s^{-1}$. This means dissipative effects occur in superconductors too for currents of these frequencies. We can understand this result very readily at the qualitative level. In a superconductor with current which changes with time we always have an electric field strength $\vec{E}$[1] (see Section 5.1.2, Equation (5-1)). This field strength $\vec{E}$ naturally acts on the unpaired electrons and imparts additional energy to them. Since these unpaired electrons can interact as usual with the lattice by means of collisions, dissipative processes occur. We have, in principle, an ohmic resistance so long as the current in the superconductor changes with time.

In the case of slow changes, i.e. low frequency alternating currents, the necessary field strength $\vec{E}$ and, hence, the dissipation are very small. In the case of commercial alternating current with a frequency $\nu = 50\ s^{-1}$ (in Europe, it would be $60\ s^{-1}$ in the USA) and a maximum amplitude of 1 A field strengths of ca. 10^{-12} V/m are necessary for the acceleration of the Cooper pairs. At this field strength a current of about 10^{-10} A would flow in a copper wire of 1 mm^2 cross section. We see, therefore, that the part of the current carried by the unpaired electrons is negligibly small at these frequencies. This statement applies to type-I superconductors. In the case of hard superconductors in a magnetic field $B > B_{c1}$ dissipative effects and, hence, resistance can occur because of movement of flux tubes about their pinning centers. We will limit our discussion here to type-I superconductors.

As the frequency is increased the contribution of the unpaired electrons to the current compared with that of the Cooper pairs also increases. So as the frequency is increased we must expect an increase in the ohmic resistance in the superconducting state. Since the numbers of unpaired electrons decrease with decreasing temperature, the resistance must also fall as the temperature sinks. Figure 112 illustrates this behavior [176].

The resistance is considerable also below T_c for frequencies $\nu > 10^{10}\ s^{-1}$. For frequencies greater than the energy gap the behavior approximates more and more closely

1 A confusion between the electric field $\vec{E}$ and the energy E is excluded by the arrow on top of the $\vec{E}$ denoting the electric field.

to that of a normal conductor. If the energy gap 2Δ is small compared with the energy of the electromagnetic quantum $E = h\nu$ then it will have little effect on the excitation processes leading to energy exchange. At considerably higher frequencies such as are found in the visible range ($\nu \approx 10^{15}\ s^{-1}$), practically no difference is observed between the superconducting and the normalconducting states. Here the quantum energies amount to several eV and, hence, are very large compared with the energy gap (a few times 10^{-3} eV).

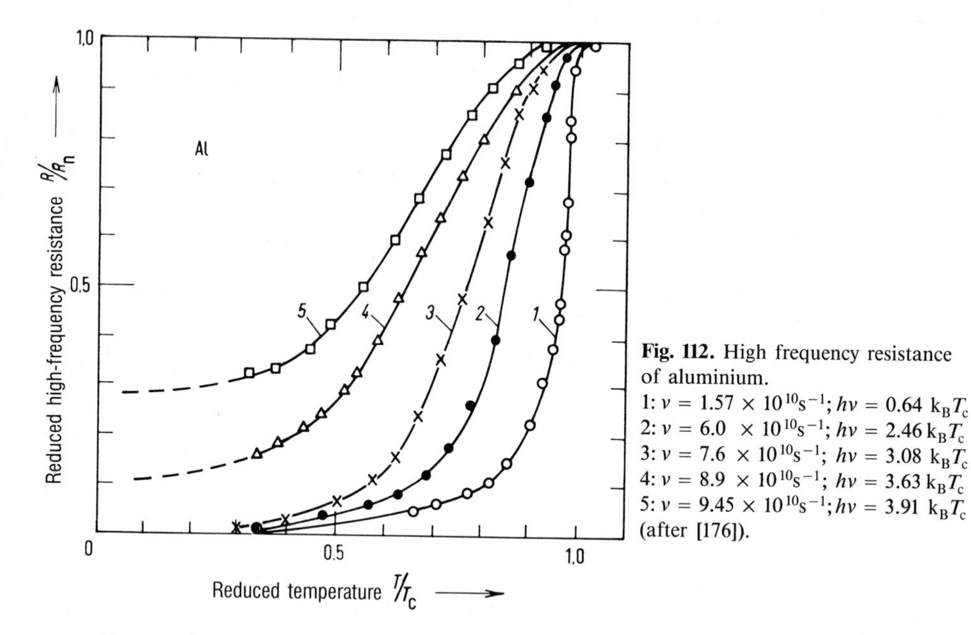

Fig. 112. High frequency resistance of aluminium.
1: $\nu = 1.57 \times 10^{10} s^{-1}$; $h\nu = 0.64\ k_B T_c$
2: $\nu = 6.0 \times 10^{10} s^{-1}$; $h\nu = 2.46\ k_B T_c$
3: $\nu = 7.6 \times 10^{10} s^{-1}$; $h\nu = 3.08\ k_B T_c$
4: $\nu = 8.9 \times 10^{10} s^{-1}$; $h\nu = 3.63\ k_B T_c$
5: $\nu = 9.45 \times 10^{10} s^{-1}$; $h\nu = 3.91\ k_B T_c$
(after [176]).

A quantitative discussion of the curves in Fig. 112 is difficult. What is known as a skin effect takes place in normal conductors at such frequencies. The high frequency currents are limited to a thin surface skin of the wire as a result of their induction effects. The skin depth decreases as the frequency is increased. On the other hand, the characteristic variable for the penetration of magnetic fields into superconductors is the penetration depth λ. The temperature and frequency dependences of these quantities are very complex in a superconductor. The determination of the energy gap 2Δ by the use of this method is illustrated, by means of the absorption of high-frequency, electromagnetic waves as an example, in Section 3.3.1. The first direct demonstration of the energy gap was made using this method [48]. This method was then pushed rather into the background as a result of the use of tunnel experiments to determine 2Δ. But it offers advantages over the tunnel effect in some cases.

It is very difficult to construct good tunnel diodes with some superconductors. Surface impurities often interfere so that one is not sure if the tunnel experiment actually reflects the behavior of the material as a whole. High frequency determination can then be very helpful. A high frequency electromagnetic wave is allowed to impinge on the sample consisting of a film with a thickness of up to ca. 10^{-5} cm and the transmitted and reflected power is determined (T_s and R_s).

Figure 113 illustrates an example of such a determination in a film of lead [177]. Both powers are referred to those determined in the normal state, T_n and R_n. The frequency of $\nu = 2.4 \times 10^{10}$ s^{-1} was chosen so that $h\nu_0 \ll 2\Delta$ (0). The lead film is normal above T_c. As the temperature is reduced the film goes into the superconducting state at ca. 7 K (crosses in Fig. 113). The condensation to Cooper pairs decouples the electron system more and more from the other degrees of freedom of the sample. This causes the reflection to increase. The transmitted radiation is reduced and becomes zero if the film is thick enough[1]. The variation of T_s/T_n (T for transmission) and of R_s/R_n reflect the temperature dependence of the energy gap. The curves correspond to a ratio $2\Delta(0)/k_B T_c = 4.5$, i.e. there was a large deviation for this film from the value predicted by the BCS theory ($2\Delta(0)/k_B T_c = 3.5$). Such superconductors are referred to as being "strongly coupling". The deviations near T_c are a result of fluctuations (Section 8.1). Such investigations have been carried out successfully on amorphous films (Section 8.5.3) [178].

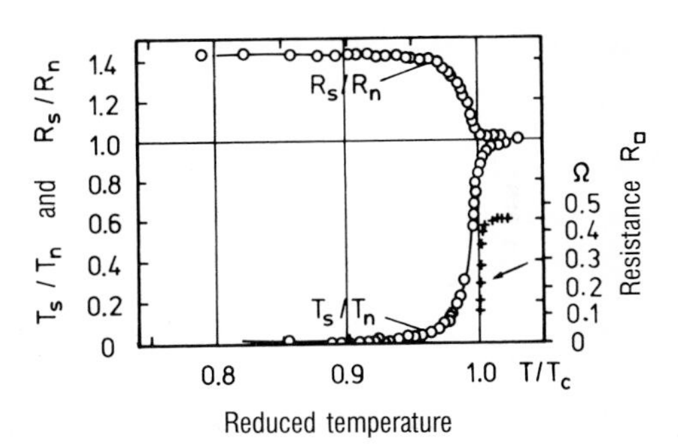

Fig. 113. Reflectance and transmittance of a lead film for electromagnetic radiation with a frequency $\nu_0 = 2.4 \times 10^{10}$. R_s or R_n = amount reflected, T_s or T_n = amount transmitted in superconducting or normalconducting state. Transition temperature $T_c = 7.05$ K, thickness $d = 14$ nm; the resistance $R_\square$ is referred to a square geometry (length l = width b) (after [177]).

Since as $T = 0$ is approached the concentration of unpaired electrons approaches zero it should be possible to make the high frequency resistance as low as desired, at least at low temperatures. This is not, however, what happens; a more or less large "residual resistance" is observed instead (see Section 9.5.2), whose occurrence is decisively dependent on the state of the surface. The details of the mechanisms which determine this residual high frequency resistance have not yet been completely explained.

8.4 The influence of interfaces

In our consideration of finite superconductors, such as thin superconducting layers, we have assumed until now that they were always bounded by interfaces with vacuum or an insulator (quartz sheet) (see Section 5.1.3). Since such an interface does not allow an exchange of electrical charge it makes sense to assume that the interface has no specific

---

1 It should be noted that the normalized presentation no longer says anything about the amount of power transmitted or reflected in the normal state. In the case of films with good electrical conductivity T_n can be very small (<1% of the incident radiation). If the experimental device is sufficiently good it is possible to follow the changes in these small power levels quantitatively as a function of temperature.

effect on the Cooper-pair density n_C, but rather that the behavior of n_C in the interior continues as far as the geometrical boundary of the superconductor [1].

This assumption completely loses its justification if we consider the interface between a normal conductor and a superconductor, say an interface between Pb and Cu. Here the electrical charges are able to diffuse across the interface and, hence, interchange practically freely. Normal conductors only contain uncorrelated (unpaired) electrons. Whereas when $T < T_c$ we also have in the superconductor a finite number of Cooper pairs. The exchange of charge will now lead to free electrons diffusing into the superconductor and Cooper pairs into the normal conductor. We must, therefore, expect that the concentration of Cooper pairs in the superconductor will be reduced in the neighborhood of the interface. On the other hand, a certain concentration of Cooper pairs will be present in the normal conductor in the neighborhood of the interface. With very great simplification it may be said that: In the region of the interface the normal conductor tends towards superconductivity, while the superconductor is, on the other hand, somewhat less favorable for pair correlation. If we now make one of the two conductors sufficiently thin the influence of the interface will alter the behavior of the whole sample to a measurable extent. The coherence length is a natural measure of the depth to which the neighboring material is detectable. There is a local variation in Cooper-pair density up to distances which are comparable with the coherence length. According to Section 8.2 it is the various relaxation times that are of importance for the processes that take place.

The first observations concerning the importance of such a junction with a normal conductor were made in the 1930s on layers of lead that had been deposited electrolytically on a constantan wire (55% Cu + 45% Ni) [179]. The transition temperature of this lead layer fell rapidly as its thickness was reduced and became immeasurably small for a thickness of ca. 350 nm. It was only much later, namely with the development of a successful microscopic theory, that such investigations were taken up again. In order to obtain quantitative results it is necessary to fulfil a range of conditions that are not easy to achieve experimentally. Thus, the contact between the layers must actually be a metallic one and must not be interfered with in any manner, such as by a thin oxide layer. On the other hand, diffusion of the two layers into each other must also be avoided, since the interface would then consist of an alloy.

These conditions can be complied with very well by evaporated layers. As an example we will discuss the behavior of a layer packet of Pb and Cu. Here we need to examine the influence of several parameters. Thus, it is necessary to vary the thickness D_{Pb} of the lead and the thickness D_{Cu} of the copper layers. As we will see the mean free paths of the electrons in the two layers are also of importance. Figure 114 illustrates the transition temperature as a function of the thickness of the lead D_{Pb} [180]. Here the Cu layer is thick with respect to the coherence length. The transition temperature drops rapidly when $D_{Pb} <$ ca. 50 nm. It is possible to extrapolate to a critical thickness of ca. 10 nm at which the transition temperature approaches zero. This critical thickness is appreciably less than that observed in previous experiments (350 nm). In order to interpret this difference it must be remembered that the results reported in Fig. 114 were obtained with a lay-

1 It is possible to demonstrate with very thin films that this assumption does not apply to the outermost layers of atoms [185].

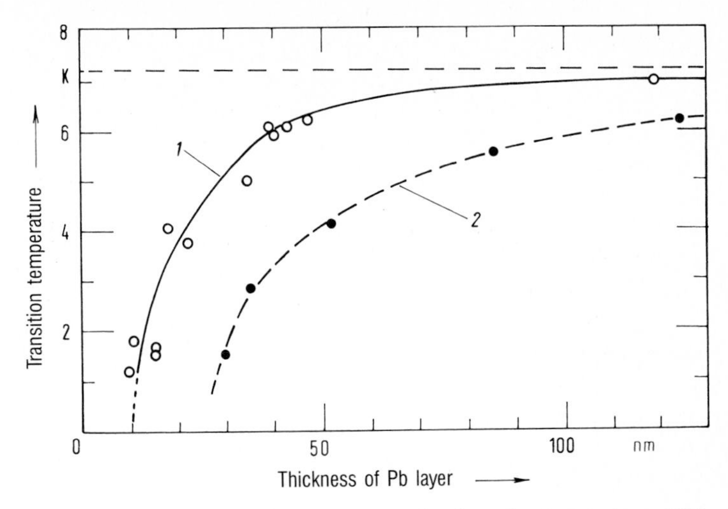

Fig. 114. Transition temperatures of double layers of lead and copper. Condensation temperature: 10 K, thickness of Cu layer $D_{Cu} \gg \xi$; free path: in lead ca. 5.5 nm, in Cu ca. 4.5 nm (Curve *1*) and 80 nm (Curve *2*) (after [180]).

er packet that was produced by quench condensation at a substrate temperature of 10 K. Both metals are then in a high state of disorder and, hence, have a low mean free path of ca. 5.5 nm in Pb [1] and ca. 4.5 nm in Cu. This means that the distance over which the Cooper-pair density is subject to local variation is also very short. The influence of the mean free path in the normal conductor is illustrated by the broken line in Fig. 114. These transition temperatures were determined on layer packets where the Cu layer was annealed before being covered with lead. This allowed an increase from an original 4.0 nm to ca. 80 nm in the mean free path. However, the Pb layer was again applied at 10 K, that is with approximately the same amount of disorder. It can readily be seen how this increases the influence of the Cu layer. The same reduction of T_c was observed for thicker layers of Pb. Alteration of the mean free path in the superconductor has a very similar effect. The shorter the free path the smaller the depth influenced by the interface since the coherence length decreases with the mean free path (Table 7, p 133).

The influence of the thickness of the normal conductor is illustrated in Fig. 115 [180]. Here the transition temperature is plotted as a function of D_{Cu} for various thicknesses of Pb layers. Here all the layer packets were condensed onto the substrate at 10 K. The transition temperature approached a limiting value at large thicknesses of normal conductor. The limiting value is given in Fig. 114.

The quantitative theoretical treatment of these effects, known as "proximity effects", involves the strength of the electron-phonon interactions in addition to the coherence length and the mean free path. For this reason there is the hope that such determinations might produce evidence concerning whether a metal such as copper could become super-

1 The disorder reduces T_c of lead. Even if this effect is small it results in a $T_c = 7$ K of the lead film (compared to 7.2 K for bulk lead) for $D_{Ph} \rightarrow \infty$.

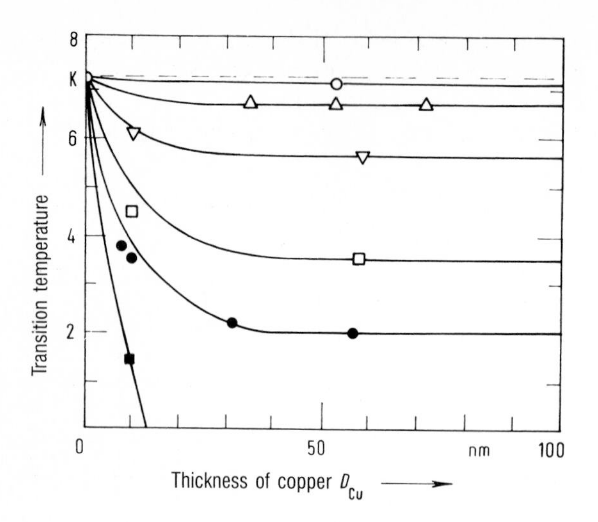

Fig. 115. Transition temperatures of double layers of lead and copper. Condensation temperature: 10 K, thickness of Pb layers: ○ 100 nm, △ 50 nm, ▽ 30 nm, □ 15 nm, • 10 nm, ■ 7 nm (after [180]).

conducting if the temperature were low enough. For this purpose it is necessary to study interfacial effects at temperatures that are as low as possible. The results of such an investigation are reported in Fig. 116 [181]. They are in agreement with the assumption of an attractive electron-electron interaction (via phonons) in copper, which would yield a T_c for copper of ca. 2×10^{-3} K. But it does not appear to be certain that the experimental accuracy and the state of the theory allow such a conclusion to be made with certainty. The extrapolation of determinations on very dilute alloys at very low temperatures (mK range) yielded an appreciably lower transition temperature for copper (Section 1.2 [8]).

Interfaces with materials having marked paramagnetic moments (e.g. Fe or Mn) have a particularly large effect on superconductivity. As we will see in Section 8.6 these moments can break Cooper pairs. This means that the Cooper-pair density rapidly approaches zero at such interfaces. Figure 117 provides an example of this [182]. Here the transition temperature of a layer packet of Pb and Ni is presented as a function of the

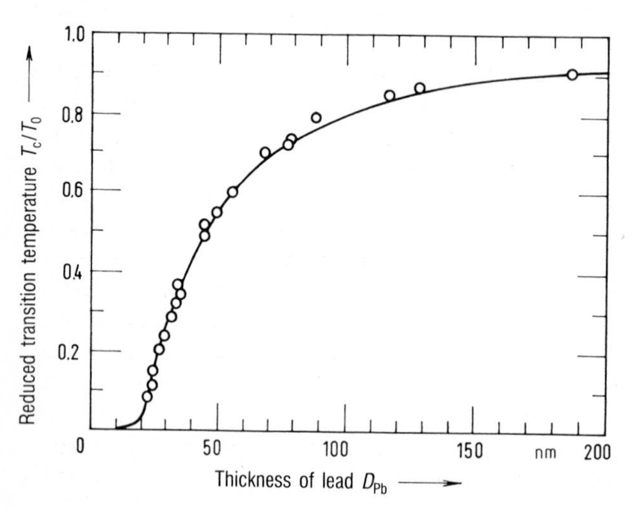

Fig. 116. Transition temperatures of double layers of lead and copper. Condensation temperature: 200 K, thickness of Cu: 320 nm. Solid curve calculated according to Moormann (after [181]).

Pb thickness. The general shape of the curve corresponds to that in Fig. 114. The extreme strength of the effect of nickel can be seen from the fact that a mean layer thickness of only 2.5 nm Ni (full circle) is sufficient to produce almost the maximum effect. Increasing the thickness of nickel to even 115 nm does not produce any additional effect (triangle).

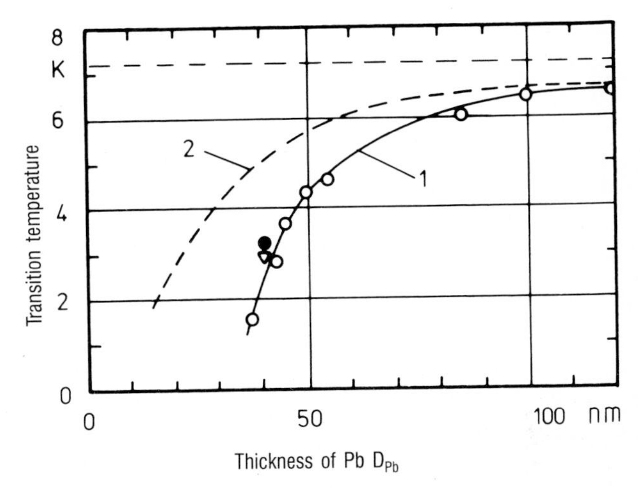

Fig. 117. Transition temperatures of double layers of Pb and Ni. Condensation temperature: ca. 80 K; thickness of Ni: ○ 5 nm, • 2.5 nm, ▽ 115 nm. Curve 1: calculated behavior taking into account the pair-breaking properties of Ni, Curve 2: calculated behavior for a normal conductor without pair-breaking with $T_c = 0$ K (after [182]).

The "pair-breaking" effect of nickel finds particular expression here. The solid curve is obtained theoretically by taking "pair breaking" into account. The broken line, on the other hand, is obtained for a normal conductor without a paramagnetic moment.

Finally we must consider the interface between two superconductors with different transition temperatures, T_1 and T_2. Figure 118 reproduces some examples of this [183]. The broken line shows the effect at a thick layer of copper for comparison. If the Cu were to become superconducting at a sufficiently low temperature, then the broken curve should be displaced in an analogous manner to the others and approach the ordinate.

The metallic contact of a superconductor I with a material II with a lower attractive or even a repulsive electron-electron interaction lowers the transition temperature of I. On the other hand, the interaction in II is increased. So that a not too thick layer of a normal conductor, e.g. a film of silver between two superconductors, will become a superconductor. This has been demonstrated in persistent current experiments. Tunnel experiments have also revealed that a normal conductor in contact with a superconductor itself becomes superconducting, i.e. contains a finite concentration of Cooper pairs.

The mutual influence of materials with differing strengths of electron-electron interaction is of particular importance for heterogeneous alloys. If the precipitates, of say a superconductor in a normal conductor, are very small (diameter comparable with the coherence length) then the superconductivity can be suppressed. A superconducting matrix can be strongly affected in its superconducting properties by normalconducting precipi-

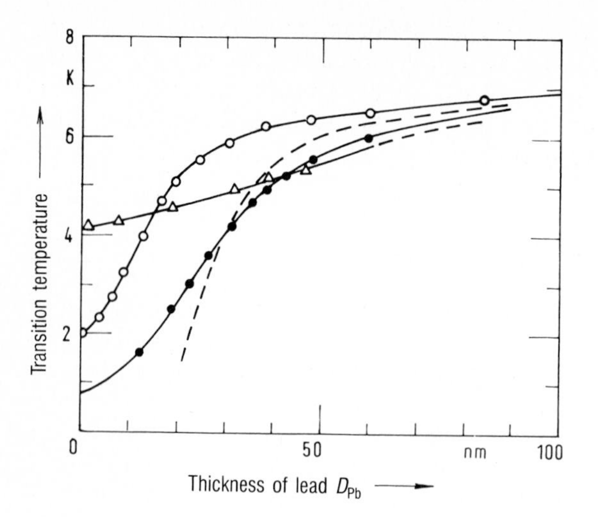

Fig. 118. Transition temperatures of double layers of Pb with another superconductor. Condensation temperature for Pb 10 K, condensation temperatures and thickness of other superconductors:
△ Hg 110 K 114 nm; ○ Al 180 K 150 nm; • Cd 180 K 150 nm;
for comparison: Cu 304 K 160 nm.
The solid curves are calculated according to a theory due to Werthamer (after [183]).

tates. However, when interpreting results in terms of interfacial effects it is necessary to take all parameters, particularly the mean free path, into account, in order to be able to estimate the correct order of magnitude of the effects.

The interface with a dielectric can also, under some circumstances, affect the transition temperature of a thin layer. If, for instance, an oxide layer forms on superconducting material what is known as an electric double layer can be produced. The strong electric fields of such a double layer can affect the electron concentration in the superconductor. Changes in the transition temperature of the order of a tenth of a degree have been observed [184].

We know very little concerning the behavior of an interface with vacuum. It is assumed that the Cooper-pair density is constant up to the geometric boundary. This can only be an approximation since every boundary undoubtedly affects a change in the characteristic parameters such as atomic distance and binding behaviour. If these changes are limited to a few layers of atoms then the influence on the superconductivity will be correspondingly small, or will only become appreciable for very thin layers. A reduction in the transition temperature has actually been observed for very thin layers [185].

The increase in T_c with decreasing thickness observed for Al films [186] is with certainty not a result of the changing geometry but of the structural changes in the thinner films. Very large numbers of defects can be frozen into very thin films under the influence of the substrate. This causes the films to have higher T_c (Section 8.5.3).

8.5 The influence of lattice defects

We will regard all deviations from the strict periodicity of the crystal lattice as lattice defects, regardless of whether these result from the incorporation of foreign atoms or from pure constructional errors, i.e. the displacement of atoms from their regular places. The primary influence of such lattice defects on the system of the conduction electrons is as scattering centers which shorten the mean free paths of electrons. As we saw (Section

5.2), this reduces the coherence length, the incorporation of defects turns a type I superconductor into a type II superconductor. This change can occur without other characteristic parameters, such as the transition temperature or the difference in the free enthalpies, being appreciably affected [1]. In general, however, lattice defects affect all properties of a superconductor. Some of the effects that occur will be discussed in this section.

8.5.1 The anisotropy effect

The correlation of the electrons to Cooper pairs takes place via the elastic vibrations of the lattice, the phonons. Now this interaction within a crystal can be directionally dependent. As a great simplification it can be said that the presence of such anisotropy makes certain crystal directions more "favorable" with respect to superconductivity than others.

This is apparent, for instance, in the fact that the energy gap Δ is of different sizes in different directions in the crystal. The degree of anisotropy is given by the mean square of a parameter α [2]. A value of $\langle \alpha^2 \rangle = 0.02$, such as was found for tin, means that the energy gap in various directions in the crystal deviates from the mean by about 14%. Such varying energy gaps have been detected in tunnel experiments (see Section 3.3.3) or measurements of the absorption of ultrasound (see Section 3.3.2).

The transition temperature T_c is very dependent on the favored direction in space, for the transition to the superconducting state occurs as the first Cooper pairs are formed in the equilibrium state [3]. If we now incorporate defects into the crystal then the electrons will be scattered at these irregularities. But this means, as will amost immediately become obvious, that the momentum of an electron is rapidly dispersed in all directions as a result of scattering. The interaction responsible for the superconductivity is averaged in this manner. The particularly favorable directions no longer express themselves so forcefully because electrons with momenta directed in these directions are rapidly scattered into other less favorable directions. This leads to a reduction in the transition temperature with increasing concentration of scattering centers.

The present theory of superconductivity makes it possible to calculate this influence of defects on T_c. Figure 119 illustrates some results obtained for tin [187]. Foreign atoms were employed as scattering centers. Since different foreign atoms possess differing scattering properties the residual resistance ratio $\varrho^* = R_n/(R_{273} - R_n)$ is employed as a measure of the mean free paths of the electrons. The transition temperature is plotted against ϱ^* in Fig. 119. The universal, linear reduction of T_c at small defect concentrations, which is in agreement with theory, is decisive for the anisotropy effect (part b of Fig. 119). When there is a large number of defects, i.e. at low mean free path, the averaging over all directions in space becomes complete, for the pure anisotropy effect T_c should take up a limiting value lying a few percent under the value obtained for the defect-free superconductor.

---

1 In Section 5.2.1 we assumed that the area under the magnetization curve (Fig. 72), which after all corresponds to the energy of transition, remains constant on transition to type II superconductivity.

2 The averaging must be made over all the electrons of the Fermi surface.

3 The transition temperature is a thermodynamic quantity for the whole system and cannot, therefore, depend on direction.

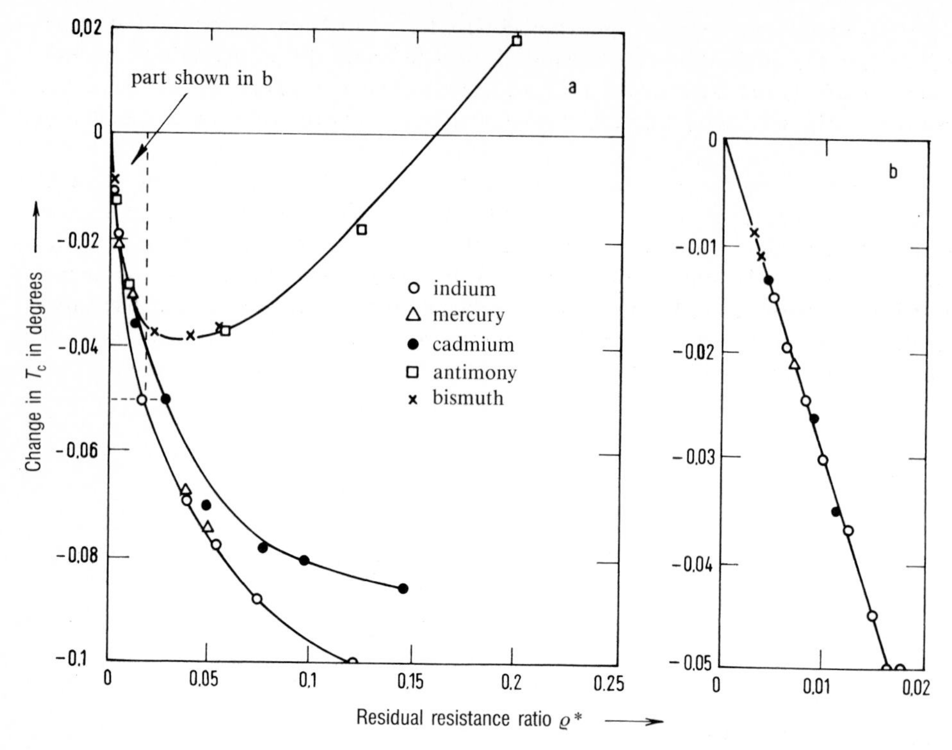

Fig. 119. Change of the transition temperature of Sn by foreign atoms (after [187].

Figure 119 shows that this behavior does not occur at large defect concentrations. Rather specific properties of the foreign substance are expressed, which leads to differing dependences of the transition temperature for different foreign atoms. It is not yet possible to understand these influences quantitatively for all systems. They are combined under the term "valence effect"[1].

8.5.2 The valence effect

The specific effect of the foreign atoms on the transition temperature of the host metal can depend on a change in the number of free charge carriers or on a change in the lattice constants. For instance, if Bi atoms, with five valence electrons each, are incorporated into tin which possesses four valence electrons then one would expect that the number of free electrons in the tin would be increased. This would, according to the model of a free electron gas, lead to an increase in the density of states $N(E_F)$ of the electrons (see Section 2.2).

1 This term was not a very good choice, since other parameters of the foreign atom are of importance in addition to its valence.

Since the impurity atoms will, in general, also have a different atomic volume to that of the atoms in the host lattice, mechanical stress fields will be produced in their neighborhood. The lattice constants of the host metal will also be changed at the same time. All these effects of defects generally result in a change in the value of T_c (see Section 4.3).

In order to separate these various effects it is necessary to choose special alloy systems in which it is possible to alter just one parameter, such as the number of valence electrons or the atomic volume, at a time. A satisfactory analysis has only been carried out in a very few cases. At the moment there is too little understanding of the quantitative interaction between superconductivity and other metal parameters (see Section 1.2). The importance of such experiments on selected alloy systems lies just in the possibility they offer of discovering quantitative relationships.

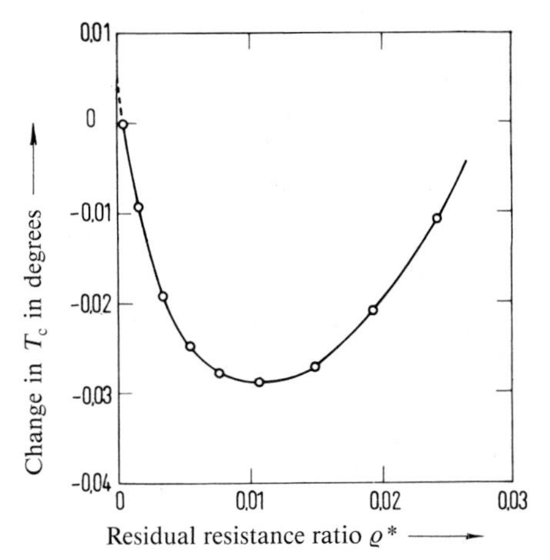

Fig. 120. Change of the transition temperature of thallium by the incorporation of lattice defects. Residual resistance ratio of the undeformed sample $\varrho_0^* = 0.4 \times 10^{-3}$ (after [188]).

Until now we have only considered defects resulting from the introduction of foreign atoms. But the purely structural defects in a crystal lattice can also have the same effects, in principle. An example of this is illustrated in Fig. 120 [188]. Here the transition temperature of a thallium wire is plotted as a function of the residual resistance ratio. In this experiment the residual resistance of the Tl wire was gradually increased by plastic deformation at liquid He temperatures. In principle, the change of T_c corresponds completely to that produced by the incorporation of foreign atoms. However, quantitative analysis here is even more difficult than is the case for alloys since plastic deformation creates not only statistically distributed atomic defect centers but also extended defects such as grain boundaries. The differing scattering mechanism brings with it additional complications for interpretation[1]. But it is understandable that the purely structural defects bring about similar changes in the material to those caused by foreign atoms. If, for instance, an atom is displaced from its regular lattice position by plastic deformation and takes up a position between the other atoms, what is known as an interstitial site, then

1 The reduction of T_c observed here cannot be explained in terms of anisotropy, since $\langle a^2 \rangle$ would, as is known from other results, yield an appreciably smaller reduction [189].

it will create a mechanical strain field there just as would a foreign atom, whose atomic volume is greater than that of the host metal.

A change in the number of free electrons would not be expected at first glance, since we do not have any atoms with differing numbers of valence electrons. But it must be remembered that the environment of an atom is altered by structural defects. This also alters the states of the electrons. So it is easy to understand that purely structural defects can have an effect on the density of states $N(E_F)$.

Until now we have basically considered the effects of the defects on the system of free electrons. Naturally the lattice vibrations, that is the phonons, can also be affected, particularly at high defect concentrations. Since these determine the interaction leading to the existence of superconductivity we must expect that these changes in the phonon system will also have a large effect on the superconductivity.

8.5.3 The electron-phonon interaction

It is possible to create particularly large numbers of defects on preparing the sample by condensation onto a very cold substrate (e.g. a quartz plate at liquid helium temperatures). This condensation process corresponds to very severe quenching [190]. The atoms, which impinge upon the substrate in a completely statistical manner, lose their energy very rapidly so that they remain frozen in the wrong sites. The amount of disorder that can be attained in this manner can have a considerable effect on the transition temperature of the superconductor.

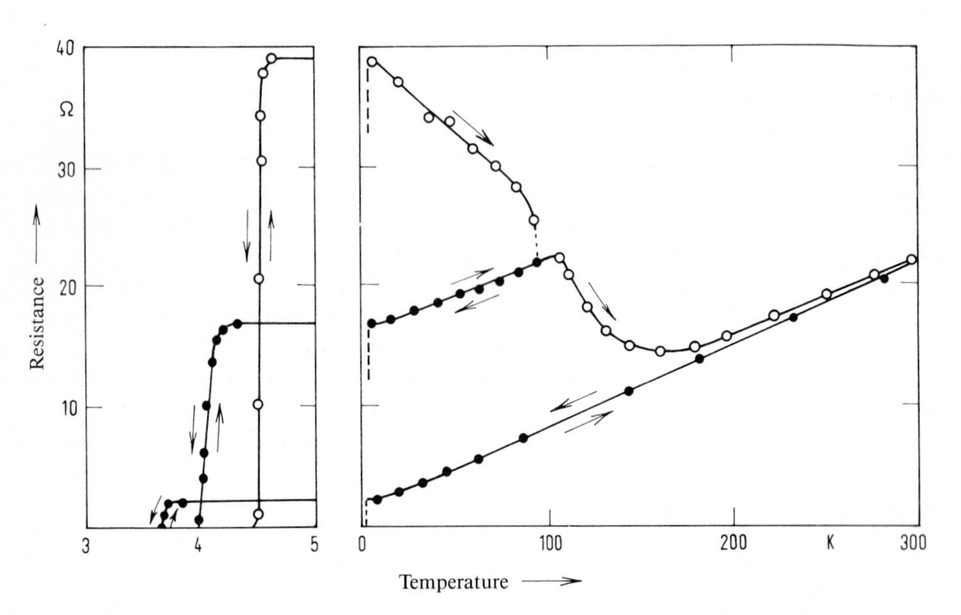

Fig. 121. Behavior of resistance of a quench-condensed film of tin. Condensation temperature: 4 K, thickness of the film 50 nm, length: 10 mm, width: 1 mm. The full circles have been observed on cooling after tempering (after [190]).

Figure 121 reproduces the behavior of condensed films of tin [190]. The left-hand side of the figure illustrates the transition curves and the right-hand side the behavior of the resistance. Immediately after condensation at 4 K the tin film has a high resistance on account of the large number of lattice defects which are frozen in. The high transition temperature of 4.6 K, that is 0.9 K or 25% higher than that for normal compact tin, is astonishing.

It can be seen from the temperature behavior that this large change in T_c is associated with lattice defects [1]. On warming, more and more atoms acquire the energy needed to adopt regular lattice positions. The disorder is reduced, the resistance decreases. At the same time the transition temperature is displaced to lower values. After warming to ca. 90 K T_c becomes ca. 4.1 K. A sufficient period of tempering at ca. 100°C yields the transition temperature of the compact material.

Electron-difraction studies of such quench-condensed layers – an example is given in Fig. 122 – reveal that these layers grow crystalline [191]. The complete diffraction pattern of the structure of white tin is present immediately after condensation at low temperatures (part b of Fig. 122). The diffraction lines are merely somewhat broadened. This broadening corresponds to an average crystallite size of ca. 10 nm.

It is possible to inhibit the crystallization even more if a substance, which does not fit into the tin lattice, is condensed at the same time as the tin, i.e. a substance that is as insoluble in the host lattice. Figure 123 illustrates electron-diffraction patterns of a tin film during whose preparation 10 atom% copper was also condensed [191, 192]. It has evidently been possible to freeze in an especially large concentration of defects in this manner. The diffraction pattern (part b of Fig. 123) immediately after condensation only exhibits a few broad rings such as are present in the diffraction patterns of liquids.

This extreme disorder has a correspondingly large effect on the superconductivity. The transition curves and temperature dependence of the resistance are reproduced in Fig. 124. The transition temperature is 7 K, that is raised by almost a factor of 2. The resistance is also raised on account of the high degree of disorder [193].

This extremely disordered state is very unstable. A very pronounced ordering process takes place even at ca. 20 K causing the resistance to fall to half its value. At the same time the transition temperature falls to ca. 4.5 K. Further tempering yields the known displacement of T_c to the value for the compact material [2].

Here we have considered just one example of the effects of very great lattice disorder in some detail. Similar and even greater changes of T_c have been found for other superconductors [194]. Thus, on the simultaneous condensation of a few atom% Cu with Al, the transition temperature is raised to 4 K (as compared with 1.2 K for compact material) [195]. When Ge is used as the defect-forming substance or by implantation of hydrogen, germanium or silicon at low temperatures, it is possible to raise the transition tempera-

---

1 A further indication that the effect is caused by lattice defects is given by the behavior of samples that have been plastically deformed at low temperatures. It is true that only lower concentrations of lattice defects can be created in this manner. But they exert the same effects [190].

2 At ca. 220 K Cu forms an intermetallic compound with Sn, which is observable in the fall in resistance. New lines associated with this compound appear in the electron-diffraction pattern. This change has very little effect on the superconductivity because the remaining almost pure tin shortcircuits the non-superconducting compound.

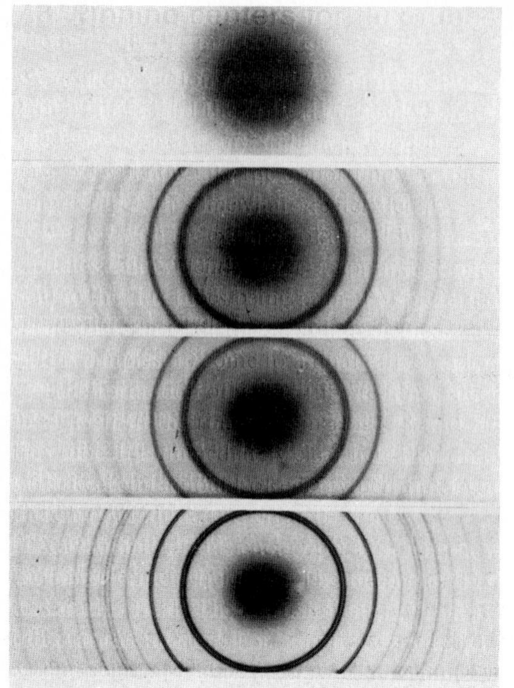

Fig. 122. Electron diffraction patterns of a pure quench-condensed tin layer.
Layer thickness: 14 nm, measurement temperature: 20 K.
a) substrate before condensation
b) Sn layer immediately after condensation at 20 K
c) after tempering at 90 K
d) after tempering at 300 K (after [191]).

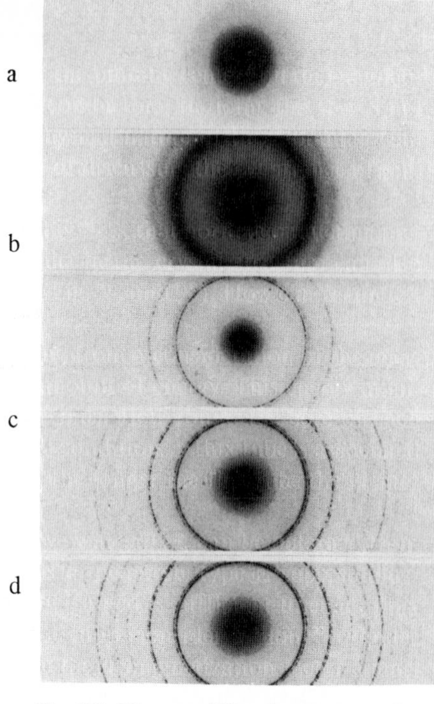

Fig. 123. Electron diffraction patterns of a quench-condensed tin layer containing 10 atom % copper as defect substance.
Layer thickness: 14 nm, measurement temperature: 20 K.
a) substrate before condensation
b) layer immediately after condensation at 12 K
c) after tempering at 30 K
d) after tempering at 90 K
e) after tempering at 300 K (after [191]).

ture of Al into the 7 to over 8 K range [196]. Quench-condensed beryllium films have a T_c of ca. 9.3 K [197), while the T_c of compact Be is now known to be ca. 0.03 K.

These large variations in T_c, some of which had already been investigated at the start of the 1950s by Hilsch et al., could not be satisfactorily explained for a long time. We now have a qualitative understanding and even the starting point for a quantitative explanation of these findings. Tunneling experiments and measurements of specific heat have shown that the bonding between the atoms in these extremely disordered films[1)] is somewhat weaker than is the case in the ordered crystal. This means that the frequencies of the lattice vibrations are reduced[2)]. The theory now predicts that the electron-phonon

1 These films are often referred to as "amorphous", without implying anything more than the extreme disorder.

2 For the same mass weakening a spring reduces the frequency.

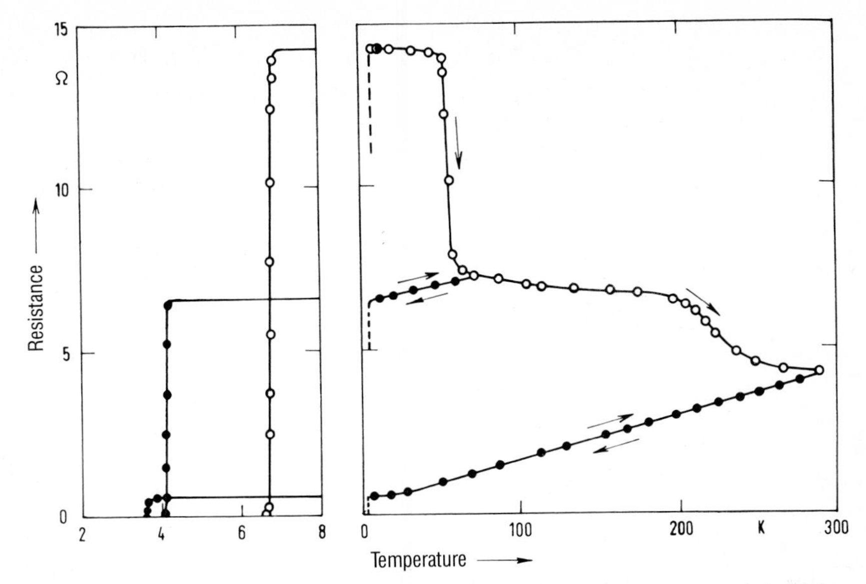

Fig. 124. Variation of resistance of a quench-condensed tin layer with 10 atom % copper as defect substance. Layer thickness: ca. 50 nm, width ca. 5 mm, length 10 mm, condensation temperature: 10 K (after [193]).

interaction will increase and that the constant λ^* (see Eq. (3-3)) will become larger. This leads to an increase in T_c. Theory also predicts that such a reduction in phonon frequency will have a greater effect on T_c the smaller the ratio T_c/Θ_D (see Eq. (3-3). So it would be understandable that in the series In, Sn, Zn, Al in which T_c/Θ_D decreases, the transition temperature is raised to increasing extents by extreme disorder[1]. Alongside this explanation of the change in T_c in terms of a reduction in the phonon frequency, it is also possible that the scattering by defects leads directly to an increased electron-phonon interaction [198]. Such a mechanism is possibly responsible for the changes in T_c with moderate numbers of defects.

At our present state of knowledge concerning the quantitative relationships between important parameters we are forced to develop theoretical models and test them on suitable substances. The investigations of disordered superconductors are particularly suited to this purpose and it is there that their importance lies.

To conclude this section on the influence of defects we will mention some results which have also been obtained with greatly disordered layers, but where the disorder is only of minor importance compared with a fundamental change in the short-range order of the atoms.

---

1 The very great changes in the case of Be may well have another, not completely understood cause. Since very beryllium-rich alloys with a cubic structure also have T_c values over 9 K a strong influence of the crystal structure may be presumed.

8.5.4 Metastable modifications

Let us consider the example of bismuth. The quench condensation of Bi vapors onto a substrate at 4 K yields an extremely disordered film [194]. Figure 125 reproduces the electron-diffraction pattern of such a film [155]. Only a few faded rings are observable immediately after condensation (*b*). Warming to 20 K leads to an ordering process, yielding the sharp rings of the normal Bi lattice.

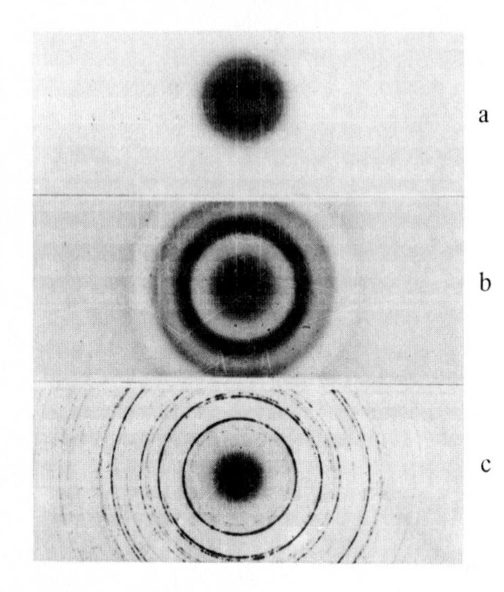

Fig. 125. Electron diffraction patterns of a quench-condensed bismuth layer.
Layer thickness: 7 nm, measurement temperature: 4.2 K.
a) substrate before condensation,
b) Bi layer immediately after condensation at 2 K,
c) after tempering at 300 K.
The crystallization takes place between 14 and 20 K (after [191]).

The disordered Bi film becomes superconducting at 6 K (Fig. 126). With the ordering process the superconductivity disappears. Compact bismuth (in the normal modification) is tested down to a few thousandths of a degree, without the observation of any superconductivity [7]. The superconductivity of the disordered Bi film is certainly not a result of the disorder alone. Rather measurements of the Hall effect [199] reveal that in the amorphous Bi film the concentration of free electrons is several powers of ten higher than in normal Bi. The very small concentration of free electrons in Bi (ca. 10^{-5}/atom) depends on the special lattice structure of Bi. The Bi lattice is a layer lattice. Each Bi atom within each layer has 3 nearest neighbors which are covalently bonded. This means that the structure of bismuth is very different from that of a typical metal, which prefers the closest or almost closest packing on crystallization. The great deviation of Bi from closest packing is also demonstrated by the fact that the density *increases* on melting.

If we condense Bi atoms onto a very cold substrate it is possible to freeze in a denser packing. The Bi atoms lose their energy so quickly that they are not able to undertake the relatively large displacements necessary for transformation to the normal Bi lattice. In this manner the atoms are frozen into another short-range order. This is the basis of the completely different behavior.

This view is supported by the results of high-pressure experiments. At pressures above ca. 27 000 bar the normal Bi lattice is replaced by a denser packing [200], this can go

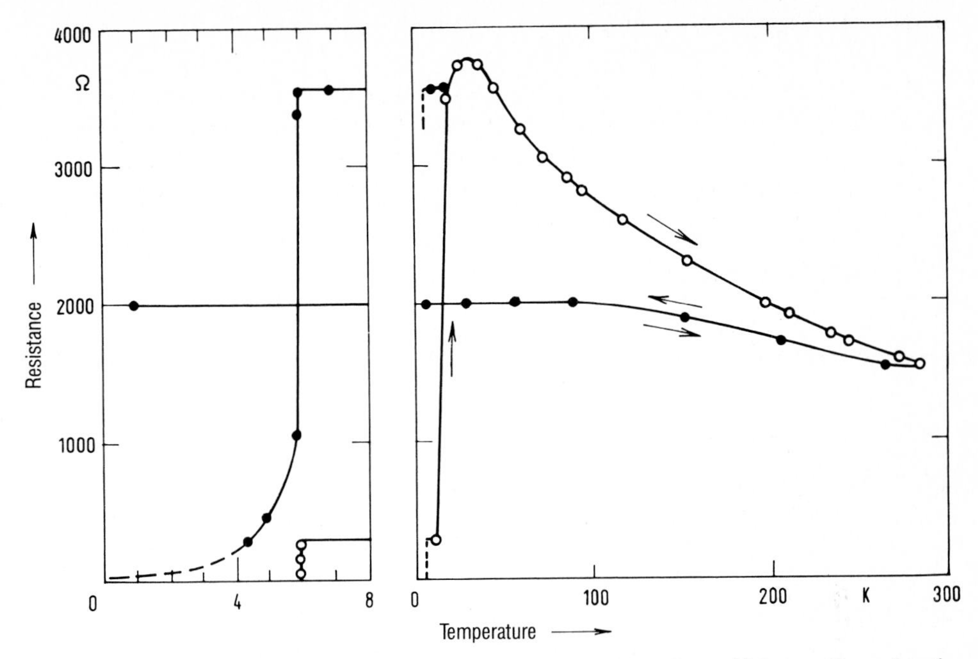

Fig. 126. Variation of resistance of a quench-condensed bismuth layer. Layer thickness: 40 nm, length 10 mm, width 1 mm, condensation temperature: 10 K (after [194]).

into the superconducting state and has a transition temperature of 7 K. Like the quench-condensed film this high-pressure modification has a much higher concentration of free electrons. The Bi films condensed at 4 K, thus, resemble a highly disordered modification of the high-pressure phase rather than of the normal Bi lattice.

Similar results have been obtained for other substances such as gallium. A very interesting similarity was found here between the investigations of condensed films and the high-pressure experiments [201].

The amorphous metals (metallic glasses) have acquired great importance over the last decades. Very rapid cooling of the melt (splat cooling) has made it possible to freeze alloys in an amorphous phase [202]. These alloys include a range of superconductors [203]. This provides a particularly good opportunity to study the low-energy excitation specific to amorphous substance on metals and at very low temperatures [204]. In the superconducting state at sufficiently low temperatures the free electrons are practically completely decoupled from the energetic processes occurring in the metal. This makes it possible to measure the excitation of the lattice alone [205].

8.6 The influence of paramagnetic ions

The incorporation of paramagnetic ions has a particularly large influence on the transition temperature of superconductivity. Alloys with paramagnetic ions occupy, therefore, a special position. Their properties will be discussed in this section.

By a paramagnetic ion we mean a foreign atom that has a fixed magnetic moment even after incorporation in the lattice. The incorporation of such moments reduces T_c very considerably. It might at first be presumed that the decisive effect was the magnetic moment. However, Matthias et al. [206] were able to demonstrate with a systematic investigation of the alloys of lanthanum with rare earths that the spin of the incorporated atom is the decisive variable for the reduction of T_c.

Figure 127 presents the change of T_c of La alloys containing just 1 atom % of added rare earth. The largest reduction was observed for gadolinium. Gd has the largest spin, but not the largest magnetic moment (Fig. 127).

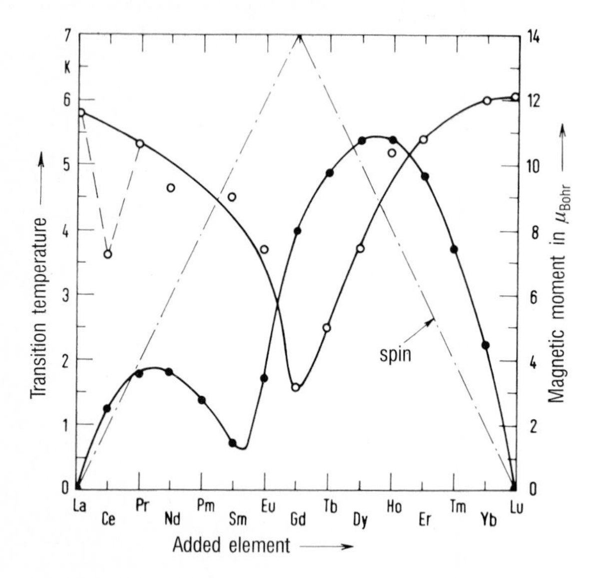

Fig. 127. Transition temperatures of lanthanum alloys containing 1 atom % of a rare earth and the effective magnetic moments of the added rare earths (solid circles) (after [206]).

This influence of ions with spin is very easy to understand, in principle. The interaction between the paramagnetic ion and the conduction electrons will lead to the environment of the ion preferring a direction of spin of the electrons. This can, depending on the type of interaction, be the parallel or the antiparallel configuration. In the superconducting state some of the conducting electrons are correlated to Cooper pairs. When this happens the spins of the two electrons making up the Cooper pair are arranged antiparallel. If one of these electrons – we are deliberately adopting an exaggaratedly corpuscular point of view, but this makes the basic situation particularly clear – comes into the field of influence of one of the paramagnetic ions, then the interaction with this ion comes into competition with the pair correlation. Hence, quite independent of whether the parallel or the antiparallel configuration is preferred, the Cooper pairs can be broken up because the electrons alter their direction of spin under the influence of the paramagnetic ion.

Thus, the incorporation of such ions will reduce the Cooper-pair correlation. This will reduce the transition temperature. This fact can be understood on the basis of the present theory [207]. This predicts a linear reduction of T_c for small concentrations of paramagnetic ions. This linear effect has been observed for a whole range of systems. Figure 128 only illustrates a few examples for lead alloys. Here the experimental difficulty appears that many superconductors have practically no solvent power for paramagnetic ions[1]. It is possible, in such cases, to force a statistical distribution of the added atoms by condensing the paramagnetic ions together with the host metal on a cold substrate. The extreme quenching conditions prevent the precipitation. The additive then precipitates out on tempering and this can be followed via the change in T_c. The large scatter in the results (Fig. 128) illustrates the difficulty of preparing such metastable alloys.

Another possibility for the preparation of metastable alloys, particularly with small concentrations of additives, is offered by ion implantation (Graph 1 in Figure 128a). Here the paramagnetic ions are created at a source and then accelerated to a few hundred kilovolts and shot at low temperature deep into the film of host metal at this energy. It is possible to obtain a very homogeneous distribution of the additive atoms by appropriate variation of the energy.

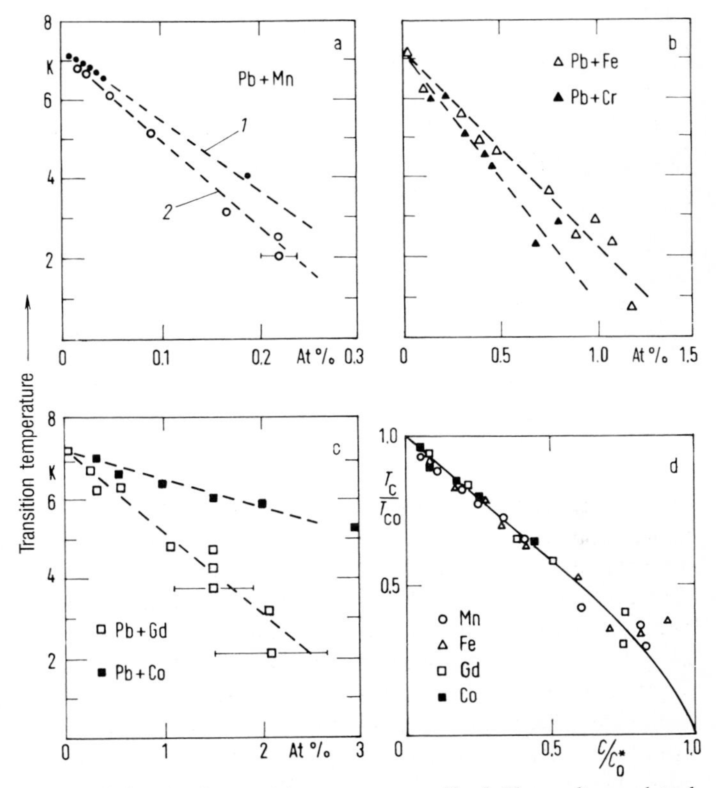

Fig. 128. The influence of paramagnetic ions on the transition temperature of lead. The results are plotted in reduced units in part d. The solid curve is that predicted by theory if the following critical concentrations are employed. Mn: 0.26 atom %, Fe: 1.2 atom %; Cr: ca. 1 atom %, Gd: 2.8 atom %, Co: 6.3 atom %. All the results, apart from Curve 1 in part a, were determined using quench- condensed films. Curve 1 in part a was obtained for alloys that were produced by ion implantation at liquid He temperatures. Pb + Fe, Pb + Cr, Pb + Co [208]; Pb + Mn [209, 210]; Pb + Gd [211].

1 The alloys of lanthanum with rare earths are particularly favorable with respect to solubility.

Table 11. Reduction in the transition temperature of some superconductors by paramagnetic ions.

Superconductor	Additive	$-dT_c/dc$ in K/atom %
Pb	Mn	21* a), 20** b)
	Cr	ca. 6* c)
	Fe	4.7* c)
	Gd	2.0* d)
	Co	0.8* c)
Sn	Mn	69* e), 14** b)
	Cr	16* e)
	Fe	1.1* e)
	Co	0.15* e)
Zn	Mn	315 f) 285* i) 343** j)
In	Mn	53* k) 50** l)
La	Gd	5,1 g), 4.5* h)

* quench – condensed films
** ion implantation at low temperatures

A review is given by E. Wassermann: Z. Physik *220*, 6 (1969)

References: a): [209]; b): [210]; c): [208]; d): [211]; e): [192]; f): [213]; g): [206]; h): [214]; i): [215]; j): [216]; k): [217]; l): [218]

It is possible to correlate the results in Fig. 128 and from similar experiments quantitatively with the theory by introducing a parameter c_0^* that characterizes the strength of the interaction (Fig. 128d). There is as yet no understanding of the variables involved in the interaction such as would allow its calculation from other solid state constants. The initial decrease in T_c as a function of the concentration c of the ions is summarized for some systems in Table 11. At high concentrations ($c > 1$ atom %) the relationships become very obscure, because the interactions between the paramagnetic ions can then become important. For example an ordering of the spins of these ions should lead to a reduction of the pair-breaking effect on the Cooper pairs.

In the case of superconductors with low transition temperatures even very low concentrations of paramagnetic impurities can lead to complete abolition of superconductivity (Section 1.2). If, at present, we cannot observe superconductivity in a whole range of metals even at very low temperatures then this could, in some cases, be a result of such impurities. The superconductivity of molybdenum, for instance, was only discovered after an extreme degree of purity had been reached [212].

It has been demonstrated in recent investigations [219] that the dependence of the transition temperature on the concentration of the paramagnetic ions can also be fundamentally different than was first calculated by Abrikosov and Gorkov. This is the case for systems where an antiparallel alignment of the electron spin to the spin of the paramagnetic ion is favored energetically. These systems are known as Kondo systems [1]. A range of special features can also be observed in these in the normal state. For instance, the electrical resistance passes through a minimum at low temperatures.

1 J. Kondo developed the first quantitative description of these systems and was, therefore, able to account for the anomalies theoretically.

Müller-Hartmann and Zittartz made for superconductors containing such paramagnetic ions the astonishing prediction that, under certain conditions, such an alloy could become superconducting at a temperature T_{c1}, and then, at a lower temperature T_{c2}, become normalconducting once more. Superconductivity was then predicted to begin once again at a very low temperature $T < T_{c3}$. The reason for this astonishing behavior lies in the temperature dependence of the interaction between the electrons and the spins of the paramagnetic ions. If this pair-breaking interaction increases very rapidly with falling temperature – which the theory predicts for certain systems – then it is qualitatively understandable that the already occurring correlation to Cooper pairs which begins at T_{c1} should be completely abolished at lower temperatures on account of the very greatly increasing pair-breaking interaction. The correlation to Cooper pairs should then begin to take place once more at very low temperatures.

The theoretical predictions concerning the existence of a temperature T_{c2} in the sense described have been spectacularly confirmed by Winzer [220] for the system $(La_{1-x}Ce_x)Al_2$. Pure $LaAl_2$ has a transition temperature of $T_c = 3.26$ K. On adding Ce a state is reached at $x = 0.63$ atom % where there are the two transition temperatures T_{c1} and T_{c2}. Inductively determined (see Section 5.1.1) transition curves are illustrated in Fig. 129. Here the magnetic susceptibility χ has been plotted against the temperature. The magnetic susceptibility χ is practically zero in the normalconducting state whereas it becomes -1 in the superconducting state. Somewhat above 1 K the sample becomes superconducting with a steep transition curve. The normalconducting state returns below 0.2 K. At this low temperature the pair-breaking effect of the Ce ions is sufficiently strong to be able to prevent superconductivity completely. A second transition to the superconducting state has been observed at ca. 50 mK in La-Y-Ce alloys of suitable composition [221]. Resistance measurements have also detected both transitions. This is a particularly beautiful example of the interaction of theory and experiment.

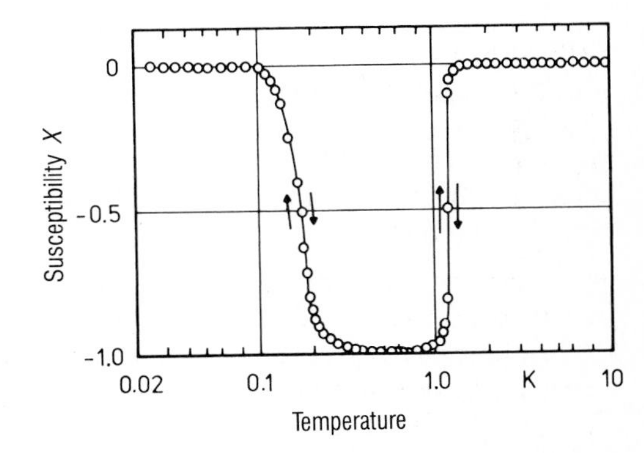

Fig. 129. Transition curves of $(La_{1-x}Ce_x)Al_2$. $x = 0.63$ atom % (after [220]).

Besides reducing transition temperatures paramagnetic ions also affect the energy gap of a superconductor. This influence is most easily understood on consideration of the lifetimes of the Cooper pairs. The possibility of the occurrence of pair breaking by the paramagnetic additive reduces these lifetimes. A finite lifetime implies an uncertainty of the energy according to $\Delta E\,\Delta t \geq \hbar$ (Δt = mean lifetime, $\hbar$ = Planck's constant/2π). This blurs the sharp edge of the energy gap. Figure 130 [222] illustrates plots of the densi-

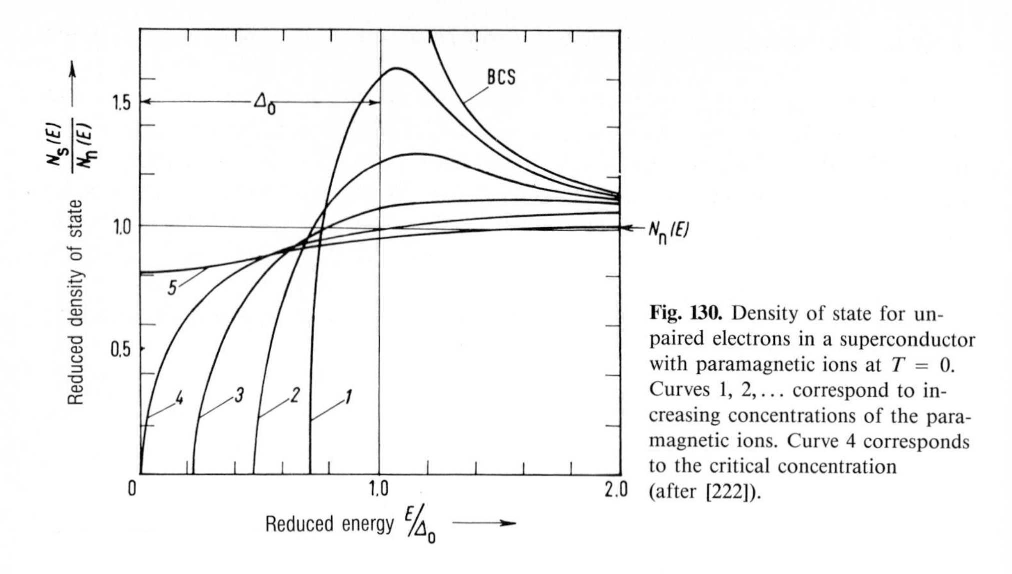

Fig. 130. Density of state for unpaired electrons in a superconductor with paramagnetic ions at $T = 0$. Curves 1, 2,... correspond to increasing concentrations of the paramagnetic ions. Curve 4 corresponds to the critical concentration (after [222]).

ty of states at the temperature $T = 0$ against the reduced energy E/Δ_0 for various concentrations of paramagnetic ions[1]. As the concentration increases the energy gap becomes smaller, while, at the same time, the blurring of its boundary becomes greater. This leads to the disappearance of the energy gap at a particular concentration c_0 even though the densities of states still deviate considerably from those of the normal state. We still have a finite number of Cooper pairs.

This state with no finite energy gap any more in the excitation spectrum, but still with correlation to Cooper pairs, is known as gapless superconductivity. In this state the superconductor still does not offer any electrical resistance if the current is not too high. The resistance (at $T = 0$) first appears at a critical concentration c^*_0 that is appreciably higher (ca. 10%) than c_0. This experimental finding indicates that the superconducting state is not determined by the presence of a gap in the excitation spectrum but by the presence of Cooper pairs.

It is difficult to understand the existence of a dissipation-free current in the absence of a finite energy gap. When there is a finite energy gap it can be argued that the interactions only then begin when the kinetic energy of the Cooper pairs is sufficiently high to be able to supply the finite energy of excitation (see Section 2.2). However, this argument evidently does not touch the core of the matter for it breaks down in the absence of a finite energy gap. We must rather assume that a certain concentration of Cooper pairs is thermodynamically stable, i.e. deviations from this equilibrium concentration are always compensated by relaxation processes (see 8.1). If the superconductor carries a current then all the Cooper pairs possess exactly the same momentum. If pairs now disappear – no matter by what process – for each one two electrons with a suitable momentum must form a Cooper pair, which has exactly the same momentum as all other Coo-

1 The employment of the reduced energy makes the results independent of the absolute size of the energy gap, so that they can be used universally for various superconductors.

per pairs. The stability is provided by the strict correlation of the Cooper pairs with one another.

It has been possible to demonstrate the effect of paramagnetic ions of the energy gap experimentally by means of tunneling experiments [223]. The more refined theories [224] make the further prediction that discrete states should appear in the energy gap of a superconductor containing paramagnetic ions. This prediction has also been confirmed experimentally. Figure 131 reproduces the normalized densities of states for lead containing very small concentrations of Mn ions. Two maxima can clearly be seen in the density of states within the energy gap. In these investigations the paramagnetic ions were implanted at low temperatures in the lead film of a previously prepared Mg-MgO-Pb tunnel diode with almost ideal behavior at low temperatures.

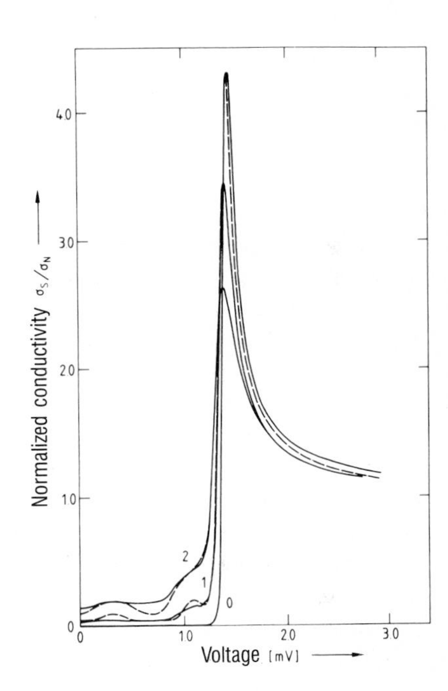

Fig. 131. Normalized conductivity σ_s/σ_N (proportional to densities of states of the electrons) of lead with added manganese (after W. Bauriedl, P. Ziemann and W. Buckel: Phys. Rev. Lett. *47*, 1163 (1981)). Measurement temperature: 200 mK; Curve 1: pure lead, Curve 2: Pb + 25 ppm Mn, Curve 3: Pb + 250 ppm Mn. The broken line was calculated with the aid of the MHZB theory.

8.7 Special superconducting materials

In Section 1.2 superconducting elements and a few examples of the vast range of superconducting alloys and compounds were listed in order to reveal that superconductivity occurs in very many materials of very different types. Patterns of any kind are scarcely discernible. We will use this section to discuss some special materials in greater detail. Even though we still have no complete quantitative understanding of the relationship between the general parameters of solids and the characteristic variables of superconductivity the last few years have, nevertheless, brought some decisive advances. Some of these will be discussed on the basis of examples of specific substances. The superconducting oxides provide us with quite a new set of problems.

8.7.1 The β-tungsten structure

A group of conductors with transition temperatures above ca. 18 K e.g. Nb_3Sn, Nb_3Ga, Nb_3Ge and others have the β-tungsten structure (A15 structure). This structure can evidently be very favorable for superconductivity [1]. The β-tungsten structure is illustrated in Fig. 132. The arrangement of the Nb atoms in chains parallel to the X, Y and Z axes is very typical of this structure. These orthogonal chains do not cross each other. The distance apart of the Nb atoms in the chains is less than in the lattice of pure niobium.

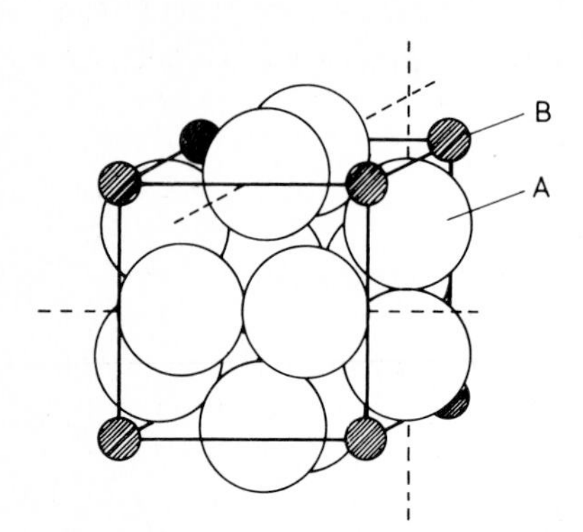

Fig. 132. Unit cell of the β-tungsten structure (A15 structure) of binary compounds of type A_3B. The broken lines give the directions of the 3 orthogonal chains of A atoms.

It is assumed that it is these chains which are responsible for the special properties. Models have been developed, based on this chain structure [225]. If the chains are regarded as "onedimensional metals" then peaks are obtained in the densities of electronic states [2]. If the Fermi energy lies in the range of such a peak in the density of states, then two effects are to be expected. Firstly the electron-phonon interaction could be particularly large and secondly the density of states at the Fermi energy should be particularly temperature-dependent since small displacements of E_F as a result of changes in occupation of states because of changes in temperature could cause a large change in N(E).

These predictions have been confirmed on the basis of the elastic behavior of some substances with high T_c, e.g. Nb_3Sn and V_3Si. In these substances there is a very large reduction in the rate of propagation of certain sound waves as the temperatrure is reduced [226]. This means that the lattice becomes very soft with respect to certain elastic deformations. This "softening" can finally lead to the structure becoming unstable [227]. A lattice transition can be observed in single crystals of Nb_3Sn at 40 K and of V_3Si at 21 K. The transition consists of a slight tetragonal distortion of the original cubic lattice.

It might be thought that the high T_c values were connected with the softening of certain phonons. Investigations of the densities of state of the phonons of the vanadium

1 It must be mentioned here that substances that are members of this group exist, which have very low T_c values (Nb_3Os: T_c = 1 K) and there are others known for which superconductivity has not yet been demonstrated (Nb_3Sb: no sc. at $T > 1$ K [7]).

2 It is clear from Fig. 9 that the density of states $N(E) \propto dk/dE$ can change in the direction of infinity for one direction in space. In the simple example in Fig. 9 the divergences lie at the edges of the bands.

compounds V_3Si, V_3Ga and V_3Ge, by means of neutron-diffraction experiments [228], appear to support this hypothesis. The change in the density of states of the phonons is highest for V_3Si with the highest $T_c = 16.9$ K of the three substances and decreases appreciably in order of decreasing T_c from V_3Ga ($T_c = 13.9$ K) to V_3Ge ($T_c = 6$ K).

Many experiments have revealed that the transition temperatures of these substances are sensitively dependent on the state of order of the chain. The highest T_c values can only be obtained after careful tempering (heating to a suitable temperature) over a long period. Care has to be taken here that no new phases are produced alongside the β-tungsten structure. The choice of a suitable temperature requires knowledge of the phase diagram of the substance involved. According to our present state of knowledge it appears confirmed that, for substances with a β-tungsten structure, the highest degree of order also brings with it the highest transition temperature [229].

In view of the theoretical understanding it is clear that the simple models [225] are insufficient for the explanation of all observations [230]. But it seems to be decisive to take into account the coupling between the orthogonal chain systems with each other and between the chain atom A and atom B in the substance A_3B [231]. Only a start has been made here. When this group of substances with high transition temperatures is considered the question naturally arises as to the highest value of T_c that can be achieved with them. As already established in Section 1.2 it is not possible to give a certain answer to this question at the moment. All we can hope to do is to extrapolate certain trends. Thus, for the series Nb_3Sn ($T_c = 18.2$ K), Nb_3Ge ($T_c = 23.2$ K) and Nb_3Si it may be predicted that the T_c for Nb_3Si will be higher simply due to the lower mass of Si. Such as estimate yielded a value of $T_c > 30$ K [232].

Another extrapolation is based on a regularity in connection with the lattice constants of many superconducting A_3B compounds of the β-tungsten structure [233]. It has been found that the transition temperatures of these substances fall when the lattice constant a increases as a result of defects of any sort. If the reduction in T_c is plotted against a then all the points lie on a rather narrow strip which indicates the universal character of this relationship. In this connection it is possible to plot the known values for a Nb_3Si sample which is very disordered. For this purpose it is necessary to know how much the lattice constants of this sample, with a large degree of disorder, differ from those of ideal Nb_3Si. This deviation can be obtained relatively readily from the available data. The prediction is then obtained that for this sample the reduction in T_c as a result of lattice expansion will be ca. 22 K. The transition temperature observed was 9 K. Thus, this consideration yields a transition temperature of ca. 30 K for ideal Nb_3Si. There is an excellent description of the properties of A15 superconductors in [234].

If it proves possible to prepare ideal samples of Nb_3Si we should know in a few years whether these extrapolations were justified. At the moment this represents a challenge to the materials scientist. A technically applicable superconductor with a T_c of ca. 30 K would be sure to stimulate very many applications.

The last paragraph has been retained from the 3rd German Edition because it very clearly illustrates how the situation with respect to higher transition temperatures was viewed in 1984. Today, only a few years later, we have superconductors with transition temperatures above 120 K.

8.7.2 Palladium hydrogen systems

The superconductivity of palladium hydrogen was discovered in 1972 by Skoskiewicz [235]. He found transition temperatures over 1 K when the H/Pd ratio was equal to ca. 0.8. When the hydrogen concentration was increased to give H/Pd ≈ 0.9 the transition temperature rose to ca. 4 K.

This discovery was so surprising because the system Pd-H had been the subject of so many investigations on account of its other interesting properties. One would scarcely have expected to find superconductivity in this part of the periodic table.

The steep increase in T_c with increasing hydrogen concentration raised the expectation that the value discovered by Skoskiewicz would not be the upper limit for T_c in this system [1]. The H concentration was, therefore, increased further by ion implantation at low temperatures. It was possible in this way to reach a maximum transition temperature $T_c = 9$ K [236]. Further increases in the H concentration then led to a gradual reduction of T_c. The maximum value of T_c was obtained at the ratio H/Pd = 1. Figure 133 illustrates these results of the ion implantation experiment [237].

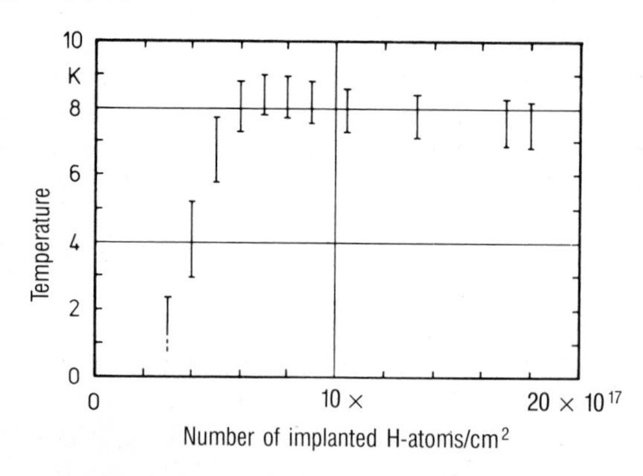

Fig. 133. Transition temperatures of Pd-H samples as a function of the numbers of implanted H atoms. The bars give the width of the transition curves for $0.1 < R/R_n < 0.9$. The Pd foil was homogeneously preloaded to a concentration of H/Pd = 0.74 in hydrogen gas at 300°C. The implantation was made at *one* implantation energy. This only further increased the H concentration in a ca. 150 nm thick layer (after [237]).

An unexpected result was achieved on changing from hydrogen to deuterium. A somewhat lower transition temperature would be expected on account of the isotope effect (Section 3.1). However, the maximum T_c was at about 11 K and occurred again at the ratio D/Pd = 1. This anomalous isotope effect, which is correlated to other anomalies on going from H to D in Pd (e.g. an increase in the diffusion constant), reveals that the incorporation of deuterium also changes the properties of the host lattice.

These results triggered off a series of experimental and theoretical investigations. It is possible to obtain H or D concentrations corresponding to a ratio H/Pd = D/Pd = 1 by electrolysis at ca. 200 K. The essential results obtained by ion implantation were completely confirmed in this manner [238].

Experiments with Pd-noble metal alloys brought a further great surprise [239]. Here transition temperatures as high as 17 K were observed on implantation of hydrogen at-

---

1 Skoskiewicz employed electrolysis at room temperature to load his palladium wires. This meant that the H concentrations achievable were limited to H/Pd = ca. 0.9.

oms. Figure 134 illustrates these results. A clear regularity can be observed on passing from the PdAu to the PdCu alloys. It has not yet been possible to provide a satisfactory explanation of these high T_c values within the framework of existing theoretical concepts.

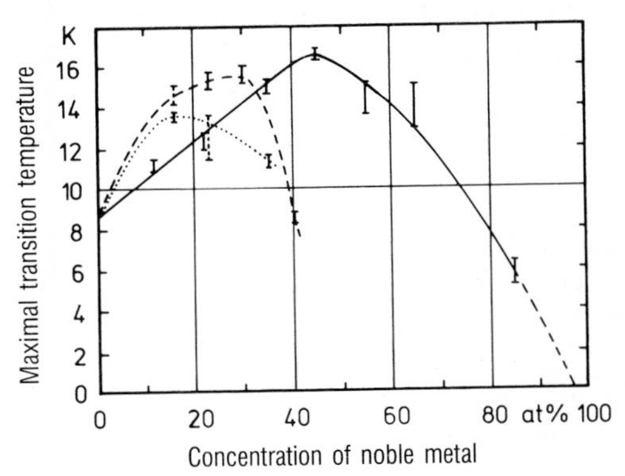

Fig. 134. Maximum transition temperatures of Pd-noble metal alloys after loading with hydrogen by ion implantation at liquid He temperatures as a function of the noble metal concentration (after [239]).

On the other hand, we now have a very good understanding of the superconductivity of Pd after loading with hydrogen. The hydrogen is incorporated interstitially in the Pd lattice. This causes the Pd lattice to expand somewhat. Basically the incorporation of hydrogen leads to the occurrence of additional lattice vibrations, simply because there are more atoms per unit volume. One can speculate that these additional lattice vibrations increase the electron-phonon interaction and, thus, favor superconductivity [240]. A prerequisite of this is naturally a sufficiently strong coupling of these vibrations with the system of electrons, a condition that is not necessarily filled.

It is now possible to determine by means of tunneling experiments whether there is this strong coupling between the hydrogen-dependent lattice vibrations and the electron system. Since this method has also been employed with great success with other superconductors too, e.g. the amorphous layers (Section 8.5.3), we will discuss it briefly for the Pd-H system.

We saw in Section 3.3.3 that the tunneling effect for single electrons can be employed to determine the density of states for these single electrons. Now, if the coupling between the lattice vibrations and the electron system is sufficiently strong, the density of states will be altered by this coupling. These changes will be evident in the tunneling characteristic and can be determined if the experimental arrangements are sensitive enough. Some computation then allows determination of the coupling constant λ^* for the attractive electron-electron interaction and the constant μ^* of the repulsive coulomb interaction (see Eq. (3-3)). In this manner, what is known as the Eliashberg function $\alpha^2F(\omega)$ is obtained, which is closely linked to λ^*. $F(\omega)$ is the density of states of the phonons. We will represent the energy here with ω in order to avoid any chance of confusion with the density of states of the electrons.

We can only determine phonons – and this is the decisive point – that couple with the electron system with tunneling experiments. On the other hand, it is possible to perform neutron-scattering experiments and determine the density of states of the phonons independent of whether they interact with the electrons. A comparison of both determinations, therefore, provides evidence concerning the amount of coupling. Figure 135 illustrates such a comparison for PdD samples [241, 242]. In spite of the greatly differing deuterium concentrations it can clearly be seen that the significant structures are visible in both sets of results. In particular the phonons at high energies, which are derived from the incorporated deuterium, are also visible in the results of the tunneling experiment, i.e. these vibrations are well coupled with the electrons. The increase in the transition temperature in the palladium-hydrogen system is comprehensible in the light of this knowledge.

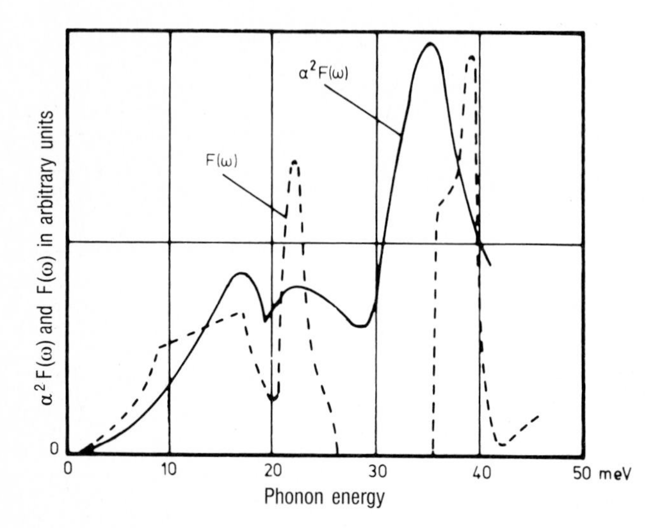

Fig. 135. $\alpha^2F(\omega)$ from tunneling results and $F(\omega)$ from neutron scattering experiments as a function of the phonon energy ω (after [241, 242]).

In view of these results it would seem obvious to search for other substances where the electron-phonon interaction is increased on incorporation of hydrogen. However, whether even higher transition temperatures would be obtained than with the palladium-noble metal alloys is a completely open question [243].

8.7.3 Organic superconductors

In 1964 W.H. Little [16] put forward the hypothesis that it ought to be possible to find organic superconductors with very high transition temperatures (p. 23). These would be long-chain molecules with conjugated double bonds along the length of the chain and suitable ligands. It has not until now proved possible to confirm this hypothesis.

The first organic superconductor, the hexafluorophosphate of TMTSF, was discovered by Jerome et al. [244] in 1980. It is necessary to subject the substance to an all-round pressure of 12 kbar to inhibit a metal-insulator transition. Under this pressure the organic conductor maintains its good metallic properties down to the lowest temperatures. The structural formula of TMTSF is illustrated in Fig. 136a.

The transition temperature $T_c = 0.9$ K is still very low. But the authors pointed out that they observed superconducting fluctuations at appreciably higher temperatures. They are, therefore, of the opinion that further efforts in the synthesis of organic materi-

Fig. 136. Structural formulae of organic superconductors. a) TMTSF, b) BEDT-TTF.

als could lead to the production of superconductors with higher transition temperatures. The compound $(TMTSF)_2ClO_4$ remains metallic down to the lowest temperatures at normal pressures and becomes superconducting at $T = 1$ K [245].

In recent years organic superconductors have in fact been discovered with appreciably higher transition temperatures. The recently discovered organic superconductor $(BEDT\text{-}TTF)_2Cu(CNS)_2$ [246] has the highest transition temperature of these. This substance becomes superconducting at 10.4 K. The structural formula of BEDT-TTF is illustrated in Fig. 136b. T_c increases to 11.2 K on replacing end hydrogen by deuterium. These compounds have no connection with Little's hypothesis.

8.7.4 *The new superconductors*

The sensational development of the superconducting oxides was referred to in Section 1.3. Here we will consider some additional facts concerning them. In doing so we will largely confine ourselves to $YBa_2Cu_3O_7$, which is, as yet, the best investigated compound and has a transition temperature between 90 K and 95 K.

The structure of $YBa_2Cu_3O_7$ is illustrated in Fig. 137. The copper atoms occupy two different types of position with respect to their oxygen environment. On the one hand, they occur at the base of a four-faced pyramid. The 6th position of the oxygen octahedron is not occupied. These pyramids are joined at the corners and, thus, form layers. The electrical conductivity parallel to these layers is considerably higher than perpendicular to them [84]. The second copper position is only surrounded by 4 oxygen atoms. Cu-O-Cu-O chains exist along the axis labelled b. But there are no oxygen atoms between the copper ions along the axis a in this plane.

The structure is orthorhombic. It is formed from a tetragonal structure on cooling samples below 700°C [247]. During this transition domains are formed even in single crystals. Because of the extremely short coherence length (Section 7.4.1) the domain boundaries of $YBa_2Cu_3O_7$ can have an extremely strong influence on its properties [248]. Thus, very thin interfacial layers can act as Josephson junctions. It was, for instance, only possible to reach the high critical current density (Section 7.4.2) when it became possible to produce almost monocrystalline layers by epitaxial growth [155].

A series of investigations, e.g. of the Hall effect [1)] [84, 249] and electron microscopy [250], have revealed that the mechanism of conductivity in the layers of Cu-O pyramids

1 The Hall effect is the name given to the occurrence of an electrical voltage between two equivalent points at the edges of a conduction band if a magnetic field is applied perpendicular to the band surface and a current is flowing through the band. The deflection of the charge carriers transporting the current in the magnetic field is the cause of this "Hall voltage" (E.H. Hall, 1789).

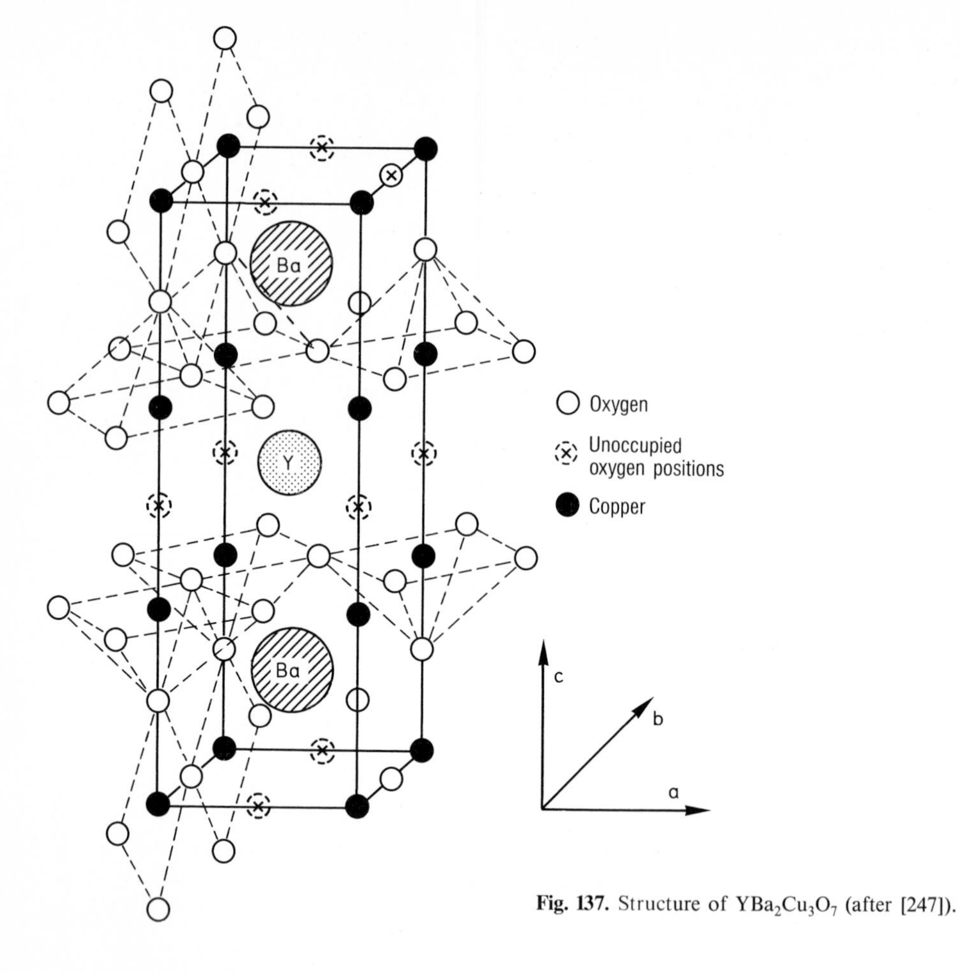

Fig. 137. Structure of $YBa_2Cu_3O_7$ (after [247]).

is carried by defect electrons [1]. According to Hall-effect investigations on single crystals it is negative charges that carry the current perpendicular to the layers [249]. There are questions that have not yet been answered here.

An important parameter for the properties of $YBa_2Cu_3O_7$ is its oxygen content and evidently also the ordering at the oxygen positions. Oxygen can be expelled from $YBa_2Cu_3O_7$ by heating in vacuum, thereby producing a substance $YBa_2Cu_3O_{7-x}$. The changes in the characteristic properties as a function of x have been studied in detail. Figure 138 illustrates the variation in the transition temperature T_c and specific electrical resistance ϱ (at 300 K) of polycrystalline $YBa_2Cu_3O_{7-x}$ as a function of the oxygen content [251]. When $x \approx 0.25$ the transition temperature is reduced to ca. 60 K. The mini-

1 If there are only a few unoccupied states to be found in a band (Section 1.1) then the mechanism of conductivity in such a band can be described in terms of these holes, which are referred to as defect electrons and which must be ascribed a positive charge $+e$.

mum in the specific resistance ϱ indicates that a new ordered phase is formed here. The temperature dependence of this phase is metallic, $d\varrho/dT > 0$.

The transition curve is broad in the concentration range around $x = 0.25$, because here statistical variations in the oxygen concentration lead to large changes in T_c in different parts of the sample. Since the oxygen was removed at temperatures < 520 K in these experiments, the sample remained orthorhombic over the whole range of concentrations investigated [251]. If oxygen is removed at higher temperatures ($T >$ ca. 700°C), then the oxygen deficit stabilizes the tetragonal phase. Some results obtained by other authors [252] are also included in Fig. 138. The scattering indicates how sensitively these investigations depend on the experimental conditions.

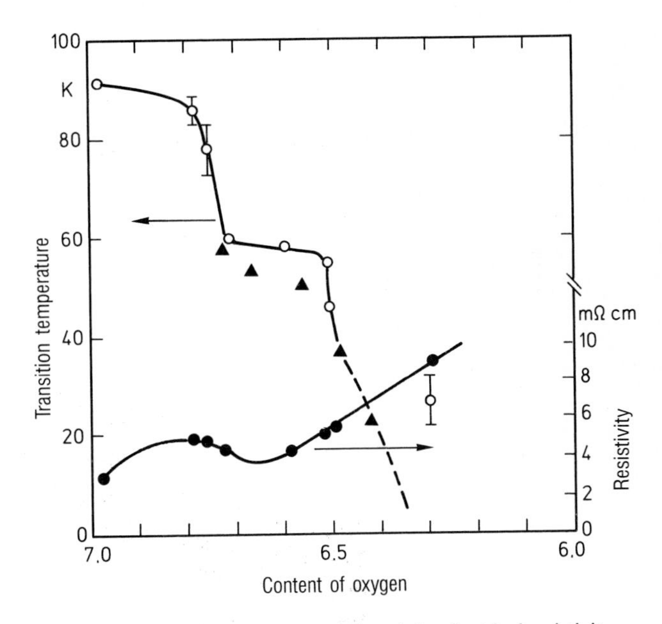

Fig. 138. Influence of the oxygen content on the transition temperature T_c and the electrical resistivity ϱ of $YBa_2Cu_3O_7$; ○○○○, ●●●● after [251]; ▲▲▲▲ after [252].

One peculiarity of the new oxides is the manner in which the specific electrical resistance rises surprisingly linearly with the temperature over a very large temperature range. There is no convincing explanation of this phenomenon, which could perhaps itself provide evidence concerning the reason for the high transition temperature.

Soon after the discovery of the high transition temperatures of the Y-Ba-Cu-O system it was found that Y could be replaced by almost any rare earth metal, when superconductors with transition temperatures about 90 K were always obtained. Only Ce, Pr and Tb did not yield compounds with such a high value of T_c. Magnetic ions, such as gadolinium, do not have an interfering effect here as they do in classical superconductors (see Section 8.6). This shows that superconductivity is limited to regions in the sample – probably the layers of the oxygen pyramids – which are a "great distance" from the rare earth ions, i.e. there is little interaction.

In spring 1988 new superconductors became known with even higher transition temperatures. Thus, transition temperatures around 110 K were found for the system Bi-Ca-Sr-Cu-O [22]. The highest transition temperature known today, of around 125 K, was observed for $Tl_2Ca_2Ba_2Cu_3O_x$ [23]. The high transition temperatures around 120 K have been confirmed [253]. It is also of great interest that with $Ba_{0.6}K_{0.4}BiO_3$ a material has been found that does not contain copper and becomes superconducting at 28 K [254]. The discovery by Bednorz and Müller has led to enormous developments in the field of materials science.

We still do not have a satisfactory theoretical understanding of the new superconductors. While the BCS theory (Section 2.1) makes all aspects of classical superconductors understandable, at least at the qualitative level, it is by no means certain – some eminent theoreticians have even positively denied – that the BCS theory can also explain the high T_c values obtained for the new substances. Numerous new models have been proposed and old ideas have been revived [255]. There is certainly a need for more experimental and theoretical efforts before we achieve a satisfactory understanding of these exciting materials. What is certain is that here too the superconducting state is supported by paired charge carriers that macroscopically occupy a quantum state. The question as to whether we can find material with even higher T_c values has already been discussed in Section 1.3.

8.7.5 Heavy fermion superconductors

At the end of the 1970s a transition to superconductivity was observed at ca. 0.5 K for the compound $CeCu_2Si_2$ [256]. The superconductivity of this compound was very surprising because this was a metallic conductor where the electrons had effective masses, which were several hundred to one thousand times greater than the masses of free electrons. These masses were calculated from the extremely high electronic density of states at the Fermi energy. The name heavy fermion materials refers to these large masses. We now know a range of such materials. These include four superconductors, namely $CeCu_2Si_2$: $T_c = 0.7$ K [1]; UBe_{13}: $T_c = 0.85$ [257]; UPt_3: $T_c = 1.5$ K [258] and URu_2Si: $T_c = 1.5$ K [258].

These superconductors are of particular interest from two points of view. On the one hand, a system of heavy fermions bears similarities to ^3He. The ^3He atoms are fermions. ^3He can take up a superfluid state below 2.6 mK. However, this state is much more complex than the superfluid phase of ^4He. On the other hand, the very large mass of the electrons is a result of very strict correlations between the electrons. We now believe that such very strong correlations are decisive for the superconductivity of the new oxides. One may guess that a better understanding of the correlations occurring in heavy fermion materials might also give us a better insight into the mechanism acting in the new superconductors.

There is some evidence that electron pairs with spin 1 (triplet pairing) may sometimes exist in the heavy fermion superconductors [259].

---

1 Since the early experiments it has been possible to improve the homogeneity of the samples considerably, with the result that somewhat higher T_c values have been found.

9. Applications of superconductivity

Thoughts concerning its technical application are as old as superconductivity itself. Kamerlingh Onnes himself hoped that it might be possible to produce very high magnetic fields very economically with the aid of electrical conductors having no resistance. We saw in the previous chapters how long the way was and how much knowledge had to be acquired before this hope of Kamerlingh Onnes could be realized.

Superconducting magnets of every desired size and geometry have now been planned and already manufactured. They are employed not just in scientific research, in high energy and solid state physics, as was the case in the 1960s. They are already in planning for controlled fusion, for superconducting motors with outputs of several thousand kilowatts, for energy storage facilities and for the magnetic suspension of trains.

However, the consideration of superconductivity for large-scale commercial applications is not limited to superconducting magnets. A great deal of effort is going into the study of superconductivity for the transmission of electric power. If the energy consumption of our society continues to rise as it has until now, particularly in the large conurbations, it could turn out in the not too distant future that superconducting cables and storage elements will offer crucial advantages.

The application of superconductors in measurement technology is no less revolutionary; here it opens up for us the opportunity to increase the sensitivity of many determinations by orders of magnitude over what could be achieved with normalconducting circuitry. Superconducting circuit and memory elements could bring decisive improvements to large computers, under certain conditions.

All these applications, most of which would have been regarded as being in the realms of "science fiction" only 20 years ago, are now either reality (e.g. very large magnets) or are being studied seriously and with a large input of effort. Within the compass of this book we can only try to provide an impression of the multiplicity of possibilities and the problems that accompany them. This is the purpose of the following sections.

The new oxides provide us with superconductors which can even be employed at the temperature of liquid nitrogen ($T_B = 77$ K). It is certain that these materials will open up new opportunities for application. It is also certain, however, that a great deal of development will be necessary before we have in our hands suitable conductors for use in say magnet construction or for microcircuitry. It is economic questions that will play a crucial role here.

We will now discuss the already existing applications of the available conductors, whose development required more than a decade from the discovery of the basic facts. The fundamental approach remains the same for the new superconductors. The outlook for the future will, thus, become clear.

9.1 Magnets

9.1.1 Economic considerations

Magnets are a particularly self-evident application for superconductivity. When the magnetic field has been set up then, in principle, no further electrical power is required to maintain it. For this purpose it is necessary, at the moment, to cool the whole coil to liquid helium temperature (a few K) and to maintain this temperature. Even a rough calculation reveals that for high fields $B > 10\ \mathrm{T} = 100\ \mathrm{kG}$ a superconducting magnet is considerably more economical than a conventional magnet with a copper coil even when the volume is small, as say in a coil with 4 cm internal diameter and 10 cm long. Moderate fields from 2 to 10 T in volumes of the order of m^3, such as are employed, nowadays, in elementary particle physics and for controlled fusion experiments can only be produced in a realistic manner with superconductors.

Some figures will make the advantages of superconducting magnets quite evident. Let us take a copper coil (Bitter magnet) with an internal diameter of 4 cm and a length of 10 cm. In order to maintain a field of 10 T it is necessary to provide at least 5 000 kW of electric power, all of which has to be completely removed by cooling water. This means that at least 1 m^3 of cooling water has to be pumped through the magnet per minute and then through a cooling tower. In contrast, the power required to maintain a superconducting magnet of the same size at its low temperature is negligibly small. A very much more detailed estimate is required for the comparison of superconducting magnets with what are known as cryomagnets, where the coil must be constructed of very pure material, e.g. aluminium[1)] and which, although they are normalconducting, have a drastically reduced resistance on cooling to very low temperatures. Figure 139 reports the approximate costs of a large magnet over the period of its use [260]. These assume a field strength of 7 T in a volume of $2 \times 2 \times 2\ m^3$. Superconducting magnets of this order of magnitude are now in operation and others are under construction [261].

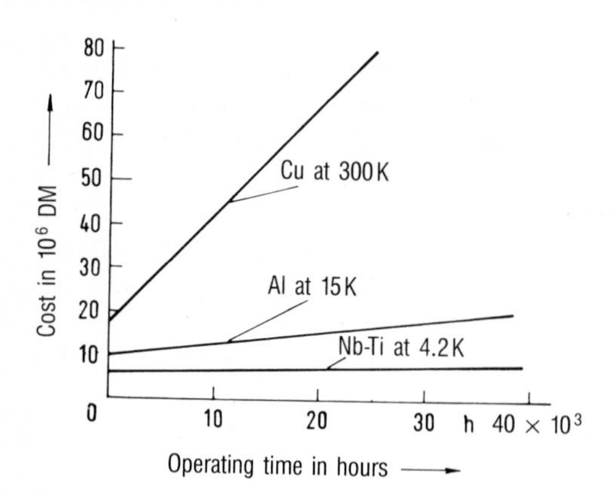

Fig. 139. Comparison of the costs of magnets with various types of coil materials as a function of the time of operation. Field strength: 7 T, field volume: ca. 8 m^3 (after [260]).

1 Since lattice defects of all types, including impurities, create an additional temperature-independent resistance only very pure samples yield the desired low resistance at very low temperatures.

All the considerations described above refer to magnets where the field is to be kept constant after the switching-on process (constant field magnets). Efforts are also being made to develop superconducting magnets whose field can be changed with time. Here losses connected with the change in the magnetic field play a crucial role. Suitable conductor structures (see next section), which have been developed, raise hopes that the losses may be reduced sufficiently for economic solutions.

Cooling with liquid nitrogen is at least 50 times cheaper than it is with liquid helium. In addition, the cryostats required are simpler and cheaper. It is to be hoped that suitable conductors constructed of the new materials will turn out to be sufficiently inexpensive for us to be able to exploit the advantages of cooling with liquid nitrogen. Great efforts are at present being made in all industrial countries to develop such conductors.

9.1.2 Stability problems

In order to make the construction of the coils as economical as possible efforts are made to employ load currents that are as high as possible for the superconducting material. The upper limit is given by the critical current of the particular material. We have plotted this critical current as a function of the magnetic field for some technically interesting superconductors in Fig. 99 (see Section 7.2) and in Fig. 106 for $YBa_2Cu_3O_7$. When attempts were made to construct superconducting coils it was unfortunately found that the levels obtained for short pieces of wire were not nearly achieved. This phenomenon was known as the degradation effect. Instabilities occur that can lead to normal conductivity even at current loads which are a factor of 2 less than the expected critical current. Particularly for large coils there is a danger of destruction, if the energy stored in the field cannot be dissipated in a suitable manner. Since it is absolutely essential that large coils operate reliably under all conditions, nothing else could be done initially but to reduce the current load and use more superconducting material to create the field. The degradation effect represented a great handicap at the start of the development of superconducting magnets, particularly for high field magnets.

It was only in 1965 that it became possible to overcome these problems to a very large degree with the aid of "stabilized" wires [262]. It had been recognized that the instabilities were the result of flux jumps in the superconducting material. Here as a result of some event, temperature changes or shocks, entire bundles of flux break away from their pinning centers and suddenly migrate through the material under the influence of the Lorentz force [1]. The rapid motion of whole bundles of flux creates a great deal of heat. If this heat cannot be removed rapidly enough the temperature will rise which can bring regions of the superconductor into the normalconducting state. The normal resistance that is created in this way causes further heating which results in the expansion of the normalconducting region. This makes the whole coil unstable so that it goes – possibly very rapidly – completely over into a normal state.

1 Several flux jumps evidently occurred on passage round the magnetization curve illustrated in Fig. 94 (see Section 7.1); in this case they made themselves evident in irregular changes in the magnetization.

The stabilization is achieved by coating the superconductor with a normal conductor with as little resistivity as possible, e.g. copper or aluminium. If the electrical contact to this normalconducting coating is good – and this is crucial – then if a region of normal conductivity occurs in the coil the current passing through it finds a normalconducting short circuit. The heating effect resulting from the normal resistivity that occurs is kept low. This gives such a conductor the opportunity to cool down again after a flux jump and to become superconducting once more. "Completely stabilized" conductors can be produced in this way if the thickness of the normal conductor is made large enough. However, it is necessary to accept an increase in the size of the coil as a result of the presence of the normalconducting coating.

This situation is illustrated schematically in Fig. 140 [144]. The solid curve *1* represents the ideal critical current density as measured for short lengths of sample (Fig. 99). The broken line represents the current density which may not be exceeded in the case of nonstabilized conductors on account of the degradation effect, without instabilities occurring.

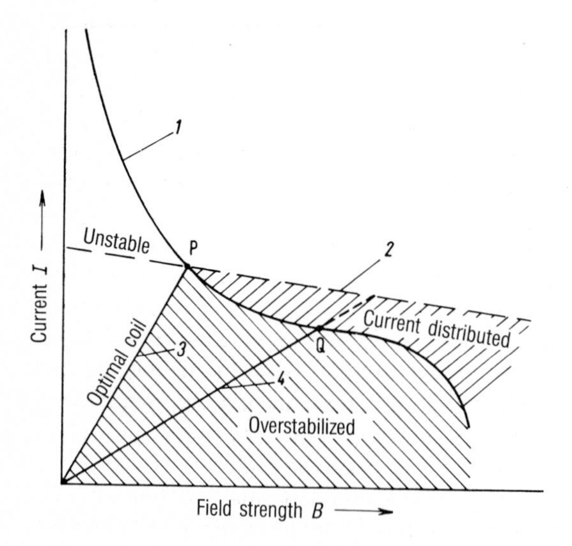

Fig. 140. Schematic representation of the stability behavior of superconducting magnet coils. Curve 1: critical current; Curve 2: normal metal limited current; P: operating point on optimization of coil geometry and complete stability, Q: operating point with overstabilization.

Full stability is achieved for a wire coated with normal conductor when the whole of the current can flow through the normal conductor and the heat so produced is transferred so well to the bath that no too temperatures occur in the coil material. In this case the superconductor will return to the superconducting state even after a very large flux jump. The maximum current laid down in such a case naturally also depends on the resistance R of the short circuit through the normal conductor and on the conduction of heat to the helium bath. The straight line *2* represents this maximum current I_r under the conditions of complete stabilization. It decreases with increasing field because the resistance of a normal conductor, in general, increases with the magnetic field. The quality of the contact between the superconductor and the normal metal is decisive for this maximum current I_r – it is also known as the "normal metal limited" or "minimum propagating" current. The various high field superconductors have to be judged technically on the feasibility of using them for the manufacture of such combined conductors having good quality. We will return to this point soon.

An optimal coil construction, where complete stabilization is just guaranteed and whose whole current will flow through the superconductor in the undisturbed state, is represented by straight line *3*. The magnetic field is proportional to the current ($B = \mu_0 nI/l$). The maximum current is represented by the intersection between curves *1* and *2*. At this point P all the current flows through the superconductor in the absence of disturbance and just with the critical current I_c. On the other hand, if a disturbance occurs the whole current can flow through the normal conductor under stable conditions.

An even larger margin of safety is to be desired for very large coils. It could, for example, happen that a movement of the not ideally fixed conductor could lead to a diminution to some extent of the contact between the superconductor and the normal conductor. This would reduce the level of I_r. In order to achieve safety in such cases the coil characteristics of very large magnets are arranged to follow something like curve *4* in Fig. 140. If the coil is operated with the critical current, point Q, then complete stabilization is achieved even if I_r is reduced. Such coils are known as overstabilized coils. They can be operated with currents between *1* and *2*. In this case there is a division of the current between the superconductor and the normal conductor. A voltage is set up in the coil, i.e. electrical power must be expended to operate the magnet. Such a mode of operation can be efficient, particularly for the production of peak fields.

The requirements of stability create quite new criteria for the technical applicability of superconducting materials. In the early days of development of superconducting magnets it was almost exclusively niobium-zirconium alloys with transition temperatures of ca. 10 K that were employed. These alloys have now been completely replaced by niobium-titanium alloys with an approximate composition Nb + 50 atom% Ti[1)] (Fig. 99). Apart from a somewhat higher critical field this alloy also has decisive metallurgical advantages. The composite conductor can be manufactured by incorporating a thick Nb-Ti rod into a copper billet of suitable size and then drawing the whole thing down to wire dimensions. This provides for excellent contact between the superconductor and the copper. Cross-sectional ratios of copper to Nb-Ti of 2:1 to 10:1 are employed. Figure 141 illustrates some types of conductor in cross section.

Envelopment with aluminium would have great advantages, in that the resistance created by the magnetic field has a saturation character and, hence, can remain appreciably lower than for copper at high magnetic fields. Figure 142 illustrates this important phenomenon for both normal conductors, copper and aluminium, with several samples of differing residual resistance in each case at zero fields, that is with differing numbers of defects [144]. The evident advantage of aluminium is counteracted by two disadvantages. On the one hand it is metallurgically difficult to manufacture a composite conductor with very good contact along its whole length. On the other hand, the lower mechanical strength of Al compared to Cu is also a disadvantage since in manufacturing large coils the very great forces involved mean that additional structures are necessary to contain

1 A slight variation of the composition brings about changes in the critical current in the sense that Ti-richer alloys have higher critical currents at low fields but smaller critical fields at high ($B > 5$ T) fields [144].

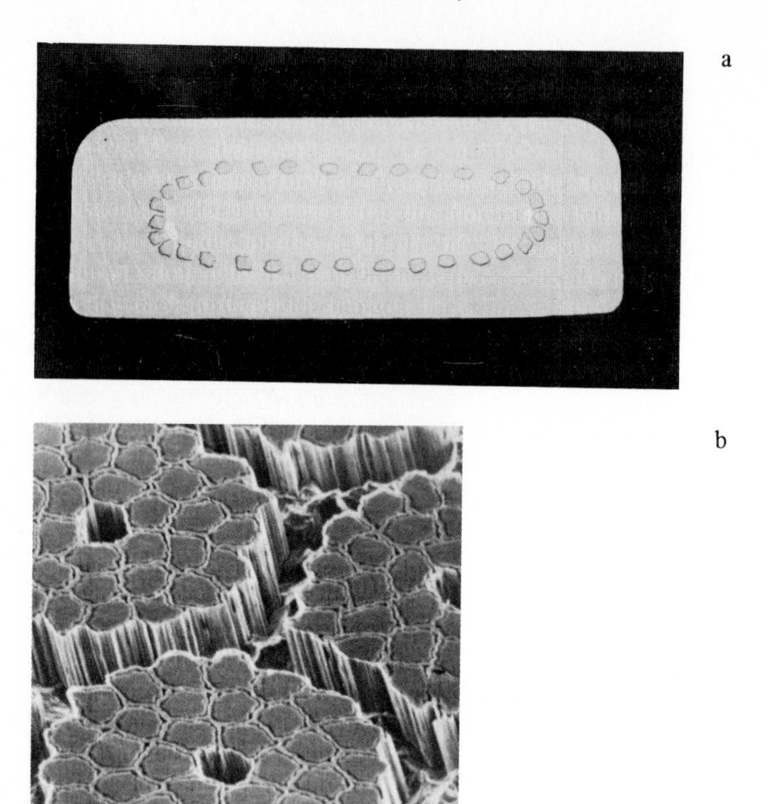

Fig. 141. a) Very overstabilized Nb-Ti conductor in Cu matrix. Seven such bands electron-welded at the short sides were employed as conductors for the bubble chamber magnets at CERN. I_c at 5 T is 110 A. b) Filament conductors of NbTi, scanning electron micrograph. The superconducting material is embedded in a matrix of Cu and CuNi. The copper had been removed for electron microscopy by deep etching. (My thanks are due to Dr. Hillmann of Vaccumschmelze GmbH for providing this micrograph. See also Ullmanns Encyklopädie der technischen Chemie, 4th Ed., Vol. 22, p. 349, Verlag Chemie, Weinheim 1982.)

them in the case of aluminium-coated conductors. A closer discussion of such questions of detail does not seem pertinent at this point, since the advances that are being made can rapidly make all quantitative information pointless.

We will, therefore, only refer to two further developments in the field of suitable conductor materials, namely to Nb_3Sn as a representative of the superconductors with particularly high transition temperatures (see Section 1.2) and particularly high critical fields (Fig. 77) and to "multicore" wires with very many fine superconductors, separated by a normalconducting matrix.

The Nb-Ti alloys have a maximum critical field of ca. 13 T at $T = 0$ and $I = 0$. In coils with a reasonable current load it is not possible to produce fields of greater than ca. 10 T. Nb_3Sn, whose maximum critical field ($T = 0$ and $I = 0$) lies well over 20 T can be employed for higher fields. It has been a magnificent technological achievement to develop wires from this brittle material that have a diameter of ca. 1 mm and contain

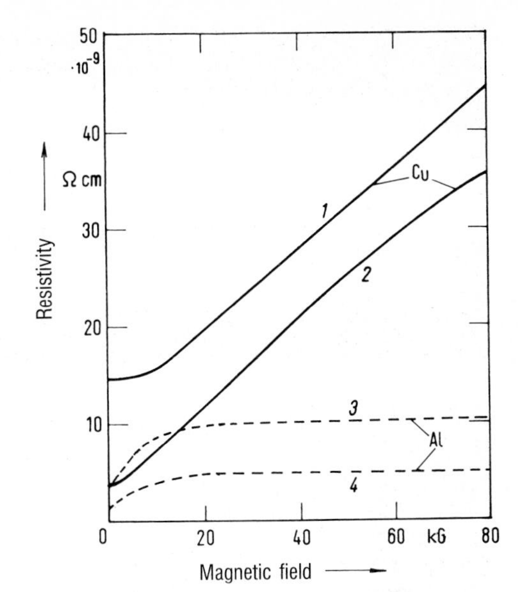

Fig. 142. Change of the specific electrical resistance of Cu and Al in a magnetic field. Cu 1: $\varrho^* = 118$; Cu 2: $\varrho^* = 462$; Al 3: $\varrho^* = 720$; Al 4: $\varrho^* = 1470$; $\varrho^* = (R_{273} - R_n/R_n$; R_n = residual resistance at 4 K (after [144]).

several thousand fine Nb_3Sn threads embedded in a metal matrix. Niobium rods are inserted into tin bronze (Cu+Sn) and the whole bundle is drawn out to the desired diameter. The Nb_3Sn is then prepared in a tempering process that follows (heating to a temperature between 700 °C and 800 °C) when the Sn from the bronze reacts with the Nb. Cross-sectional pictures of this "multicore" conductor are very similar to that illustrated in Fig. 141b for the NbTi conductor. V_3Ga (Fig. 77) also possesses very good properties.

The impressive success achieved in the development of these conductors with Nb_3Sn and V_3Ga makes it very likely that suitable processes will also be developed for the manufacture of useful conductors from the new oxides. It should be remembered in this context that the first usable Nb_3Sn conductors consisted of layers which were deposited on Nb or steel bands and covered with normal conductors. Figure 143 illustrates such a band

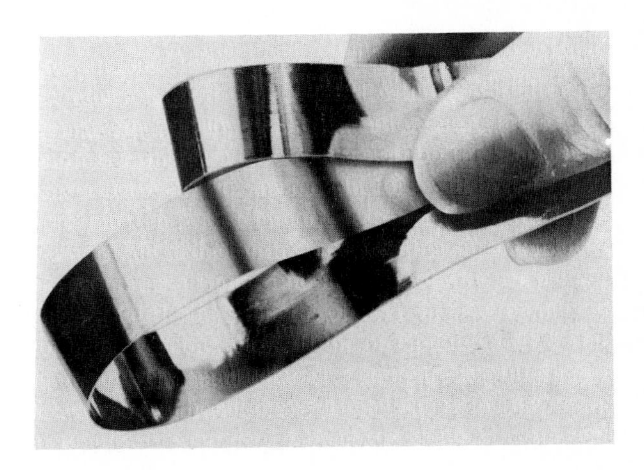

Fig. 143. Nb_3Sn on steel or niobium substrate (reproduced by kind permission of General Electric, USA).

and gives an impression of its flexibility. The first generation of conductors incorporating the new superconductors could also take the form of such bands.

The more advanced conductor configurations are made up of multicore conductors, since these possess "internal" protection against flux jumps. Systematic investigations have revealed that the tendency to flux jumps decreases as the diameter of the superconducting wire is reduced. So that wire bundles, where each of the separate superconductors is very thin, possess this "intrinsic" stability which appears to be very desirable for magnets where the field is to be altered with time. A slight twisting of the multicore wire allows a further improvement of its properties in a variable field. Conductors based on NbN (T_c = 15–17 K [XVII], [7]) are under development at the present time. Here the NbN layer is deposited on a carbon fiber of great breaking strength [263]. The individual fibers can be assembled into a cable. Since the NbN layers also have very high critical magnetic fields excellent conducting material is produced in this way. To summarize this consideration of stability, it should be emphasized once more that it is not just the properties of the superconductor but also the efficiency of cooling that has a decisive effect on the stability. The mechanical strength of the windings is another important factor. The large forces that are set up in the windings at high fields can lead to small displacements. Experience has shown that windings that are not well immobilized mechanically are a cause of instability and, hence, of a large degradation effect.

9.1.3 Coil protection

Even when stabilized materials are employed and the coil is properly constructed events can occur (e.g. gas leakage into the cryostat vacuum) which lead to the coil becoming normalconducting. The magnetic field then collapses and the whole of the energy stored in the magnet is converted into heat. This energy is considerable in the case of a large coil. A magnetic field of 50 kG in a volume of 1 m^3 contains a stored energy of 10^7 Ws (ca, 2.8 kWh). If this energy is transferred into heat in an uncontrolled manner when normal conductivity begins then the coil will be completely destroyed.

Various processes can occur to bring this about. On the one hand, instantaneous heating can lead to local melting of the material of the coil because the induction processes when the field collapses produce very large currents. It may also be mentioned in passing that the induction processes can also destroy the whole cryostat. For if large eddy currents are set up in the walls of the cryostat then the large forces that occur may exceed the allowable mechanical stress. All these points must be given careful consideration in the design of superconducting magnets.

If the electrical resistance of the coil is very large in the normal state the current that is produced will remain small so that destruction by ohmic heating will be avoided. But, in this case, the collapse of the field will create large voltages that can lead to electrical breakdown between the windings of the coil.

In order to avoid such catastrophic results of an unforeseen transition of the coil to normal conductivity it is necessary to incorporate protective devices, particularly in the case of large coils. The purpose of such protective devices is to decouple the stored energy from the coil as rapidly as possible. This can, in principle, be achieved in various ways. Figure 144 illustrates three possibilities on the basis of equivalent circuit diagrams [260].

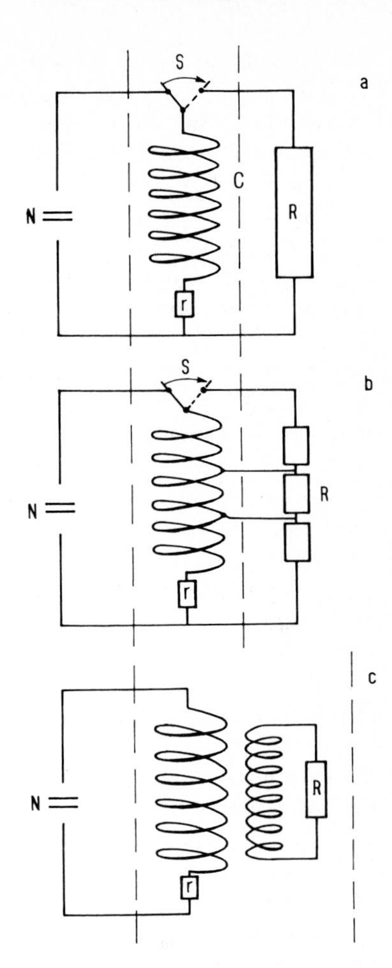

Fig. 144. Examples of electrical protection of coils. N = current supply, r = internal resistance of the coil, R = protective resistance, S = switch, which is automatically activated when the breakdown occurs. The broken line surrounds the components in the He bath.

An obvious method is shown in Fig. 144a. The magnetic coil is connected to an external resistance R. If this resistance R is large compared with the internal resistance r of the coil then in the case of a field collapse the greater part of the stored energy E, namely the proportion $ER/(R + r)$ will be converted to heat in the external resistance. This method has the advantage that only a little liquid helium is evaporated in the cryostat. Its disadvantage lies in the high voltages which occur during the field collapse because R has to be large compared with r. Coils that are protected by this method have to be particularly carefully protected against electrical breakdown between the windings.

An improved modification of this method is illustrated in Fig. 144b. Here the coil is split up into separate regions, each of which is connected to its own protective resistance. This means that the voltages that occur are also divided up. This avoids extreme peak voltages.

In special cases it can be best to decouple all the stored energy from the coil inductively (Fig. 144c). To do this it is necessary to surround the coil with a closed conductor of low induction. Then during the field collapse the majority of the energy conversion will take place in this conductor, perhaps a cylinder of copper. However, in order to provide good induction coupling it is necessary to bring this conductor very close to the coil, i.e. in the helium bath. The disadvantage is now that all the energy is dissipated in the helium bath which leads to very vigorous boiling. Under normal bath conditions about 250 liter helium gas are produced per kWs which corresponds to the evaporation of ca. 0.35 l liquid. Such induction decoupling is not employed in the case of large coils storing thousands of kWs energy. The advantage of the method is that large voltages are not produced in the windings.

We will restrict ourselves to these few examples, since the aim here is merely to illustrate the principles involved. Every large coil has to be individually optimized together with its cryostat as a whole.

The large forces which occur in coils for high magnetic fields constitute a considerable problem. Thus, the forces that occur between the two coils of the bubble chamber magnet that was built for the Argonne National Laboratory amount to 450 tonnes at the full field strength of 1,8 T. This force amounts to 9000 tonnes for the bubble chamber magnets (Fig. 148) of the European Organization for Nuclear Research (CERN) in Geneva. This makes very massive support structures necessary and these must be so designed that no supercritical forces, e.g. because of eddy currents, are induced if the field collapses.

The forces that force the windings of cylindrical coils outwards are also very large for high magnetic fields. This radial tension is ca. 300 kp/cm^2 when the internal diameter of the coil is ca. 50 cm and the magnetic field 5 T. The coil construction must be able to resist these forces and, on the other hand, it must not affect the magnetic field. If it is possible to accommodate the mechanical supports within the cryostat then a constructional solution is not too difficult. But considerable problems must be solved if the purpose for which the magnet is required demands that the support elements be outside the cryostat. The very massive structures necessary on account of the large forces involved result in a large heat flow into the cryostat. The designer attempts to arrange the support elements so that the materials are primarily stressed in tension in order that their cross sections may be kept as small as possible.

These few examples are only intended to provide an impression of the design problems that occur and must be solved in the construction of reliable superconducting magnets with high field strengths.

9.1.4 Applications of superconducting magnets

9.1.4.1 Research magnets

A large number of small superconducting magnets are already produced commercially as standard products with field strengths up to 8 T. Some of these coils are illustrated in Fig. 145[1)]. They find many applications, particularly in solid state physics. We will discuss, by means of just one example, what the advantages of high field magnets are.

Nuclear magnetic resonance investigations are an important method of structure determination in organic chemistry. Here the substance being investigated is brought into a

1 In the recent past superconducting coils for fields up to 20 T (!) have been offered commercially.

Fig. 145. Laboratory magnets with superconducting coils (reproduced by kind permission of Siemens, Research Laboratories, Erlangen).

magnetic field. Because of directional quantization the nuclei of hydrogen atoms are only able to take up two orientations in the magnetic field, namely parallel and antiparallel. These two orientations differ in their potential energies. The parallel orientation has a lower potential energy, i.e. work has to be performed to displace the moments from their parallel orientation. The energy difference between the two orientations is proportional to the magnetic field:

$$\Delta E = 2\mu B \qquad (9\text{-}1)$$

μ = magnetic moment[1]; B = magnetic field.

The hydrogen atoms of an organic molecule are distributed between these two orientations. The energetically more favorable energy level is somewhat more heavily occupied. Figure 146 illustrates this distribution schematically [264]. If this system is now irradiated with electromagnetic radiation with the quantum energy $h\nu = \Delta E$ then this radiation will excite transitions between the levels with energies E_1 and E_2. Since the lower level is more densely occupied more transitions will take place from the lower to the upper level than in the reverse direction, i.e. radiation will be absorbed. If the magnetic field is maintained constant and the frequency ν of the radiation varied then a typical absorption signal is obtained at the resonance frequency $\nu_R = \Delta E/h$ (Fig. 146b).

The crucial factor is that the hydrogen nucleus does not just see the external field B_e but also its environment. This causes the resonance frequencies of the hydrogen nucleus in different chemical environments, e.g. as CH_3 or CH_30 groups of an organic molecule,

1 More precisely the component of the magnetic moment in the direction of *B*.

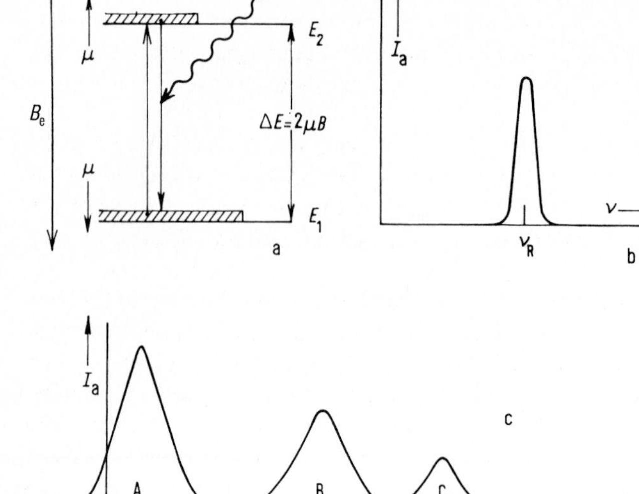

Fig. 146. Schematic representation of nuclear spin resonance. a) Term scheme, b) absorption signal, c) spectrum of ethyl alcohol C_2H_6O. Lines A, B and C represent the protons of the CH_3, CH_2 and OH groups of the C_2H_6O molecule. The splitting shown was obtained at 30.5 MHz with a resolution of ca. 10^6 (after [264]).

to differ somewhat. This frequency displacement is known as the "chemical shift". Thus, it is possible to identify the environments of hydrogen nuclei using nuclear magnetic resonance methods. This method is of great importance for the determination of the structures of organic molecules.

Now the resolution of the lines is difficult or impossible when the breadth of the lines is greater than their distance apart on the frequency axis. All that is then observed is a large, unresolved signal. Apart from the instrumental constants (e.g. the homogeneity of the magnetic field) the width of the lines is determined by the interaction of the nucleus being observed with more distant groups in the molecules. This interaction leads to the absorption line for a particular group of protons being split up into a number of narrower lines and, hence, having a certain half-width.

High magnetic fields can bring about an appreciable improvement in the resolution here. The chemical shift is proportional to the external magnetic field, i.e. the lines of the various proton groups move further apart as the magnetic field increases. The splitting as a result of interaction with neighboring groups is independent of the external field B_e, i.e. the half-width associated with these interactions remains unaffected even at high fields. This makes it possible to separate the lines of differing groups at high field and to determine their chemical shifts.

The high fields offered by superconducting magnets, therefore, offer advantages. However, it is necessary in the case of high resolution nuclear magnetic resonance spectrometers to maintain the magnetic field very homogeneous in the region of the sample. Every field inhomogeneity increases the half-width of the line. R. Richard of the University of Oxford in cooperation with the Oxford Instruments company have succeeded in constructing superconducting magnets for 7.5 T whose field inhomogeneity in the sample volume (a sphere of 1 cm diameter) is less than 2×10^{-7} [265]. With these magnets it is possible to resolve nuclear resonance lines whose frequencies only differ by ν = 0.05 s^{-1}, with a measurement frequency of 2.7×10^8 s^{-1}. This results in the enormous resolution of 5×10^9 *). The fact that it is now possible to construct such magnets so that they consume less than 1 liter liquid helium daily for cooling purposes means that they have virtually completely displaced conventional iron magnets in the field of high resolution spectroscopy.

Nuclear magnetic resonance has found a very promising application in the field of medicine in recent years in the form of NMR tomography (NMR for *N*uclear *M*agnetic *R*esonance) (see Phys. Blätter *39*, 84 (1983)). Here the nuclear spin resonance described in Fig. 146a is employed for the investigation of human tissue. This provides entirely new diagnostic evidence.

While the high resolution nuclear spin resonance basically exploits the shift in the resonance frequencies of protons (other nuclei can also be investigated) as a result of their differing chemical environments (chemical shift) (Fig. 146b), in NMR tomography it is the strengths of the resonance signals that are employed as a measurement of the density of the protons. In addition, it is possible to choose the relaxation times of the nuclear spins as the parameter to be measured. A relaxation process was described in Section 8.2. The relaxation times [1)] are determined by the environment of the protons and can be employed diagnostically.

In order to localize the nuclear spin signals, i.e. to produce a map of the signal height or the relaxation time, the constant magnetic field B_e is overlaid with a locally varying magnetic field, e.g. a field that increases in the X direction. Depending on the resonance frequency it is then possible to localize the position of the nuclei being detected on the X axis. Appropriate variation of the direction of the axis allows compilation of "pictures" of the proton density and the relaxation times. A considerable computational effort is naturally required as it is in X-ray tomography.

Superconducting coils are employed for the creation of the required magnetic fields. This has opened up an entirely new field of application for superconducting magnets. Instruments are being built today which can investigate the whole body of a patient. NMR tomography is of particular importance for the investigation of pathological changes in the brain.

This diagnostic method is just at the beginning of its career. But it is already particularly evident that tumors have a significantly different relaxation time. Since, in contrast to X-ray tomography, this method does not involve ionizing radiation the investigations can be repeated frequently without causing harm to the patient and it is, hence, possible to follow the development stages. A very impressive description of this new process is to be found in a publication by Siemens AG, Bereich Medizinische Technik, Erlangen, Order No. M-RA5/1624-01. The new superconductors have considerable application potential here [266].

---

* The high resolution was demonstrated with a 10% solution of *o*-dichlorobenzene in CCl_4.

1 A distinction is made between two relaxation times τ_1 and τ_2. τ_1 is a measure of the time that a system of nuclear moments requires to achieve equilibrium orientation in a magnetic field. τ_2 is related to the phase of the preceding nuclear moment. Both relaxation times can be employed diagnostically.

A further important possibility for the application of superconducting magnets in solid state research will only be mentioned here. The best methods for the determination of the quantum states of electrons in metals and semiconductors employ magnetic fields. In these fields the electrons rotate in circular paths as a result of the Lorentz force. A measurement of the frequency of this movement provides information concerning the characteristic parameters of the electronic system, e.g. concerning the effective mass of the electrons [1]. However, this rotational frequency can only be measured if, put in a simplified manner, the electrons make at least one rotation within a mean free path, or, in other words, in the time between two collisions. Since the circular paths are large in small fields it is only possible to carry out such experiments on very pure samples with large mean free paths. The application of the high fields made possible by superconductors shortens the circular paths and makes it possible now to investigate substance with smaller mean free paths. This also makes it possible to introduce scattering centers deliberately and then to study their effects on the electron system. Superconducting magnets are being increasingly employed for this purpose too.

Until now we have spoken of possibilities within solid state research. High energy physics opens up a wide field of applications for superconducting magnets, in particular, for very large ones. In this field particles are now accelerated to several 100 GeV (1 GeV = 10^9 eV). These highly energetic charged particles, e.g. protons or deuterons, have to be kept in their tracks by means of suitable magnetic fields. To simplify the matter somewhat, this is all the easier the stronger the magnetic field which is available. In particular, increasing magnetic fields make it possible to reduce the diameter of circular accelerators while maintaining the energy constant. Or, on the other hand, the diameter can be kept constant and the energy of the particles increased by increasing the strength of the magnetic field. The large accelerator at Geneva has once again been equipped with conventional magnets. But it appears certain that later generations of circular accelerators will have to employ superconducting magnets to guide the particles. A large accelerator equipped with superconducting magnets is already in operation at the Fermi laboratory in Chicago. The deviating magnets of the new accelerator ring HERA at DESY in Hamburg are also constructed with superconducting coils.

Superconducting magnets can also be employed to guide the particles after they have left the accelerator. In this case it is not necessary to change the field with time. Such a magnet is illustrated in Fig. 147. Magnets of this type are known as quadrupole magnets; in them two extended pairs of coils (ca. 1 m long) generate cross magnetic fields. The maximum field is 3.5 T which achieves the magnetic field gradient of 0.37 T/cm required to focus the particles. The coils are immersed in a cryostat which has a thermal opening of 12 cm diameter for the steel tube containing the beam. This magnet illustrates very vividly that superconducting magnets are built today even for special purposes.

In view of this development in the construction of magnets the question arises as to whether the cooling, which, after all, always requires helium, does not constitute a basic

---

1 A quantum condition for the motion of the electrons in a magnetic field B allows only certain paths and also leads to characteristic oscillations of the magnetic susceptibility and the electrical resistance as a function of $1/B$. These oscillations are known as the DeHass-van Alphen effect and the Shubnikov-DeHaas effect.

Fig. 147. Superconducting quadrupole magnet with Nb_3Sn winding. Length 1 m, warm diameter 12 cm, maximum field 3.5 T, field gradient 0.37 T/cm (reproduced by kind permission of Siemens, Research Laboratories, Erlangen).

obstacle. This can be answered with no. The employment of very small refrigeration aggregates to screen the He bath allows the construction of magnets that only require to be topped up with helium every 15 months. The ability to cool with nitrogen would naturally bring advantages with it.

Bubble chamber magnets are another example of the application of superconducting magnets in high energy physics. In a bubble chamber the path of a high energy particle is made visible by the formation of bubbles along its path through a liquid, as a result of local heating caused by the ionization process [1)]. An applied magnetic field deflects the path of the charged particles as a result of the Lorentz force. The amount of deflection can be used to calculate both the charge of the particle as well as the momentum of the particle if its charge is known. Very high magnetic fields are required to produce a measurable deflection of the path of the particle at the very high momenta of the highly energetic particles.

A very large superconducting bubble chamber magnet has been built for the European Organization for Nuclear Research in Geneva, CERN. Figure 148 illustrates one of the

---

1 The effective cross section for ionization of a highly energetic particle decreases greatly with increasing energy. It is, therefore, necessary to employ liquid chambers in order, as a result of the higher density, to produce sufficient ionization along the tracks for them to be visible.

Fig. 148. Pancake coil of the superconducting bubble chamber magnet for the European Nuclear Research Institute in Geneva. Details in text.

40 pancake coils during construction. The magnet has an internal diameter of 4.7 m. The two coils each made up of 20 pancake coils have a length of 1.5 m each and are mounted 1.05 m apart. The magnetic field is 3.5 T and the force produced between the coils is 9000 tonnes. The magnets are driven by a current of 5700 A. The weight of the overstabilized conductor having a length of 60 km is 100 tonnes, of which 3 tonnes are superconductor. These few figures provide a very good idea of the size of the project.

Finally, it must be mentioned that hybrid magnets have been developed for the generation of constant magnetic fields of very high intensity. Here a field of ca. 15 T is generated by means of a large superconducting coil. A conventional copper coil (electromagnet) can be mounted inside the superconducting coil to provide a further 15 or 20 T. This makes it possible to provide fields of 30 T without too much consumption of electrical power [267]. It is possible to reach magnetic fields of almost 20 T with superconducting hybrid coils (outer part of coil NbTi, inner part of coil Nb_3Sn) [328].

9.1.4.2 Magnets for energy transformation

Very great efforts are being put into nuclear fusion, i.e. the fusion of 4 hydrogen nuclei to a helium nucleus, such as takes place spontaneously in the hydrogen bomb, so that it takes place in a controlled manner. The energy involved in this process is very large.

Since hydrogen is available on Earth in virtually unlimited quantities this thermonuclear process could be a future energy source. At the moment the investigations of fusion are all in the field of basic research. The difficulties that must be overcome to achieve controlled fusion are enormous. In order to set up a process that will yield excess energy it is necessary to heat the hydrogen gas to at least a few ten million degrees. This hot plasma, that practically only consists of hydrogen nuclei[1] and electrons, can naturally not be enclosed in material vessels. However, since they are charged particles their paths can be deflected in a magnetic field. For this reason it is possible, in spite of their high velocity, to hold the particles in a reaction volume by the use of sufficiently high magnetic fields of suitable geometry. The magnetic fields required are so high that it is certain that they can only be generated economically with superconducting magnets.

The "magnetic bottle" is a particularly simple construction. The principle is illustrated in Fig. 149. The crucial point is that the magnetic field at the ends of the coil increases greatly. This is achieved by supplementary coils. The paths of the particles are corkscrew-shaped, the diameter becomes ever smaller as the ends of the coil are approached. In this path the energy of the motion along the length of the coil decreases and that perpendicular to the axis of the coil increases. This can be understood in the following manner: The ring current, that passes through the field lines of the rotating particles, has a magnetic moment that is antiparallel to the magnetic field. This follows from Lenz's law. The ring current, therefore, behaves like a diamagnet, i.e. in an inhomogeneous field it experiences a force in the direction of regions of lower field strength. Since the field of the magnetic bottle increases towards the end of the coil the electrons are repelled from these ends, their motion is braked and finally reversed into a motion in the direction of the middle of the coil. We speak of a magnetic mirror. The larger the field that can be reached and particularly the field gradient at the end of the coil the better is the enclosing effect. Only superconducting magnets offer the hope of producing the necessary stable field strengths and gradients.

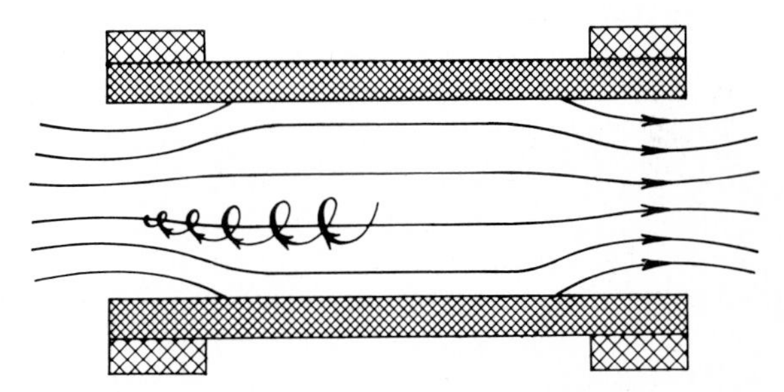

Fig. 149. Schematic representation of a magnetic bottle. The thickly drawn curve gives an impression of the path of a charged particle.

Unfortunately the simple geometry of the arrangement illustrated in Fig. 149 does not yield a stable enclosing effect. It is true that we can hold the charged particles back from the ends of the coil because of the increased field strength. However, this field configuration is not stable towards sideways movements. A configuration has to be found that yields an increasing field strength in every direction. This can be done, with a greater or

---

1 The isotope tritium with a nucleus made up of a proton and two neutrons is employed.

lesser degree of accuracy, by the use of additional windings. Various different possibilities have been put forward. Here we will only describe one, in which a *single* superconducting winding is used. It is a conductor configuration whose geometry is equivalent to that of the seam of a tennis ball. It is, therefore, known as a "baseball" coil [268]. Figure 150 illustrates the geometry of the current path. This consists of a stabilized Nb-Ti conductor. A magnetic field of ca. 2 T is generated at the magnetic walls with a current of 2400 A. In the vicinity of the conductor the field has a strength of 7.5 T. The force that is set up between the loops amounts to 450 tonnes. This force has to be counteracted by the support construction. Figure 150b gives an impression of the size and compactness of this baseball magnet. Since the enclosure conditions for hydrogen atoms are very favorable for this geometry this suggestion is a very original contribution to the solution of the stability question. The most highly developed geometry is the one for a toroidal plasma, what is known as the Tokamak design. For this purpose six very large magnetic coils for a field of ca. 7 T have been built in various parts of the world (Large Coil Task Project) and then assembled at the National Laboratory in Oak Ridge, USA, for torus geometry [261]. The aim is to gather the experience required for the construction of a large fusion reactor with superconducting coils.

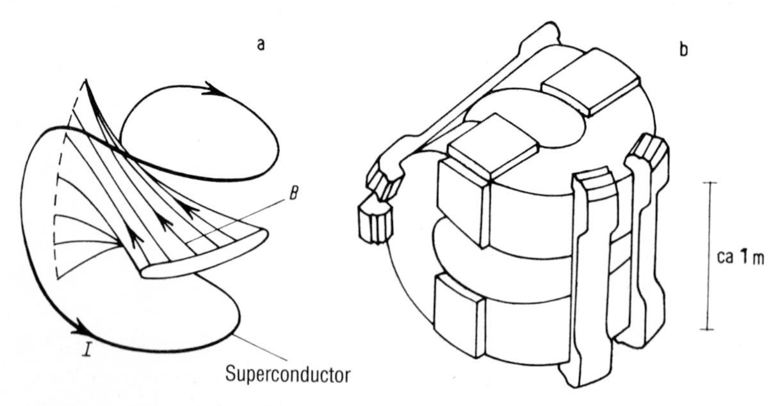

Fig. 150. Baseball coil and support construction (schematic) from the Lawrence Radiation Laboratory, Livermore, California (after [268]).

Another potentiality for the conversion of energy which requires high magnetic fields over large volumes is what are known as magnetohydrodynamic generators (MHD generators). Here gas is heated by burning a fuel or by means of a high temperature reactor and forced through a jet. Alkali metal is added in small quantities in order to increase the concentration of ions. This jet of hot gas, containing a concentration of ions and electrons which depends on its temperature, now passes through a volume in which as high a magnetic field as possible is maintained. This causes the ions and electrons to be deflected and creates a potential difference between the collecting electrodes which is completely analogous to the Hall voltage, which is well-known in metal and semiconductor physics.

If the separation of the collection electrodes D is 0.1 m and the velocity of the ions 400 m/s then a voltage of 200 V is created for a field of 5 T. The electrical power which can be tapped naturally depends on the volume and density of the gas stream. Calculations have shown that at a field of 5 T over a length of 6 m and with a diameter of 1 m it is possible to generate ca. 40 MW (= 4×10^4 kW). Such MHD generators could be employed as peak load power stations. They have the advantage of a very rapid start-up. How far superconducting magnets, which would naturally be preferable on account of their high field strength, could be employed in practice is a question of economics.

Another advantage of MHD generators with superconducting magnets is the relatively low weight per MW output. This advantage can be decisive for particular applications, e.g. for aerospace applications. For this reason very detailed discussions are in progress concerning MHD generators for such purposes, with outputs of 3–5 MW [260].

9.1.4.3 Magnetic suspension and shielding

The levitated magnet in Fig. 7 is a simple example of magnetic suspension, here of a permanent magnet, with the aid of a superconductor. It is naturally also possible to make a superconductor hover in a suitable magnetic field. The diamagnetic behavior [1] of the superconductor is the crucial factor for finding a stable configuration. The field exclusion effect in the Meissner phase means that every surface of a superconductor, with an external field B_e parallel to it, experiences a pressure p directed to the interior of the superconductor, which is given by [2]:

$$p = \frac{B_e^2}{2\mu_0} \tag{9-2}$$

p = pressure in N/m^2; B_e = external field in $V\ s/m^2$; $\mu_0 = 4\pi 10^{-7}$ V s/A m.

An external field of 0.1 T creates a pressure of 0.4 N/cm^2.

We will first limit ourselves to the consideration of devices where the superconductor is in the Meissner phase, i.e. where the exclusion effect is complete. Such devices are particularly suitable for bearings where the magnetically suspended body is to reach high speeds of rotation. For frozen-in fluxes must be avoided here as they would lead to losses during rotation. If magnetically frozen-in fluxes do not occur an extremely friction-free suspension can be achieved up to the very highest rotation velocities. The upper limiting factor for the frequency of rotation then becomes merely the mechanical strength of the rotating material.

The basic principle of such suspensions will become evident from the geometrically simple example of a planar plate. Such an arrangement is illustrated in Fig. 151. The

---

1 Normal conductors can also be held in suspended state. Here, however, it is necessary to employ a high frequency field. The eddy currents set up by the fields displace the external high frequency field as an induction current according to Lenz's law, that is they cause diamagnetic behavior. This levitation method is employed in the crucible-free melting of metals.

2 A small displacement Δx of the surface S perpendicular to the field B_e yields the displacement energy $\Delta E_B = S \Delta x\, B_e^2/2\mu_0$. $B_e^2/2\mu_0$, thus giving the force with which the field B_e attacks the superconductor. This, however, is just the pressure which is applied.

superconducting plate A is lowered towards the superconducting coil C in which a persistent current has been set up. The superconducting coil C maintains the magnetic flux Φ_d which passes through the opening in the coil. In the ideal case this flux must entirely pass through the ring-shaped gap between the plate and the upper surface of the coil. There it will create a field B_d, which is almost homogeneous when the distance d is small and when $R \gg b$, and which is given by:

$$B_d = \Phi_d / 2\pi R d . \tag{9-3}$$

The force on the superconducting plate will, therefore, become:

$$F = \frac{B^2}{2\mu_0} S \simeq \frac{\Phi_d^2 b}{2\mu_0 2\pi R} \frac{1}{d^2} . \tag{9-4}$$

As the distance d between the plate and the coil is reduced the repulsive force increases in proportion to $1/d^2$. It is limited by the requirement that $B_d < B_c$ for the plate, since this would go over into the intermediate state for $B_d > B_c$ (see Section 5.1.4).

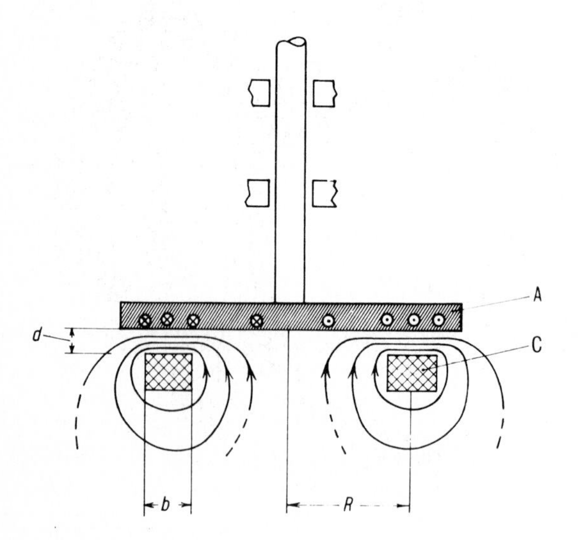

Fig. 151. Schematic representation of a magnetically supported superconducting plate. A = superconducting plate, C = superconducting coil with frozen-in flux.

Using this principle it is possible to discover several designs which yield stable suspension in 1, 2 or 3 dimensions. Figure 152 only illustrates two of these possibilities [269]. The device in Fig. 152a yields stable suspension with respect to one dimension. Figure 152b shows the generalization into three axes. The suspended rotor is now a sphere [270].

In the investigations of the stability of hot plasmas, such as are necessary for controlled fusion, devices have been built which are known as spherators, in which a torus-shaped plasma is suspended magnetically from a superconducting ring carrying a persistent current of several 10^5 A.

Magnetically suspended trains are also being studied seriously, particularly in Japan and the Federal Republic of Germany [271]. The following principle is employed: Mag-

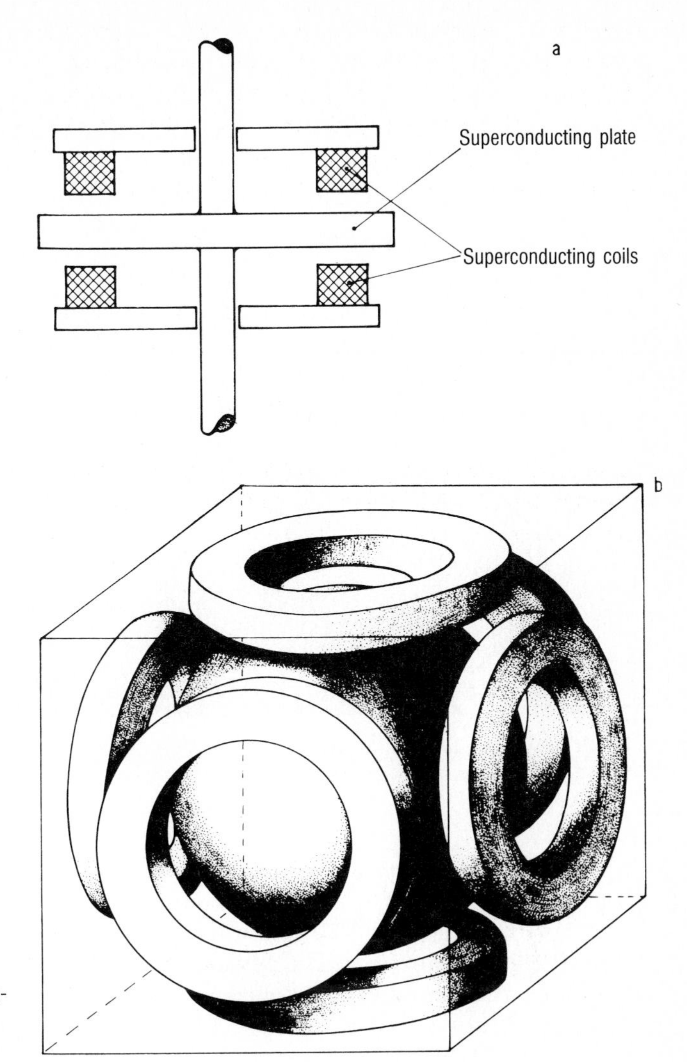

Fig. 152. Two examples of magnetic support using superconductors (after [269, 270]).

nets producing a sufficiently strong downwardly directed magnetic field are mounted in the individual cars of the train. The train track is made up of rows of loops of a good conductor, e.g. Al wire. There are no repulsive forces between the track and the train when it is at rest.

It is necessary at first to impart a certain velocity to the train by conventional means. When in motion there are repulsive forces between the magnets and the conducting loops. Eddy currents are induced in the conducting loops which generate a magnetic field according to Lenz's law, which opposes the primary field – here that of the train magnets. This field produces a repulsive force. In analogy with the previous discussion we can say that the conductor loops of the track behave diamagnetically on account of the eddy currents. Figure 153 illustrates the research train of the Siemens company on the

Fig. 153. Research vehicle of the magnetically levitated train of Siemens AG on the closed circuit track at Erlangen. (Kindly supplied by Dipl. Ing. C. Albrecht with the permission of Siemens.)

loop of track constructed for research purposes in the grounds of the Siemens AG research laboratories in Erlangen. The car contains superconducting magnets which maintain it hovering over the metal track when it reaches the required velocity.

This magnetic suspension has appreciable control advantages over the hovercraft supported on a cushion of air. Hovercraft would also get into difficulties trying to pass through tunnels. It is certainly possible to adopt this technique using conventional magnets since the fields necessary amount only to ca. 10 kG. Project studies involving this possibility have reached a very advanced state. How far superconducting magnets would be economic is a matter for detailed calculation [264]. The decision has been made in the Federal Republic of Germany to follow the conventional path.

What is certain is that the need for transportation in heavily populated regions, such as Europe, Japan or parts of the United States, will make very necessary the provision of rapid means of transportation with speeds in the 400 to 500 km/h range. Magnetically suspended trains could be this means of transport.

The magnetic separation of mixtures is a further possibility for the commercial application of superconducting magnets. The selection of ferromagnetic materials from ores with the aid of strong magnetic fields is a well-known technique. Superconducting magnets would allow very much higher field strengths and very high field gradients. This would make it possible to separate effectively materials which only vary slightly in their magnetic properties. This technique has great potential, particularly with the new superconductors [266].

Finally a project will be mentioned here where it is intended to employ the high fields created by superconductors to shield space ships from the high energy radiation of space [272]. The paths of the charged particles, protons and electrons would be deflected in the magnetic field. Naturally, for a given field, the deflection is the less the higher the momentum of the particle. For particles of the same type we can also say the higher the energy. For this reason the magnetic field of a short coil can only shield a certain volume, which extends, like a tube, round the windings of the coil and which becomes ever smaller as the energy of the particles increases. Should extended space voyages be undertaken, even if only close to the earth, then this screening possibility is likely to play an important role.

9.1.4.4 Motors, generators and energy storage

Coils with iron cores are employed to generate the necessary magnetic fields in conventional electric motors and generators. This limits the practical maximum field strength that can be employed. The intensity of the magnetic field determines, in its turn, the output that can be obtained from a machine of a given volume.

Superconducting magnets allow the generation of very much higher magnetic fields. This means that superconducting machines can be made much smaller for the same output. This can be of crucial importance for certain applications. The cryoscopic complication must naturally be taken into account in every estimate of economy. However, careful calculations have revealed that the low temperatures necessary do not have a serious effect on the competitiveness of rotating machines even today.

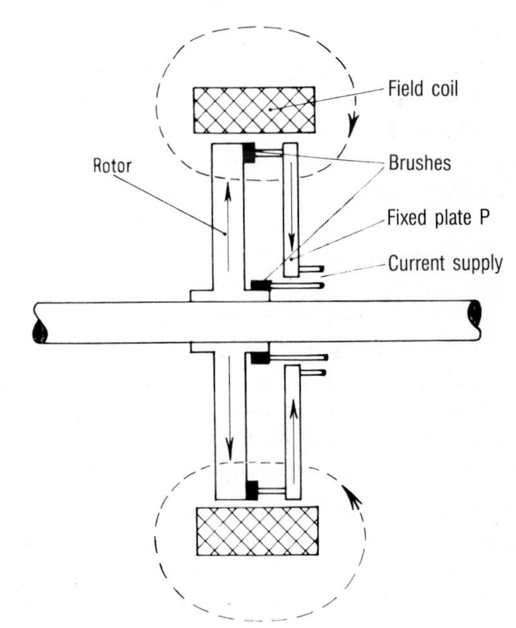

Fig. 154. Schematic representation of a unipolar machine.

An electric motor with an output of ca. 2500 kW (3250 h.p.) has been completed and tested. The motor rotates at 200 r.p.m. It is a unipolar machine with only the field-producing coils superconducting. The principle of such a motor is illustrated in Fig. 154. The rotor is made up of wedge-shaped segments which can carry a radial current. Contact brushes serve for electrical contact. The radial current experiences a Lorentz force in the magnetic field. This force provides the necessary turning moment for the rotor.

This simple principle of motor construction suffers initially from the difficulty that it requires large currents at low voltages for operation. A rapid estimate can reveal this. For a rotor 2 m in diameter and rotating at 200 r.p.m. in a homogeneous magnetic field of 3.5 T the potential drop between the axis and the periphery of the rotor is only ca. 36 volts. So that, in order to produce the desired output of 2500 kW, a current of ca. 6×10^4 A would be necessary and this would be difficult to transport via frictional brushes. However, the division of the rotor into segments makes it possible to switch the individual segments in series and, thus, to sum the potential drops. It is also possible to include several rotors and to connect them in series. The current is then carried back to the axis from the periphery by the fixed disc P (or segments). This method of construction has, at the same time, the advantage that the forces are not carried by the field coil but by the fixed plate P [273].

Figure 155 illustrates the cryostat housing for the field coil of the 2500 kW motor. The coil is made up of a fully stabilized Nb-Ti-Cu band and is designed for a field strength

Fig. 155. Cryostat housing of the field coil of a superconducting motor. Output: 2500 kW (after [273]). (Reproduced with the kind permission of Dr. Appleton.)

of 3.5 T. The motor was tested at full power in November 1969 [265]. It is employed to drive a cooling water pump at Fawley power station, England.

Superconducting motors with higher outputs are under consideration for ship's drives. The decisive factor here is that such motors can be built with a weight per kW output that is about three times less than that of conventional motors.

The situation is different as far as generators are concerned because high speeds of rotation are desired here. Detailed studies are also under way here to investigate methods of realizing this. Such generators, in combination with power transmission via superconducting cables, could open up new perspectives. It may be said, in general, that the exploitation of superconductivity in energy generation and transmission must be considered and evaluated in terms of complete systems and not just in terms of the individual components. The new multicore conductors with their high stability to flux jumps have made the subject of alternating current motors and generators topical again. Generators with outputs of up to 100 MW are under construction or being planned in several places. In the Federal Republic of Germany a detailed study is under way for a 3000 MW generator. A 100 MW generator with helium-cooled rotor has been built by Siemens as a component of an substantially larger aggregate [274].

Large superconducting coils have been proposed for energy storage [275]. Such stores must be compared for their utility with batteries of capacitors, on the one hand, and fuel cells, on the other. When this is done it appears that, at the moment, superconducting energy stores have great advantages only in the case of a few special applications. For instance, the weight per unit of energy stored is appreciably less than for batteries of capacitors but appreciably greater than for fuel cells. On the other hand, the stored energy can be released in a very short time as it is with batteries of capacitors. Applications of this type require very careful calculation for each individual case.

9.2 Flux pumps

Superconducting magnets are frequently driven by large currents of up to several thousand amperes. The supply lines for such currents cause very large heat inputs into the helium bath. In order to avoid these losses methods have been developed of inducing persistent currents in the coils for the generation of the field. The devices employed, which are generally known as flux pumps, do not have any particular technological importance but they are very instructive as examples of induction phenomena in conducting circuits without any ohmic resistance. Some variants of these flux pumps will, therefore, be discussed in this section.

We will begin with an arrangement that allows amplification of the magnetic flux by compression into a smaller area [276]. Figure 156 shows the principle of the method. A magnetic flux Φ is frozen into the complete opening of the superconducting cylinder by say cooling it in a magnetic field. The field strength B is given by:

$$B_1 = \Phi/A_t \quad (9\text{-}5)$$

where $A_g = A_1 + A_2$ is the total area of the hole. Now so long as the field does not reach a supercritical value the flux in the hole will remain constant. So if it is possible

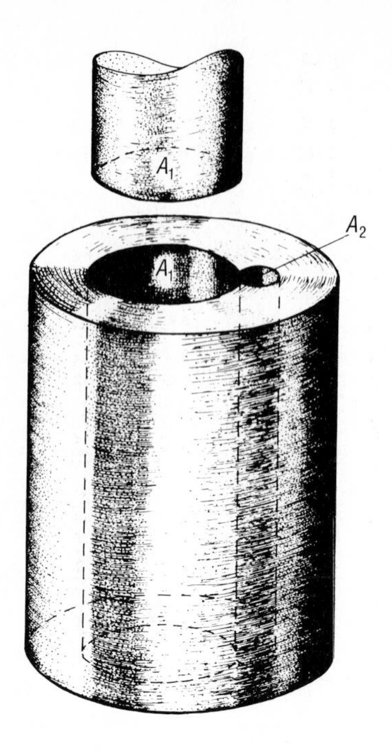

Fig. 156. Device for a single flux compression. Cylinder and piston are superconducting. The fields produced may not exceed the critical values for flux conservation.

to reduce A_g to A_2, then the field B_1 will rise to the value B_2 where B_2 is given by:

$$B_2 A_2 = B_1 (A_1 + A_2) \tag{9-6}$$

or

$$B_2 = B_1 \frac{A_2 + A_1}{A_2} . \tag{9-7}$$

It is necessary to employ hard superconductors in order to achieve really high fields. As we saw in Section 7.1 the magnetic flux penetrates the superconductor and can migrate through it if the pinning capacity is exceeded. Flux constancy does not apply any longer in this case. The magnetic flux can escape from the cylinder. This single flux compression in superconducting cylinders has not acquired any technical importance[1]. But it leads us immediately to one possibility for the flux pump. For we can repeat the compression if we employ two separate openings 1 and 2 whose topology (single or two-fold connecting) can be changed by suitable "switches" [277].

---

1 Flux compression in normalconducting cylinders is employed for the creation of large magnetic fields. A cylinder, which is as good a conductor as possible, e.g. made of copper, is placed in a magnetic field B and, thus, a flux Φ produced by an external coil. The conducting cylinder is then rapidly reduced to a much smaller diameter by means of an explosion. The magnetic flux cannot escape on account of the eddy currents created in the conductor during the brief period of compression. It is possible to create magnetic fields of several 100 T for a short period (a few times 10^{-6} s).

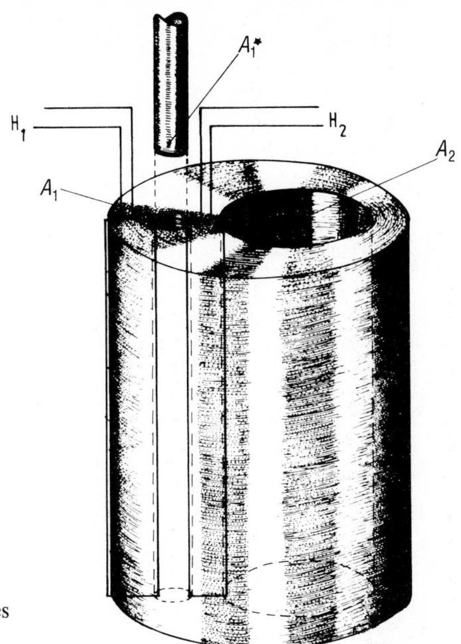

Fig. 157. Arrangement of a flux pump for repeated flux compression. H_1 and H_2 are heating coils which allow interruption of the multiple connection of holes 1 and 2.

This principle is illustrated in Fig. 157. We wish to fill the large opening 2 with magnetic flux by repeated flux compression; the two heating windings H_1 and H_2 can abolish the two-fold connection for openings 1 and 2 with the areas A_1 and A_2, by making a part of the mantle normalconducting. These flux pumps now work in the following way. We freeze a field B_0 into the two holes by cooling in an external field B_0. Then in the first stage a superconducting piston is inserted into opening 1, whose cross section A_1^* is somewhat smaller than A_1. The flux is compressed into the annulus and the field, therefore, increased. Now H_2 is switched on so that openings 1 and 2 are connected. The magnetic flux "expands" into A_2. Switching off H_2 sets up the two-fold connection for 2 again. This freezes in the new flux. Now H_1 is switched on to abolish the two-fold connection of 1. The piston can be removed and a field B_0 generated externally in opening 1. This flux is captured by switching off H_1 and setting up the two-fold connection again. The flux in 1 can be compressed once again and expanded into 2. A saturation is reached in this simple process when the flux in 2 equals the compressed flux in 1 after introduction of the piston. Such flux pumps can be employed to create a field of ca. 2.2 T over a volume of several cm^3 [277].

We have dealt with this simple, almost self-evident flux pump in some detail because its principle is employed in a whole range of flux pumps. Figure 158 illustrates the construction of a rather simple example schematically [278]. The similarity to the version illustrated in Fig. 157 will be evident immediately. Here we have two superconducting circuits 1 and 2 with switches S_1 and S_2. In circuit 2 there is a coil with a large self-induction L, whereas circuit 1 only has a small self-induction $l \ll L$.

A magnetic flux (indicated by the magnet) is created in circuit 1 with switch S_1 open. Then S_1 is closed. The conservation law for magnetic flux now applies to the completely

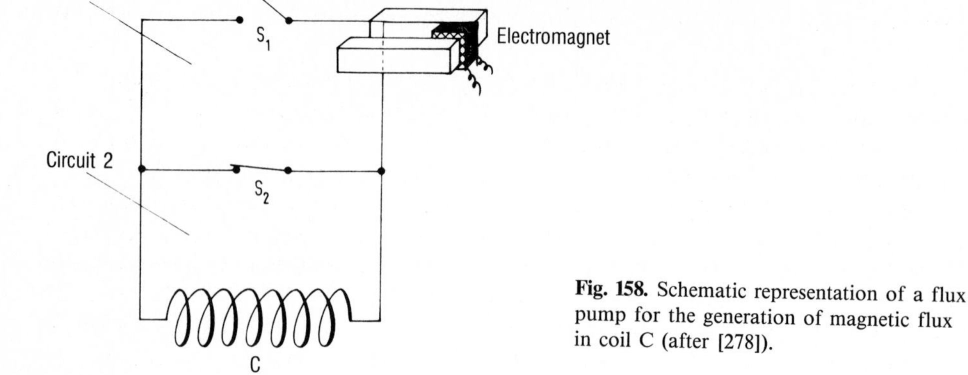

Fig. 158. Schematic representation of a flux pump for the generation of magnetic flux in coil C (after [278]).

superconducting circuit 1. When the magnet is removed a supercurrent i will be set up which will just keep the flux constant. The flux in 1 with current i is given by:

$$\varphi = li\,. \tag{9-8}$$

If switch S_2 is now opened the current has to flow through circuits 1 and 2; you could say that the flux has "expanded" into circuit 2. The flux maintenance, which still applies since a completely superconducting circuit encloses the flux, yields in this first step a current I in circuit 2:

$$I_1\,(L + l) = li\,. \tag{9-9}$$

We designate the quantities for circuits 1 and 2 respectively by means of small and capital letters.

We now close S_2 and, thus, freeze in the flux $\Phi_{(1)} = I_{(1)}L$. When S_1 is open we can generate the flux φ once more in 1 and freeze it in by closing S_1. Expansion into 2 then yields:

$$I_{(2)}\,(L + l) = li + LI_{(1)}\,. \tag{9-10}$$

From this it becomes immediately evident what the current I is at the nth step. For:

$$I_{(n)}\,(L + l) = li + LI_{(n-1)}\,. \tag{9-11}$$

It can be seen directly that I can never be greater than i, since when $I = i$ there is no increase in I when S_2 is opened. The current I approaches the value i as $n \to \infty$ asymptotically. The flux in circuit 2 can then be very much larger than that in circuit 1, namely:

$$\left(\frac{\Phi}{\varphi}\right)_{n\to\infty} = \frac{L}{l}\,. \tag{9-12}$$

We can, therefore, "pump" flux into the coil (circuit 2) by means of such an arrangement. For a given flux φ the value of Φ_{max} will be the greater, the larger the value of L/l.

The other group of flux pumps replaces switches S_1 and S_2 by a superconducting plate, in which a normalconducting region can be set up by the pump magnet [279]. This region can be used like the switches to pump the flux into a completely superconducting circuit. Figure 159 illustrates such a device semischematically.

The bar magnet is introduced from the left underneath and very close to the superconducting plate P. The plate must be constructed of a superconductor whose critical field is smaller than the field at the face of the magnet. The magnet will then create a normalconducting region, through which the flux passes, while the rest of the plate remains completely superconducting. In this manner a magnetic flux is brought into the circuit C_s which contains the coil C. If the magnet is now simply withdrawn downwards then the flux remains in the completely superconducting circuit. A corresponding supercurrent is set up in C_s. The magnetic flux is now virtually limited to the inside of the coil. The opening of circuit C_s and the interior of the coil are related to each other topologically. When the magnet is removed the magnetic flux remains constant but it is displaced to the coil because it is here that the large inductivity L is to be found.

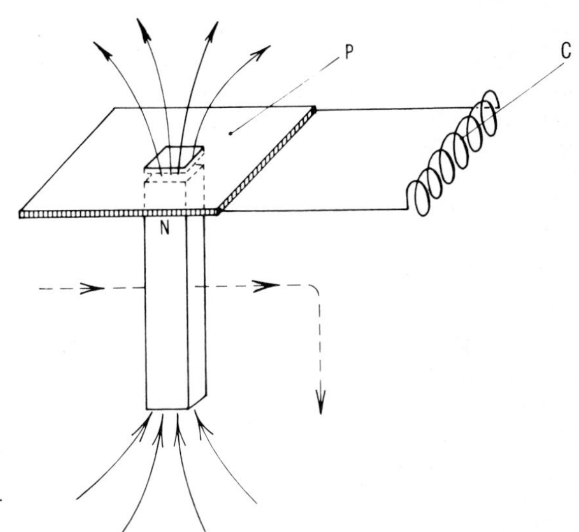

Fig. 159. Schematic representation of a flux pump with superconducting plate instead of the switches. P = superconducting plate (small B_c), C = coil.

But the magnet can also be moved out over the edge of C_s, when the edge will remain fully superconducting, i.e. the wire loop is made up of material with a sufficiently high critical field. This possibility leads to an arrangement where the magnet rotates. The flux in C_s is increased with each rotation. This flux pump too has an upper limit for the flux through C_s. The current that creates the flux in C_s flows through plate P. When passing through the normalconducting region this current has to be divided. At first it flows completely in front of the normally conducting region and after passing through it has to flow completely through the rear. This division of the current path through the plate reduces the current by a certain fraction. So the current can only increase so long as this weakening is smaller than the increase caused by removing the magnet [280]. Here, too, the current approaches its limit asymptotically. Experiments with such flux pumps have confirmed this behavior very well [281]. Technical versions of rotating flux pumps can achieve performances of 10^3 T cm^2/min.

The migrating magnetic field can be created by a fixed coils through which alternating current flows with a displacement in time in order to avoid immersing rotating parts in liquid helium. The same principle is employed as that which is used to create the rotating fields of electric motors. Such flux pumps have also been developed and tested.

Until now the use of flux pumps has been limited to individual cases because the pumping times, especially for large coils, where the application of such pumps would be most attractive, are simply too long. If one were forced for technological reasons, say involved in the manufacture of the superconductor itself, to produce the coils for high fields with only a few or even only *one* turn (hollow cylinder) then it would certainly be impossible to introduce the required current of millions of amperes from outside. Flux pumps would then have to be used to generate the ring current.

9.3 Cables for power transmission

The energy consumption of mankind will continue to increase in the future. Great efforts are being made to find new sources of power for this very reason. We discussed the importance of superconductivity for controlled fusion in Section 9.1.4.2.

As the amount of power consumed increases so do the problems of power transmission. In order to obtain higher efficiencies power stations are being built with ever greater outputs per unit. This power output of the order of 10^3 MW (10^9 watt) then has to be transmitted to the consumer. Here the interconnection often has to extend over very large distances. This is achieved today largely with the use of over-head, high-voltage lines which work at a voltage of 380 kV and more. The subsidiary distribution is then carried out at voltages from 110 kV down to 20 kV.

The increasing consumption naturally makes it necessary to extend the transmission grid. Here it is becoming virtually impossible, particularly in the great conurbations, to continually add over-head power lines. The effect on the landscape and particularly on land use of high voltage transmission lines makes it appear ever more necessary to transfer power transmission to underground cables. A great deal of development work and investment will be required in this field for the future.

Naturally one thinks too of the exploitation of superconductivity for this purpose. A conductor with no ohmic resistance appears, at first glance, to be virtually ideal for power transmission. There is no doubt that superconductivity basically offers a new and attractive possibility. Economic considerations affect this application much more than they do in the case of superconducting magnets. We already have a well-functioning network of over-head power lines, whose capacity could undoubtedly be raised by increasing the transmission voltage while retaining the same number of routes. When it becomes necessary, as it will in the future in many locations, to turn to underground cables as an alternative we have other alternatives, such as oil- or gas-cooled cables or cryocables that contain a normal conductor and are cooled to ca. 80 K with liquid nitrogen. It will be necessary over the next few years to carry out development work specifically aimed at discovering whether the employment of superconducting cables constitutes an economically competitive method for one or more of the voltage levels employed.

Here there are many parameters which require to be varied and subsidiary conditions like the required overload properties which need to be taken into account. Many of these requirements have an immediate effect on the choice of superconducting material. Here we can only refer briefly to some of the questions involved and in this way give at least some impression of the problems involved.

Direct current transmission would undoubtedly be the method of choice from the point of view of loss-free transmission. Today in hard superconductors of relatively low cross section we can transport currents of many thousands of amperes without losses of any sort (see Section 7.2). Such a system of transmission would require transformation from alternating to direct current and vice versa at its input and its output respectively. This demonstrates immediately that such power lines will probably only be economical in the foreseeable future for large distances and very high power ratings. Here, however, they will have to compete with the relatively cheap over-head power lines.

In contrast it is certain that in the future underground cables will have to be resorted to more and more in the large conurbations. A superconducting cable could become very much more readily competitive under these circumstances. But then it would at least initially have to be an alternating current cable in order to allow its connection to the existing grid. Here with alternating current cable the aim must be to make the losses, which intrinsically occur in superconductors too, as small as possible.

All these questions are at present under active discussion in various parts of the world. Constructional suggestions and cost estimates have already been made. It is not yet possible to make a final decision on the question of whether a superconducting cable could be employed economically or not. Too much fundamental information is missing and has yet to be produced. Probably in this question too, as with most radical changes, it will require the testing of a prototype line under normal operating conditions to provide sufficiently precise data for a soundly based economic consideration of the question. The new superconductors could have considerable potential in this field [266].

We will only consider a few examples of the many designs for cable construction that were developed and investigated in the 1960s.

9.3.1 Direct current cables

When direct current cables are employed the transmission can take place with practically no losses. Very small losses occur merely as a result of the load fluctuations and as a result of any residual ac voltage. Gauster, Freemann and Long [282] suggested in 1962 a cable made up of two separate strands which was intended to transmit power levels of 10^4 MW over 1600 km at $\pm$ 75 kV and a current strength of 67 kA per strand. The structure of this cable is very simple; it is made up of concentric tubes (Fig. 160). The conductor suggested was niobium. A conductor thickness of only 0.32 mm would be required to carry the assumed current with a conductor diameter of 67 mm. The use of two strands would provide the reliability required for power transmission. This cable would involve some difficulties in assembly because it would have to be constructed from short units ca. 20 m long and the connections would pose particular problems. All cables involving rigid tubes present such difficulties.

The use of extremely hard superconductors, e.g. Nb_3Sn would allow the current strength to be increased appreciably. So that for a cable of the same dimensions it would be possible to lower the voltage or increase the power transmission. In 1967 Garvin and Martisoo [283] discussed a cable for a load of 10^5 MW at $\pm$ 100 kV and 500 kA over a distance of 1000 km. When it is remembered that the total output capacity of the Federal Republic of Germany is at the moment ca. 6×10^4 MW then projects involving such performance levels acquire a somewhat utopian aspect. They are only intended, at the moment, to demonstrate that should power transmission of this order of magnitude ever be required then superconductivity would offer very large advantages. As a rough measure it may be said that direct-current superconducting cables would begin to be economically competitive at loads of more than 2×10^3 MW and at distances of more than 500 km.

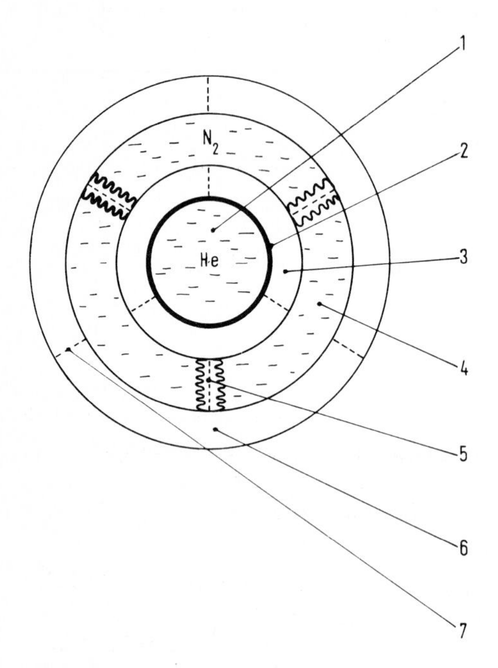

Fig. 160. Schematic representation of the cross section of a superconducting cable. 1 liquid He; 2 Nb conductor on a carrier tube; 3 vacuum insulation; 4 liquid nitrogen; electrically insulating supports; 6 vacuum insulation; 7 support construction with low thermal conductivity (after [282]).

The coolant circulation is naturally of decisive importance for every cable that has to operate at liquid helium temperatures. Detailed suggestions have also been made here. Astonishing as it might seem in view of the extremely low temperatures of the order of 4 K, it has been established that the cooling would not involve any serious technical problems. Cooling with liquid nitrogen would undoubtedly bring advantages here too.

9.3.2 Alternating current cables

Certain residual losses have to be accepted for alternating current cables. They result from the interaction of the electrons, that are not correlated into Cooper pairs, with the alternating electrical field. The heat that is produced must be transported away by the coolant at 4 K. This requires 500 times as much electrical power as the power loss involved. Every effort must, therefore, be made to keep these losses in the superconductor to a minimum. Type I superconductors are very suitable for this on account of their very large screening effects (see Section 5.1.1). Figure 161 illustrates the power losses per m^2 surface as a function of the magnetic field at the surface [284]. If maximum losses of 0.025 W/m^2 are accepted then a critical current load for a conductor 10 cm in diameter is 5500 A for Pb and 16000 A for Nb[1]. As expected the hard superconductors do not offer any particular advantages.

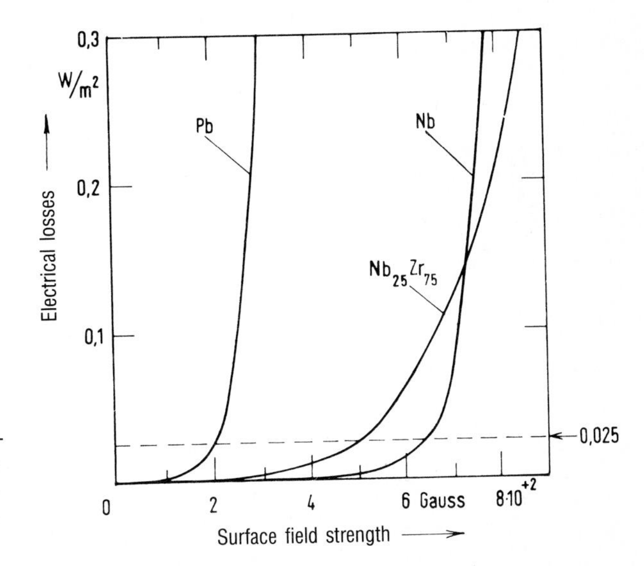

Fig. 161. Electrical losses for alternating current in superconductors as a function of the surface magnetic field strength of the current load (after [284]).

Probably the first suggestion published for a superconducting cable was that of McFee [285] in 1962. He discussed a single-phase alternating current cable for a load of 750 MW at 200 kV. Lead was suggested as the conductor. The division of the conductor into many parallel strands, running within a helium-cooled tube, allowed the current strength to be increased so that the voltage could be reduced to the generator voltage without affecting the load. This would mean that there would be no necessity for transfomation to high voltages to allow power transmission.

Many constructional suggestions involve the use of rigid tubes, which can only be manufactured in relatively short lengths and which have to be joined in a vacuum-tight manner with good contact between the superconductors. A suggestion by Klaudy [286] using corrugated tubes instead of rigid tubes avoids these constructional difficulties.

1 It is only necessary for the superconductor to form a thin surface film.

Such corrugated tubes can be constructed in lengths of up to 1500 m and wound onto cable drums. Distancing pieces of poor thermally conducting materials guarantee the necessary centering. Such a construction is illustrated schematically in Fig. 162. The innermost tube through which liquid helium flows contains many three-phase conductors suitably arranged. The space between tubes 2 and 3 is cooled with liquid nitrogen. The spaces between tubes 1 and 2 and tubes 3 and 4 are for thermal insulation.

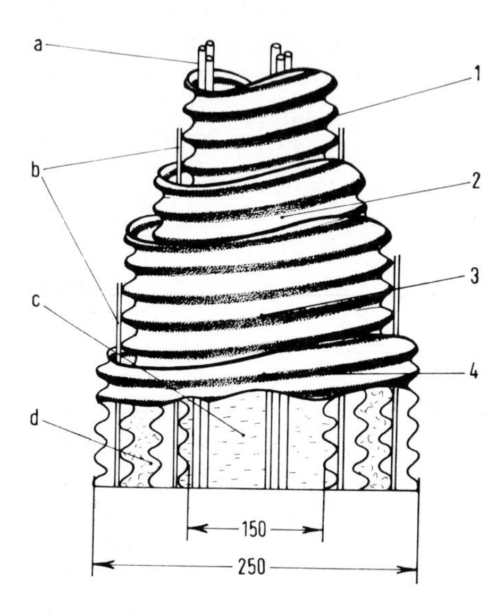

Fig. 162. Schematic representation of a superconducting cable of corrugated tubes. 1, 2, 3 and 4 corrugated tubes; a) three-phase conductor, superconductor Nb or Pb; b) distance pieces of poorly thermally conducting materials; c) liquid He d) liquid nitrogen (after [284]). P. A. Klaudy, who has made many contributions concerning superconducting cables, was one of the first to have actively followed up this idea of modern power transmission.

A cable for a load of 370 MW having a length of 200 km was the subject of a design exercise on these principles. The minimum internal diameter of the innermost tube was 150 mm and the external diameter of the outermost tube 250 mm. The losses at liquid helium temperatures were given as 0.21 W/m. The distance apart of the necessary refrigerating stations was 8 km. The total cost, including refrigerating stations, was calculated at 1.5 million DM per km. This would mean that such a cable would definitely be competitive with a conventional cable to carry the same load, where the costs would be ca. 1.3 – 1.4 million DM per km. P. Klaudy built a 50 m length of such a flexible cable and tested it with great success in the public grid [287].

In addition to all these questions of economics, which naturally play a very great role in cable projects, there is also the fact that conventional constructions can be based on many years of experience while investigations of superconducting cables only began about 10 years ago. The efforts being made at the moment concerning this modern method of power transmission will undoubtedly bring further improvements. Then the testing of prototypes will have to show whether the use of superconducting cables is feasible in view of the many subsidiary conditions involved. Consideration must also be given to whether it would not also be advantageous to operate other components of the power supply system at low temperatures.

9.4 Superconducting switching and storage elements

The phase transition from the normalconducting to the superconducting state can be effected by means of an external magnetic field. Since the transition is accompanied by a change in the electrical resistance it is possible to "steer" currents in this manner. The basic possibility of employing superconductivity for switching elements has been recognized for a considerable time. The first intensive investigations of this possibility were only made at the end of the 1950s as attempts were made to use such circuits in large computer units. The name "cryotron" for such elements comes from this period; it was suggested by D.A. Buck [288] in 1956 and rapidly found general application.

There was initially great optimism concerning the possible application of cryotrons in large computers. It is a particular advantage of such superconducting circuits that they are readily combined with superconducting memory components.

The principle of storing binary information with the aid of superconductors is very simple. A persistent current can be induced in a completely superconducting circuit (Fig. 5). Depending on the direction of flow of this persistent current it can be employed to designate either "0" or "1". It is also possible to assign "1" or "0" to rings carrying or not carrying a persistent current. Now superconducting elements are characterized by particularly low power consumption requirements. Since only supercurrents flow in the stationary state heat is only generated during the switching process. This power consumption can also be kept very low. If 10^6 such elements are each switched 10^6 times per second then the power consumption would not amount to 1 watt. This extremely small heat output allows such elements to be packed very tightly together. This small volume requirement together with the feasibility of constructing the switching elements for the logical connection in the same manner as the memory units offers convincing advantages especially for the construction of very large rapid computers. An improvement in performance (calculating speed and capacity) of a factor of 10 – 100 over that of the largest computers available today would appear possible.

The difficulties lie in the technological field. It was at first impossible to construct suitable switching and storage elements in very large numbers with sufficient reproducibility of the important variables (critical magnetic field, critical current). Since, in contrast, the microminiaturization of semiconductor elements made great strides the interest in superconducting elements was greatly diminished for a period.

It then became clear that the Josephson junction can be designed appreciably more favorably with respect to the critical variables. This led to great efforts in the USA and Japan to develop large superconducting computers. The difficulties involved in the manufacture of highly integrated circuitry with many thousands of individual elements remained so large, however, that this development work was ended in the USA. However, the problem continues to be attacked intensively in Japan. The question of whether the new superconductors will bring advantages for large computers remains, at the moment, open. The cooling costs are not important here. A first application could be the incorporation of superconducting components, e.g. conductors, into large computers that already operated at the temperature of liquid nitrogen. Such computers are commercially available today.

The basic possibility of realizing both logic circuits and memories with superconductors is of such great interest that at least the most important aspects should be discussed here. We will consider just a few of the many possible modifications.

9.4.1 Circuit elements

A wire cryotron was suggested in 1956 by D.A. Buck [288]. It consists of a tantalum wire inserted through a coil of niobium wire (Fig. 163). This cryotron can be operated at 4.2 K, the boiling point of liquid helium. Tantalum has a transition temperature of 4.4 K. The critical field is 40 G at the working temperature, which is only 0.2 K below the transition temperature. The Ta wire can be made normalconducting by means of the magnetic field of the control current in the Nb coil. The Nb coil remains completely superconducting on account of its high transition temperature of 9.3 K.

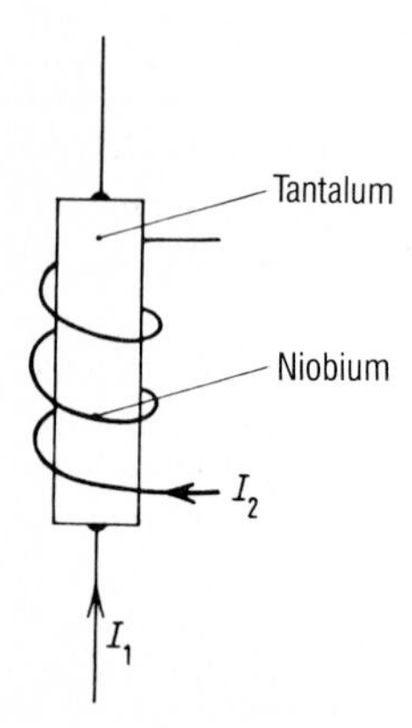

Fig. 163. Wire cryotron. I_1 cryotron current; I_2 control current (after [288]).

A wide range of logic gates can be constructed with such a cryotron. The basic principle is always the same. Several parallel paths containing cryotrons are available for the supercurrent. The desired current path is chosen by making the cryotrons in the other paths normalconducting for a short period. Figure 164 illustrates this principle using a particularly simple example. The two current paths 0 and 1 are superconducting. The distribution of the supercurrent between the two paths is decided by their self-inductivity. If the cryotron Cr_1 is briefly made to conduct normally then the whole of the current I_s will flow through branch 1. Since no potential drop is involved this configuration remains stable until cryotron Cr_2 is made normalconducting as a result of a pulse in its control coil. Such a pulse throws the current I_s to branch zero. It can be seen very clearly that the cryotron fulfils for supercurrent the function that a relay fulfils for a normalconducting circuit. The superconducting or normalconducting cryotron corresponds to the closed or open relay.

It is now possible to "read" the path taken by the current by means of cryotrons Cr_3 and Cr_4. The cryotron that is controlled by the current flowing through this branch is normalconducting and, therefore, gives a voltage signal when a read pulse is applied. The read cryotron in the other branch remains superconducting. It is naturally possible to

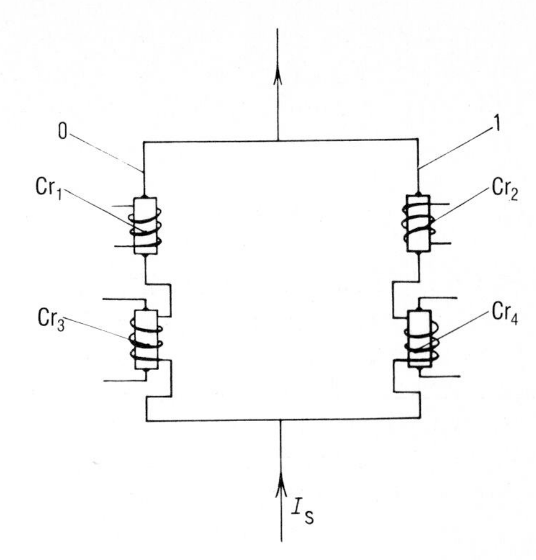

Fig. 164. Current branching with switching and read cryotrons. Cr_1 and Cr_2 switching cryotrons; Cr_3 and Cr_4 read cryotrons.

omit one of the read cryotrons, since, in any case, when there is a current I_s all of the information will be obtained by interrogating one branch. But it is sometimes useful to have a completely symmetrical circuit. We have also seen here that the current through a cryotron can be used to control one or more other cryotrons. This combination of cryotrons provides for great versatility of operation.

However, for this control of one cryotron by means of the current of another it is necessary that the critical current I_c of the tantalum wire be greater than the control current I_{ct} necessary to make a cryotron normalconducting. The two currents I_c and I_{ct} are determined by the critical field B_c of the Ta wire. This is:

$$I_c = 2\pi r \frac{B_c}{\mu_0}; \qquad I_{ct} = \frac{B_c}{\mu_0 n} \tag{9-13}$$

where r = radius of Ta wire, n = number of turns on the control coil and μ_0 = $4\pi \times 10^{-7}$ Vs/(Am).

So that

$$I_c/I_{ct} = 2\pi r n . \tag{9-14}$$

This ratio can be made greater than 1 by suitable design of the cryotron.

Above the broken line Fig. 165 illustrates a flipflop, where currents in one branch alternately increase the resistance in the other by making a cryotron normalconducting. This feedback increases the stability towards accidental changes in current distribution. Information can be written in and read out by a cryotron circuit completely analogous to that in Fig. 164. Logical networks can clearly also be realized with such cryotrons.

The relatively slow switching time is the major disadvantage of wire cryotrons. If the current in the control coil of an element is switched off in such a manner that the Ta wire of another element becomes normalconducting, then the characteristic time τ for this decay process equals L/R, where L is the inductivity of the control coil and R the ohmic resistance of the Ta wire.

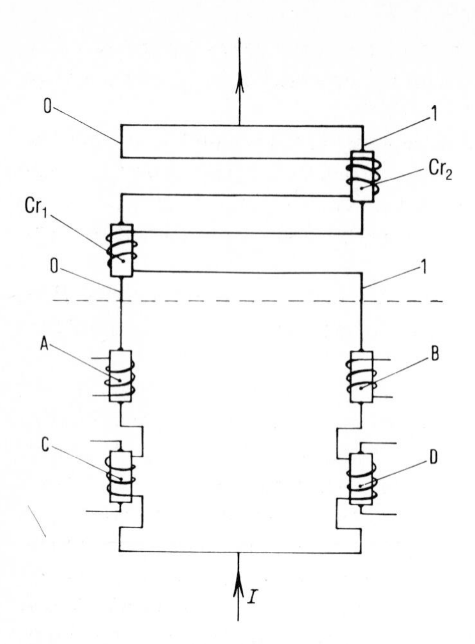

Fig. 165. Current branching with stabilizing feedback. Cr_1 and Cr_2 feedback cryotrons; A and B switching cryotrons; C and D read cryotrons.

The sizes of L and R are determined by the geometry and the material of the wire (here Ta). They are[1]:

$$L \simeq \mu_0 n^2 \pi r^2 l \tag{9-15}$$

$$R = \varrho \frac{l}{\pi r^2} \tag{9-16}$$

where n = the number of turns per unit length, r, l and ϱ represent the radius, length and specific resistance of the wire material.

We, therefore, get for τ:

$$\tau \simeq \frac{\mu_0 n^2 \pi^2 r^4}{\varrho} . \tag{9-17}$$

If we now substitute the expression for the current amplification $K = 2\pi r n$ (see Eq. (9-14)) then τ is given by:

$$\tau \simeq \mu_0 \frac{r^2 K^2}{4\varrho} . \tag{9-18}$$

The wire cryotron first described by Buck had a characteristic switching time of 40×10^{-6} s (40 μs). If it is remembered that 3 to 4 relaxation times τ are required for

1 For the sake of simplicity we will approximate by using the expression for a long linear coil.

a complete switching process, then this makes the switching time several hundred μs. The greater the current amplification K the greater will be the switching time, since a large K requires a coil with many turns, that is a large L.

Switching times of several hundred μs are unacceptable for large modern computers. Here elements with switching times of 10^{-7} to 10^{-8} s, which are several orders of magnitude shorter, have long been routine. Wire cryotrons are, thus, unsuitable for use as switching elements in computers. They find some applications in measurement techniques (see Section 9.5.1).

In order to shorten the switching time it is necessary to make L very small and R as large as possible. It is possible to do this by constructing the cryotron with thin sputtered layers instead of with wires [289].

Figure 166 illustrates a cryotron of this type. It is made up of a layer of tin, which corresponds to the Ta wire and a layer of lead that serves as the control line. The Sn layer is only a few times 10^{-5} cm thick. This ensures a large resistance R for the element in the normalconducting state. The self-inductivity of the two conductor layers and of all the sputtered connections can be sharply reduced by using a superconducting underlayer made of say Pb. The various layers of metal are separated electrically by insulating layers, usually composed of SiO. Every current in the conductors sets up an antiparallel superconducting "mirror" current in the underlayer. This means that the magnetic field is largely limited to the small gap between the conductor and the underlayer so that the self-inductivity becomes small.

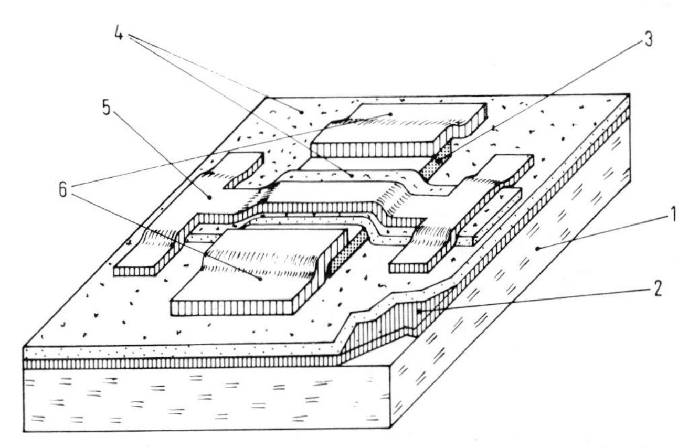

Fig. 166. Layer cryotron (semischematic). 1 base; 2 superconducting layer (Pb); 3 cryotron layer (Sn); 4 insulating layer (SiO); 5 control lead (Pb); 6 electrodes at the Sn layer (Pb). The layer thicknesses are greatly exaggerated for clarity.

The restriction of the magnetic field also brings about another important advantage in that the mutual interaction of the switching elements is kept small. The elements can, therefore, be packed very closely together. Finally the superconducting base has a further advantage. The mirror currents screen the vertical components of the magnetic field from the base plate itself and, thus, create a magnetic field parallel to the base plate and the conductors. This means that the current in the conductor layers is distributed homogeneously across their width. This increases the critical current for the tin layer which is very much to be desired from the point of view of current amplification. Switching times of less than 10^{-7} s have been achieved for such layer cryotrons. The circuit elements can be kept very small [I, p. 1315].

In spite of considerable efforts it did not prove possible in the 1960s to develop methods that allowed the manufacture of such layer cryotrons with the required reproducibility. The development of these circuit elements was stopped for this reason.

Nowadays Josephson junctions offer excellent possibilities for extremely fast, low-power switching elements. It was Juri Matisoo (IBM) in 1965 who first pointed out this application of Josephson junctions. The arrangement of the films corresponds largely to that in Fig. 166. The only difference is that the oxide film between the layers, e.g. films of Nb, is made sufficiently thin to allow the passage of a Josephson direct current. Such Josephson junctions are placed as in the cryotrons in Fig. 164, in the branch for the supercurrent I_s. The switching process consists in passing a current pulse through the control conductor which passes over the Josephson junction – but completely electrically isolated from it. The magnetic field of this current suppresses the Josephson current (Fig. 38). Since even very small fields can suffice to "switch off" the Josephson current it is possible to use very small control currents. These elements can be used to carry out all the functions described for the cryotron. The switching times attainable with Josephson junctions lie below 10^{-10} s. This makes them at least equal in this respect to the best semiconductor transistors. They are very much superior to the transistors, however, with respect to their power requirements, since Josephson junctions require several orders of magnitude less power than transitors of comparable switching time.

Since Josephson junctions are made up of thin layers which have to be separated by a very thin ($d < 3$ nm) oxide layer, the question must be asked as to whether it is possible to manufacture such junctions in large numbers, particularly in the miniaturized form necessary for integrated circuits. This question was at first answered very optimistically. It proved possible to define the thickness of the sensitive core of the element, the oxide layer, exactly to 0.1 nm. Attempts were made to microminiaturize Josephson junctions based on Nb (T_c = 9.3 K), on NbN (T_c = 14 K to 17 K) and even on Nb_3Ge (T_c = 23 K) [290]. However, there has been no success as yet in the sufficiently reliable manufacture of these elements in large numbers as would be required for integrated circuits. The question is still open as to whether superconducting computers will become important in the future. The Josephson elements are superior to the best transistors with respect to switching time and power consumption [291].

9.4.2 Storage elements

The arrangement of cryotrons illustrated in Fig. 164 can, in principle, store the information "0" or "1". A memory ought to be able to preserve the information stored in it without the supply of current from outside. This is indeed possible as we will see by exploiting a persistent current set up in the loop formed by branches 0 and 1 in Fig. 164. The information "1" or "0" is represented by the presence or absence of a persistent current in the loop. The two possible directions of the persistent current could also be defined as representing "1" and "0".

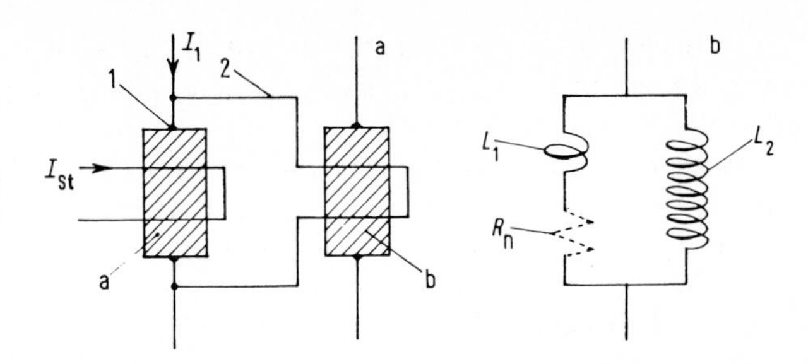

Fig. 167. a) Memory cell with read cryotron; b) equivalent circuit diagram; L_1 and L_2 self-inductions; R_n normal resistance of cryotron.

Figures 167a and b represent such a memory element and its equivalent circuit diagram. A supercurrent I_1 is distributed to branches 1 and 2 according to their inductivities L_1 and L_2. Cryotron a is superconducting. Since $L_1 \ll L_2$ almost the complete current flows through branch 1. If you now drive cryotron a into the normal state by a trigger input I_{ct} then the whole supercurrent is transferred to branch 2 and stays like this even when cryotron a becomes superconducting again. We have already discussed the stability of supercurrents in superconducting grids in connection with the circuits illustrated in Fig. 164.

If current I_1 is now switched off a persistent current I_s remains in the loop. This is not difficult to understand. While the cryotron is normalconducting the current I_1 which has been completely displaced to branch 2 creates magnetic flux through the loop which is not now completely superconducting. When I_1 is switched off this flux is maintained by a persistent current in the now completely superconducting loop. It is possible in this way to use suitable current pulses I_1 and I_{St} to store a "1". A further pulse I_{St} destroys the persistent current and, thus, sets up a "0". Cryotron b can be employed to read information or control further cells.

Figure 168 illustrates a memory element which is constructed of Josephson junctions. The Josephson junctions G_1 and G_2 (G stands for "gate") are arranged completely symmetrically in a loop. The control line, with which a "1" or a "0" can be written, runs over both junctions but is completely electrically insulated from them. Here in this device a very interesting possibility is exploited. The control current I_{St} can be so adjusted that it only generates a magnetic field of sufficient strength to switch off the Josephson current when it flows in the same direction as the current in the loop branch. When the control current and the loop current flow in opposite directions the two magnetic fields mutually weaken each other. The junction is unaffected, i.e. the Josephson current is not "switched off".

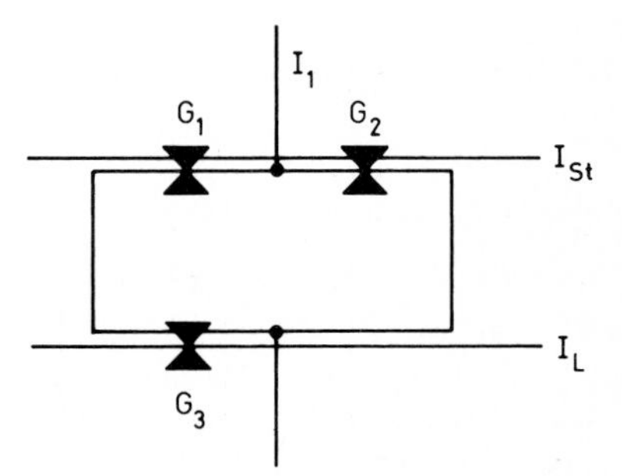

Fig. 168. Superconducting memory cell with Josephson junction (after [292]).

In our device (Fig. 168) this means that a control pulse from left to right only switches junction G_2. In contrast when the direction of the current pulse is from right to left it is junction G_1 that is switched. When, for example, junction G_1 is switched the total current I_1 flows in the loop with G_2 and creates flux Φ in the loop. When the control current I_{St} is switched off the current I_1 in the now completely superconducting loop will maintain this flux because of the persistent current. A persistent current is set up in a clockwise or counterclockwise direction depending on whether the current is suppressed in G_1 or G_2 – that is the information "1" or "0" is stored. Now we have to be able to interrogate, "to read" this information. A further junction G_3 is employed for this purpose; this is located over but insulated from a branch of the loop. Depending on the direction of the persistent current this junction will allow an interrogation current to pass, say from left to right, or suppress it. The asymmetry which results by superimposition of the junction current and the external current, in this case the reading current, is exploited here too [1)].

For a large memory it is naturally necessary to arrange many such elements in as simple a manner as possible, so that every individual cell can be interrogated. Figure 169 illustrates a method of interrogating every cell without altering the information it has stored in it [293]. We will describe the method of operation for cell I as an example. Lines A and B serve to write information into the loop abcd. The information is written in by the method described above, by means of pulses of current through lines A and B. Cryotrons K_2 and K_3 are employed to read out the information. For this purpose the reading current is passed through line C. The desired cell of those associated with line C is selected by means of line D, which also carries a supercurrent. This current makes

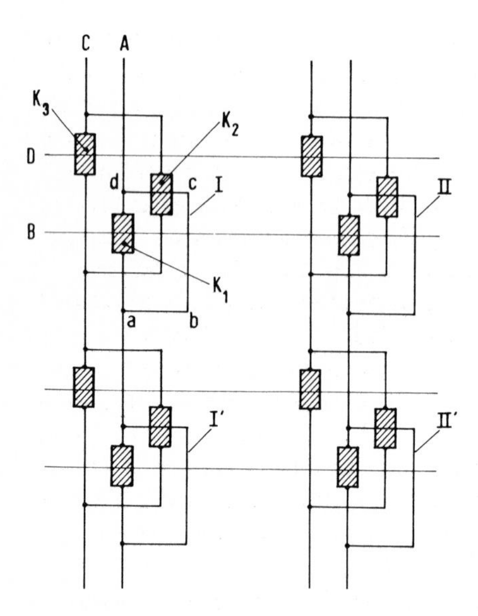

Fig. 169. Device consisting of 4 superconducting cells in a memory with read-in and "nondestructive" out possibilities for each cell. A and B read-in lines; C and D out lines.

1 It should be mentioned that it is very easy to set up logical networking with several separate control conductors over one contact. The logical configuration *and* can be realized by choosing the size of the current pulse so that only the sum of the currents produces a sufficiently large magnetic field to abolish the Josephson current in the junction.

cryotron K_3 normalconducting. Cryotron K_2 will be superconducting or normalconducting depending on whether the information stored in abcd is "0" or "1". If it stays superconducting (information "0") then there is no ohmic resistance in line C. If, on the other hand, it is a "1" that is stored then K_2 is normalconducting and a supercurrent cannot flow through C. A potential drop occurs in line C, which can now be employed further. It makes a good exercise to create the same memory arrangement with Josephson junctions.

In recent years there have been great advances in the manufacture of Josephson junctions in microminiature circuits [291]. Nevertheless, it has not yet proved possible to construct a prototype for a large superconducting computer. The requirements for the homogeneity of many thousands of Josephson junctions on one chip are extraordinarily difficult to fulfil.

A decisive advance has been made recently in the development of an active element with amplifier properties [294]. This element is known as the quiteron (quasiparticle injection tunneling effect). It is made up of two tunnel diodes arranged on top of each other (see Fig. 44). The tunneling current of one diode can be controlled by the injection of electrons from the other diode (see also Section 8.2. p. 190).

There is no question that the Josephson circuit and memory elements will have a considerable importance in the future at the very least for special functions such as measurement technology at low temperatures.

9.5 Applications in measurement technology

It is understandable that the unique electrical and magnetic properties exhibited by superconductors should also have specific applications in measurment technology. A superconducting galvanometer was built very early on; this had a very high sensitivity (ca. 10^{-11}V) on account of its extremely low internal resistance. The enormous temperature dependence of the electrical resistance on the transition curve from the normalconducting to the superconducting state allows the construction of enormously sensitive radiation meters (bolometers). Superconducting amplifiers and modulators can be constructed with the aid of cryotrons and the considerable difference in the thermal conductivity in the normalconducting and superconducting states at temperatures well below T_c (Figs. 54 and 55) allows the construction of superconducting "thermal valves".

However, the most remarkable advance has been made in recent years with devices which apply, on the one hand, flux quantization (see Section 3.2) and, on the other hand, the Josephson effects (see Section 3.4) for the purposes of measurement. What is virtually a breakthrough in measurement technology has taken place here in that it has proved possible to improve the accuracy of measurement of the intensity of magnetic fields by at least four orders of magnitude in one single step. This has opened up a new field for the determination of the magnetic properties of materials.

We will end our treatment of the applications of superconductivity by considering this fascinating method and with it, at the same time, our path through a macroscopic quantum phenomenon.

9.5.1 Amplifiers and modulators

The basic possibility of using a cryotron to achieve current amplification has already been discussed in Section 9.4.1 (see Eq. (9–14)). Figure 170 illustrates the basic unit of a cryotron amplifier designed by Newhouse and Edwards [295]. The two cryotrons a and b are brought to the desired operating point [1] by currents I_0 and I_1. This is a bridge circuit. A signal current I_{Sig} in the direction shown reduces the control current and with it

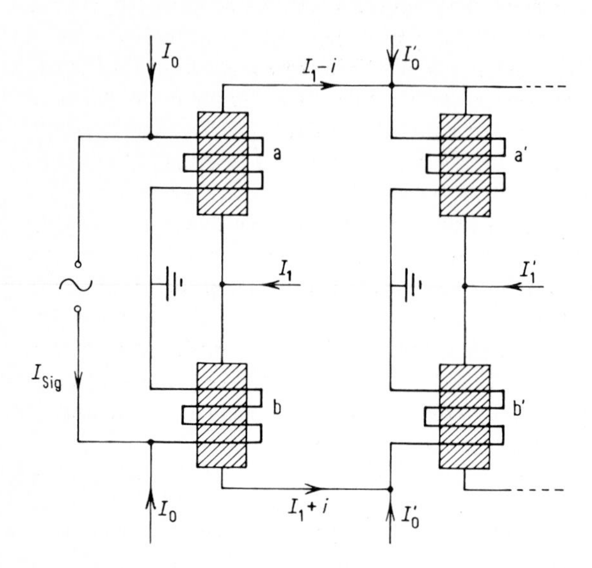

Fig. 170. One stage of a cryotron amplifier. I_0, I_0', ... control currents in the control line; I_1, I_1', ... control currents in the main line; I_{Sig} signal current; i current increase as a result of amplification (after [295]).

the resistance R_a of the cryotron a, while the control current and resistance of cryotron b are increased. This imbalance of the bridge can now be fed to a normal amplifier. The amplification for one step is then:

$$K^* = \frac{\partial I_1}{\partial I_0} . \tag{9-19}$$

This amplification effect can be made large if the control line is passed backwards and forwards over the layer many times (multicrossover cryotron). Then, however, the device has a large self-inductivity, as with the wire cryostat with a large number of turns in its control coil.

Such basic elements can be arranged one after the other in a cascade-like manner, in that the imbalance of the first stage is fed to the control of the second and so on.

At low frequencies ($\nu < 10^3\ \mathrm{s}^{-1}$) these amplifiers exhibit unusually large fluctuation phenomena. The large fluctuations result from the formation of bubbles of helium gas which are produced as a result of the ohmic heating of the cryotrons. The fluctuations

---

1 Both cryotrons must have a resistance. That is the operating point must be on the transition curve of the cryotron.

can be filtered out by a low pass feedback which can only be passed by low frequencies and reducing the amplification to one in this range[1]. The upper frequency for amplifiers of this type is determined by the joule heat. It is about $10^6\ s^{-1}$.

It is very frequently advantageous in low temperature investigations to chop small direct current signals at liquid He temperatures and to transform the voltage still at low temperature to such a level that it can be detected without effort at room temperature. Such a device employing a cryotron and designed by Templeton [296] is illustrated in Fig. 171. With the aid of the direct current the cryotron is maintained at an operating level so that a superimposed alternating current makes the cryotron normalconducting (by addition of the currents) once every cycle. This alters the current generated by the signal voltage. This modulated current is then fed to a transformer installed in the He bath whose output voltage is measured by means of a phase-sensitive amplifier. The cross talk of the alternating voltage, required for control of the cryotron with the signal, limits the sensitivity of this simple device. It is possible to remove this in several ways. For instance, a reversing switch, made up of cryotrons, can be incorporated in the signal circuit [297]. This makes it possible to change the phase of the signal without changing the control current. The phase-sensitive amplifier, thus, separates the signal from the interference. Another possibility is to use the alternating current alone in the control line of the cryotron, without any direct current component, and to make it so large that the cryotron exhibits resistance twice per cycle [298]. A signal with double the frequency can then be separated readily from the interference. It is possible to achieve voltage sensitivities of 10^{-11} V with time constants of 1 s with such arrangements.

The important prerequisite for the high voltage sensitivity is naturally the low resistance $< 10^{-5}\ \Omega$ of the measurement circuit, which allows enough current to flow to be detected even at very small voltages. Insofar as a sufficiently small resistance of the input circuit can be obtained by using superconductors a wide range of sensitive detection methods can be employed to measure the current. We will return to this for the voltage measurement methods involving the Josephson effect.

We will discuss one of these methods briefly here. Instead of the cryotron in Fig. 171 a ring core made of a ferromagnetic material is used. Figure 172 illustrates the principle of such a device [299]. Winding W_1 represents the input circuit for the signal and is

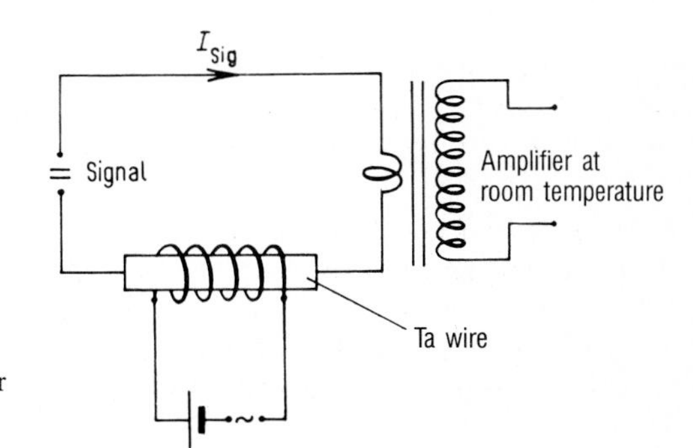

Fig. 171. Cryotron modulator (schematic) (after [296]).

1 Experiments in which the cryotrons were packed in Al foil and only cooled by gas demonstrated that the effect actually was due to the influence of gas bubbles, for the large fluctuations then ceased.

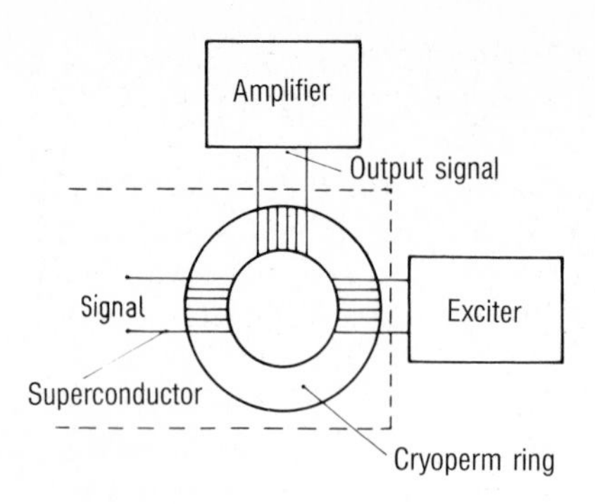

Fig. 172. Magnetic amplifier with superconducting input circuit (after [299]).

made up of superconducting wire to produce a very low resistance. One winding is employed for sine wave excitation whose amplitude should be so large that the ring is brought to magnetic saturation. As long as no direct current is flowing in the input coil the induction signals in the output circuit are fully symmetrical about the zero line. However, they become asymmetrical as soon as a signal current flows. The most convenient way of detecting the induced signal is by means of the first harmonic of the exciter frequency using a phase-sensitive amplifier. It is proportional to the signal current.

A device of this sort employing Cryoperm (Vacuumschmelze AG, Hanau) as the magnet material can yield sensitivities of 10^{-12} to 10^{-15} V with an input resistance of 10^{-5} to 10^{-8} Ω. For this purpose two magnetic rings are employed in order to compensate the basic signal to zero by excitation in opposite senses. The first harmonic (and all even-numbered multiples of the fundamental frequency) are amplified in this manner.

A cryotron adjusted to an operating point near the transition curve is an extremely nonlinear element for an alternating current in the control line. Such elements are suitable for the construction of rectifiers. Various superconducting rectifiers have been suggested. They can, for example, be employed to compensate for the losses in superconducting magnet coils [300]. Such rectifiers basically fulfil similar functions as the flux pumps described in Section 9.2. Like the flux pumps their use is likely to be restricted to specific problems.

9.5.2 Radiation meters, thermal valves and electromagnetic cavity resonators

Superconducting radiation meters are of the bolometer type. These are instruments where it is the change in resistance of an electrical conductor under the influence of the *heat* of radiation that is employed for detection. Such radiation meters are especially employed for the "far infrared" wavelength range, from 50 μm to 1 mm. They have the advantage of a very simple construction compared with other instruments sensitive in this region. The electrical conductor usually just consists of a thin, free-standing foil or a layer sputtered onto a thin support film. The sensitivity of such a detector depends ultimately on the thermal noise of the device. The time constant is determined by the thermal capacity of the detector strip and its thermal coupling to the surroundings.

Superconducting bolometers [301] offer three very material advantages. Thermal fluctuations are very low indeed since they work at low temperature. The noise level of a bolometer at 4 K with an area of 1 mm^2 and a detection time of 1 s is ca. 10^{-16} W. This level corresponds to the fluctuations at radiative equilibrium with the surroundings (at 4 K). It is possible to make other sources of noise small compared with this.

Furthermore, the heat capacity of the measurement system is also very much smaller than it would be at room temperature, i.e. the same temperature change occurs with a substantially lower level of radiation. This also makes the time constant small.

Finally – and this is the crucial advantage – the change in electrical resistance with temperature is very large on the transition curve. Levels of 1000 Ω/K are achieved without difficulty. However, the sharpness of the transition curve makes it necessary to control the temperature that sets the working range very accurately (to better than 10^{-4} K).

A superconducting bolometer is made up of a superconducting film or layer, that is suspended in a measurement cell cooled with liquid He or H_2. It is possible to adjust the coupling to the bath temperature to the desired level by gas contact at a sufficiently low pressure. For very low time constants the coupling can be achieved with a good thermal conductor, e.g. a copper sheet. The short relaxation time is naturally bought at the expense of a loss of sensitivity. The measurement cell must contain a window to admit the radiation that is transparent over the desired wavelength range.

Such bolometers are constructed of tantalum (T_c = 4.3 K), niobium nitride (T_c = 14.3 K) and tin (T_c = 3.7 K). Liquid hydrogen provides adequate cooling in the case of NbN. These bolometers allow the detection of a constant radiation level of ca. 5×10^{-12} W and a short pulse of radiation of $\geq 2 \times 10^{-13}$ Ws.

A particularly sensitive instrument was developed by Martin and Bloor [302]. The detector element consisted of a layer of tin sputtered onto a mica film 3 μm thick. The resistance change was measured in an AC bridge at 800 Hz. If the radiation to be measured is modulated to 10 Hz by means of a rotating slit and fed to a resonance amplifier it is then possible to detect power levels of 10^{-12} W. The sensitive area of this bolometer is 2×3 mm^2. The time constant is 1.2 s. The constancy of the temperature is maintained very precisely by using the output voltage of the bolometer to regulate it.

Superconducting radiation detectors can also be used to detect α-particles or other highly energetic particles. The very short dead time of such an instrument, if its operating point is chosen to be very close to the transition curve, can be an advantage. The energy that the particle gives up to the film results in a zone, perpendicular to the current, becoming normalconducting so that a voltage signal is produced. This then disappears in a period of the order of 10^{-7} s.

We saw in Section 4.4 (Figs. 54 and 55) that below the transition temperature the thermal conductivities are different in the normalconducting and superconducting states since, as the temperature falls, more and more electrons are correlated to Cooper pairs and simultaneously "condense" into a single quantum state where they are decoupled from the thermal processes in the superconductor. Since the thermal conductivity of good metals is largely transported by the conduction electrons the thermal conductivity can fall by orders of magnitude when these electrons are not available for the transport of heat. Below 1 K the ratio of the thermal conductivities $\varkappa_n/\varkappa_s$ for lead is greater than 200; at 0.3 and 0.1 K it becomes 500 and 5000 respectively. In the case of lead the transition can be brought about at relatively small fields of 800 – 900 G. So that the thermal resistance of a lead wire at 0.1 K can be reduced by a factor of 5000 by switching on a magnetic field. This yields a very efficiently operating thermal valve. Such devices have been employed in a range of experiments where low temperatures ($T < 0.5$ K) were generated by adiabatic demagnetization. A periodically operating refrigerator employing superconducting thermal valves has been developed for 0.2 K [303]. The working substance, a paramagnetic salt, is thermally coupled to the helium bath during magnetization and to the sample during demagnetization. Such apparatus has been completely replaced by the ^{3}He-^{4}He dilution cryostats, where the evaporation of ^{3}He into liquid ^{4}He (instead of into a vacuum) allows the continuous maintenance of temperatures down to ca. 10^{-2} K. Superconducting thermal valves are indispensable components of systems which penetrate down to temperatures below 1 mK. Aluminium has proved very successful as switch material at very low temperatures [304].

An electromagnetic cavity is a closed cavity of simple geometry with conducting walls. Electromagnetic oscillations can be generated in such a cavity when alternating magnetic fields are set up in the cavity and high frequency alternating currents in its walls. The frequency depends on the geometric dimensions of the cavity and on the spacial structure of the alternating fields and currents. It is possible to excite various oscillation modes in a cavity. For frequencies of the order of magnitude of $10^{10}\,\mathrm{s}^{-1}$ (cm waves) such cavity resonators acquire all the functions of electrical oscillation circuits, such as are built up out of capacitors and coils for lower frequency oscillations.

The losses, i.e. the energy which is converted into heat from electromagnetic oscillations and is lost, is determined by the high frequency resistance of the cavity walls. As we saw in Section 8.3, high frequency resistances for frequencies $\nu \ll 2\Delta/h$ become very small in the superconducting state and should, like the numbers of unpaired electrons, diminish with falling temperature. Basically it should, therefore, be possible to construct cavity resonators with as little damping as desired by using a superconductor as the material for the walls and cooling to a sufficiently low temperature.

Resonators with very small damping offer many advantages. The lower the damping, the sharper the resonance curve and the better the fundamental frequency of the resonator is specified. Very weakly damped resonators make excellent frequency standards.

The damping is often expressed in terms of the quantity Q. Q/π is the number of complete oscillations undergone in a freely oscillating cavity before the amplitude has fallen to the eth fraction. The lower the damping the smaller the energy losses per oscillation and the greater the value of Q. Superconducting cavities (internal wall of the cavity Pb or Nb) can reach values greater than 10^9.

This value is limited by the residual high frequency resistance due to the properties of the surface (roughness, oxide formation etc.). But it can be made very small by careful surface treatment. According to the BCS theory the frequency and temperature dependence of the high frequency resistance for $T <$ ca. $0.5\,T_c$ is given by [305]:

$$R = C\omega^{-n}\,\mathrm{e}^{-A\frac{T_c}{T}} + R_{\mathrm{res}} \qquad (9\text{–}20)$$

where ω = frequency, $n \approx 2$, $A = \Delta/(k_B T_c)$ with Δ = energy gap and k_B = Boltzmann's constant, T_c = transition temperature, R_{res} = high frequency residual resistance, C = constant.

Figure 173 reproduces a determination of the high frequency resistance of Pb below 4.2 K. If the residual resistances in Fig. 173 are subtracted from the determined values, then the exponential temperature dependence (broken line), predicted by the BCS theory, is very closely followed [306]. A summary of the opportunites presented by superconducting components in electronics is to be found in [XIX]. The applications in the microwave field will also be dealt with there.

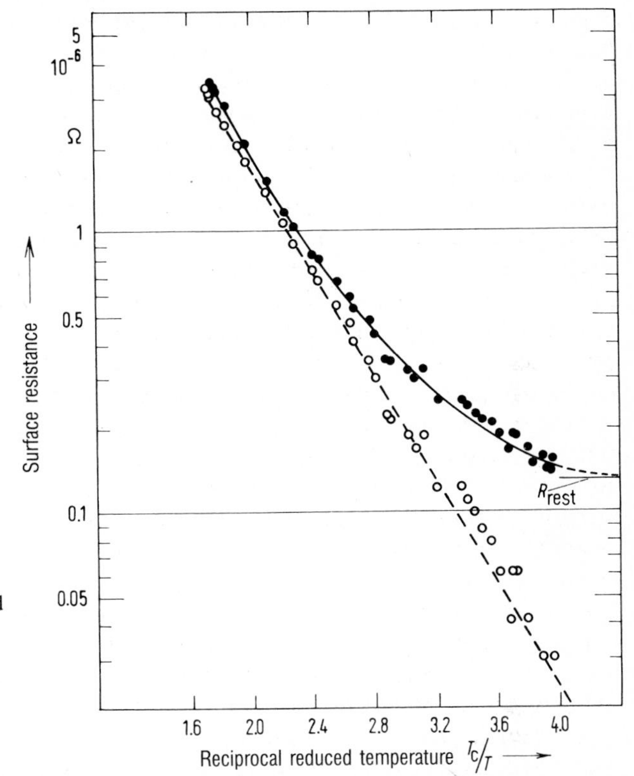

Fig. 173. Surface resistance of lead at 2.8×10^9 s^{-1}, $T_c = 7.2$ K; $R_{res} = 0.13 \times 10^{-6}$ Ω. • values measured; ○ values after subtraction of R_{res} (after [306]).

Cavity resonators with very small intrinsic losses are of great importance in measurement technology because they make it possible to study the very small losses of samples introduced into the cavity.

Interest in superconducting cavities with extremely high Q values has been stimulated by the possibility of utilizing them in linear accelerators. In such accelerators the particles pass through a series of oscillating cavities one after the other, the oscillations of these cavities are so tuned to each other that the particles are always accelerated by the electrical fields produced. Since it is alternating fields that are concerned in each cycle every particle can be accelerated that enters the field in the favorable half cycle. In practice the presently available linear accelerators[1] equipped with normalconducting cavities can

1 The largest of these accelarators has been built in Stanford (USA). It is 3.2 km long and has a final energy of 20 GeV (1 GeV = 10^9 eV).

only be operated in pulses with long dead times in between (duty cicle ca. 1/1000), since the required high frequency performance cannot be maintained continuously on account of the losses that occur.

Superconducting cavities would offer advantages here on account of their very small losses. It ought to be possible to operate a superconducting linear accelerator continuously, i.e. to accelerate particles during every high frequency cycle [305].

Such an accelerator would generate a particle current modulated only with the high frequency, while the instruments available until now use very high particle densities in the pulse to achieve the desired average current. The high particle densities are associated with difficulties of a statistical nature. High energy experiments are used to study rare events by measuring the tracks made by the particles. If very many particles arrive within a short period of time then there is a large probability that purely accidental combinations of results are produced which could counterfeit the desired result. This statistical falsification is a very serious matter in the case of very rare events. In superconducting accelerators a much lower particle density would be achieved for the same average current and this would reduce the chances of accidental events.

A further advantage would be that accelerators with superconducting cavities would allow a greater accelerating voltage per unit length, thus making the whole accelerator shorter. Calculations have shown that, even allowing for the costs of cooling to 1.8 K, a superconducting accelerator could be appreciably cheaper as a result of savings in high frequency power consumption. These advantages are so great that work is taking place at the moment in preparation for the construction of prototypes of such a superconducting accelerator unit. A superconducting accelerator unit developed at the Nuclear Research Center Karlsruhe has been tested with full success in the German electron synchrotron (DESY) in Hamburg [307].

Other applications of superconducting cavities in high energy physics, e.g. as particle separators [308], have already been exploited. High voltage electron microscopes are an interesting possible application. Very well-defined voltages of several MeV with large currents, such as are required by these microscopes, can be achieved with very small and, therefore, easily handled high frequency units.

9.5.3 Tunnel junctions as transmitters and receivers of very high frequency phonons and as radiation detectors for microwaves

Tunnel junctions, such as were described in Section 3.3.3, can be used as transmitters and receivers of phonons (sound quanta). Tunnel junctions differ from other ultrasound sources in that they allow the production of extremely high frequency phonons [309]. The production of high frequency phonons can be easily understood if the possible states of excitation of free electrons in a superconductor are remembered. These excitations are illustrated again in Fig. 174. This figure corresponds to Fig. 36b. Here, however, for simplicity we have depicted a tunnel junction with identical superconductors (e.g. Sn-SnO-Sn). In addition we are interested in the behavior at very low temperatures ($T \to 0$), i.e. in conditions where there are practically no unpaired electrons remaining free at equilibrium. The tunneling current increases steeply at a voltage of $U = 2\Delta/e$ (*cf.* Fig. 31c). Figure 174 represents a state where $U > 2\Delta/e$. A considerable tunneling current flows which increases the concentration of unpaired electrons in superconductor II (right-hand side) over the equilibrium concentration (see also Section 8.2). Almost without exception these electrons now suffer, as follows, in superconductor II. They first fall from energy $E = eU - \Delta$ to the lower limit of the excitation spectrum, i.e. to energy $E' = \Delta$, by transferring energy to the lattice (i.e. by the generation of phonons). This process takes place relatively rapidly, namely within 10^{-9} s. The result is the production of a continuous spectrum of phonons with, however, a very sharp edge at energy $h\nu = eU - \Delta$. This sharp edge comes about because the very high density of states at the lower edge of the excitation means that the number of transitions from $E = eU - \Delta$ to $E' = \Delta$ in *one* step is particularly large. This results in the production of phonons with the energy $eU - 2\Delta$.

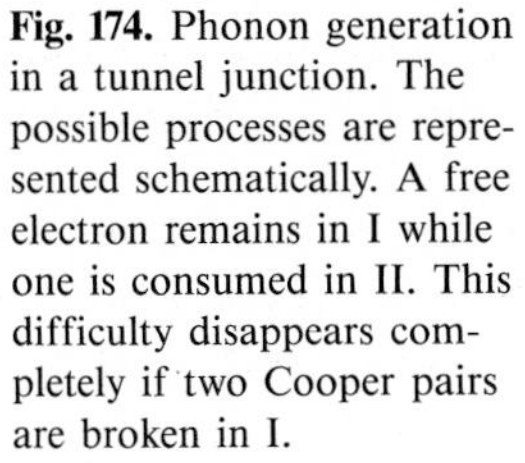
Fig. 174. Phonon generation in a tunnel junction. The possible processes are represented schematically. A free electron remains in I while one is consumed in II. This difficulty disappears completely if two Cooper pairs are broken in I.

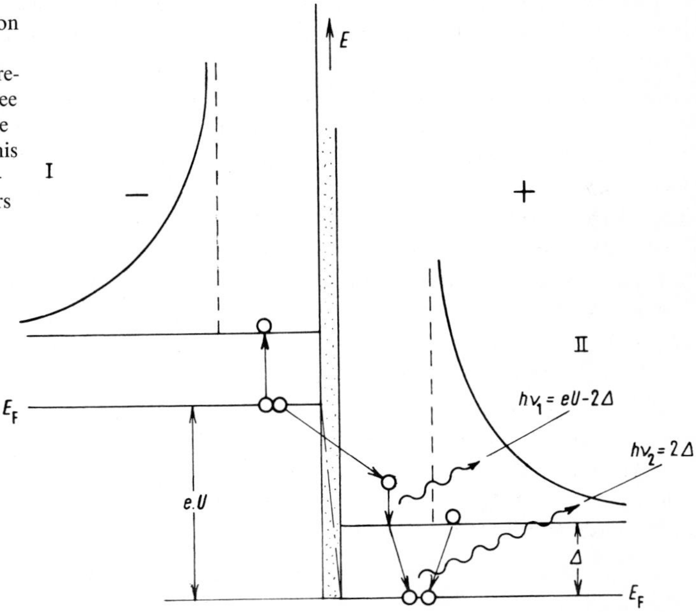

The next step, whose time constant is much larger at low temperatures ($\tau > 10^{-7}$ s) is the recombination of the unpaired electrons to Cooper pairs. This yields phonons with the energy 2Δ. If the energy gap of tin is assumed to be $2\Delta_{Sn} = 1.1 \times 10^{-3}$ eV (Fig. 20) then the frequency obtained for these recombinations is:

$$\nu_{2\Delta(Sn)} = \frac{2\,\Delta_{Sn}}{h} \approx 2.8 \times 10^{11}\,\mathrm{s}^{-1}\,. \tag{9-21}$$

Superconductors with larger energy gaps allow the generation of phonons with even higher frequencies. It should be mentioned here that the frequency of these monochromatic phonons is at least an order of magnitude higher than it is possible to achieve with any other method of generating monochromatic phonons as yet discovered.

Such high frequency phonons are also detected using superconducting tunnel diodes. Here an operating point is chosen for the diode such that $U < 2\Delta/e$ (see Fig. 31c for the case $\Delta_I = \Delta_{II}$). The tunneling current here is very small because it is only carried by the unpaired electrons present at the temperature chosen. However, if phonon radiation is encountered with an energy 2Δ then these phonons are able to break up Cooper pairs. This causes the tunneling current to increase. The increase in the current is proportional to the phonon current, i.e. to the numbers of phonons absorbed per second by the tunnel junction.

In addition to the monochromatic phonons of energy $h\nu = 2\Delta$ it is also possible, in principle, to use the continuous spectrum for spectroscopic purposes on account of the sharp edge [310] at $h\nu = eU - 2\Delta$. It is possible to displace this edge by altering the voltage U across the tunnel junction. If, when this is done, an excitation energy is passed over then there will be a particularly high absorption of phonons in the narrow range of voltage U. This spectroscopy with extremely high phonon frequencies and superconducting tunnel junctions has been developed over recent years [311].

This has resulted in the discovery of an even more efficient source of monochromatic phonons of very high frequency [312]. It was found that a Josephson junction with an applied voltage (see Section 3.4.2) not only radiates electromagnetic quanta (see p. 83 ff) but also monochromatic phonons with a frequency of $\nu = 2eU/h$. In order to increase the generation of phonons the junction, which constitutes a small cavity for electomagnetic radiation, is operated in resonance. The resonance conditions can be adjusted by means of a suitable magnetic field parallel to the surface of the contact [67]. This new source of very high frequency phonons is tuneable. Its frequency is very readily changed by adjusting the voltage across the junction. One new application for high frequency phonons involves the construction of a "phonon microscope" (D. Rugar: "High Resolution Microscopy Using Coherent Phonons in Liquid Helium" IVth Int. Conf. on Phonon Scattering in Condensed Matter, Stuttgart, 1983).

We have already seen in our consideration of the Josephson alternating current in Section 3.4.2 that a high frequency, electromagnetic alternating field brings a characteristic structure to the current-voltage characteristic of a tunnel junction. Giaever exploited this effect of a microwave field on the tunnel junction to demonstrate the Josephson alternating current (see Section 3.4.2). Such junctions can also be used to detect any external microwave radiation. Some of the devices that have been developed can detect a radiation level of 3×10^{-13} W at a frequency of $7 \times 10^{10}\ s^{-1}$ [313].

Finally it should be mentioned here that superconducting tunnel junctions can be used to investigate the solid state using polarized electrons of extremely low energy. A tunnel junction – we are considering the tunneling of single electrons – is placed in a magnetic field parallel to the junction surface, this separates energetically the states for single electrons with magnetic moments parallel and antiparallel to the magnetic field. If such a junction between a normal conductor and a superconductor in a magnetic field is subjected to a voltage then first the electrons of one direction with respect to the magnetic field, of one spin direction, will be able to tunnel. This can be used, for instance, for a very direct investigation of the orientation of electron moments in ferromagnets [314, 315].

9.5.4 Superconducting magnetometers

The fact that all Cooper pairs lie in the same quantum mechanical state and, therefore, possess one well-defined phase – we described this property as "phase coherence" in Section 3.4 – makes it possible to construct magnetometers of very high sensitivity. We met flux quantization and the Josephson effects in Sections 3.2 and 3.4 as particularly characteristic consequences of this phase coherence. Both these phenomena can be exploited for the highly sensitive determination of magnetic fields. We will now go on to discuss just a few of the possibilities in order to explain the fundamentals of these instruments by the way.

Figure 43 illustrates the critical current through a Josephson double junction as a function of the magnetic field. The distance between the maxima or minima corresponds to a change in the magnetic flux through the area of the double junction [42] by one flux quantum $\Phi_0 \approx 2 \times 10^{-11}$ T cm^2. It is clear that such devices allow a very sensitive determination of the magnetic field.

It is not entirely simple to prepare Josephson junctions with suitable oxide layers. Attempts have, therefore, been made to replace the very thin oxide layers by other devices. The crucial function of a Josephson junction, as far as the following applications are concerned, is that two superconductors should be joined by a *weak* link, that is a region containing very few Cooper pairs. This region should only be able to allow a very limited but nevertheless finite exchange of Cooper pairs and, thus, set up the desired weak link.

This function can be performed by normalconducting layers, double layers and point contacts as well as by oxide layers. These various possibilities are illustrated schematically in Fig. 175. Oxide barriers have to be very thin to allow any exchange of Cooper pairs

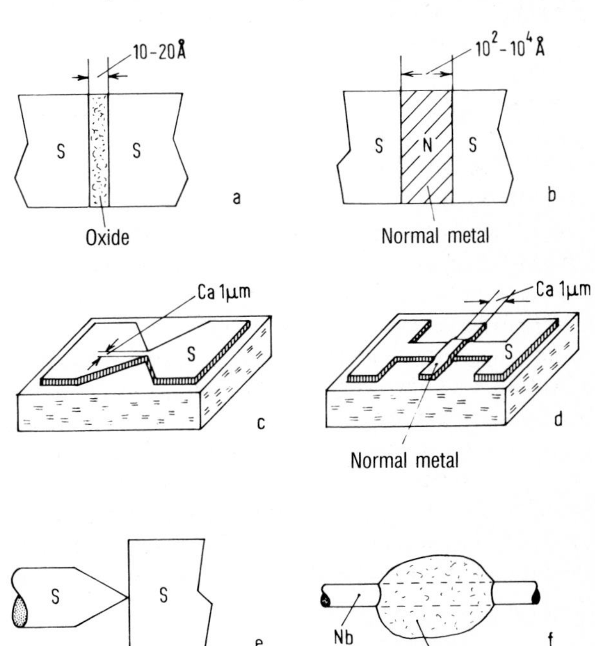

Fig. 175. Schematic representation of methods of producing a weak link between two superconductors. a) Oxide film; b) SNS junction; c) film bridge; d) double layer contact; e) point contact; f) multiple point contact.

at all. Difficulties are involved in the preparation of oxide layers with homogeneous thicknesses of only 1 – 2 nm and without microscopic holes. The SNS junctions (Fig. 175b) can, in contrast, operate with much thicker layers of the normal conductor; this is simply because Cooper pairs can penetrate normalconducting metals to a much greater depth than they can oxide layers. The decay depth of the Cooper-pair concentration (see Section 8.4) depends on the mean free paths of the electrons. When the mean free paths are very long (few defects) it is possible to use normalconducting layers up to several times 100 nm in thickness. The important difference between oxide and normalconducting junctions lies in their normal resistances. In the case of an oxide junction the resistance is usually between 0.1 and 1 Ω, in the case of SNS junctions this falls to ca. 10^{-6} Ω. Hence the electrical voltage across a junction when the critical current is exceeded (ca. 1 mA) is 10^{-4} to 10^{-3} V in one case and 10^{-9} V in the other. Even these tiny voltages can be determined without difficulty with superconducting galvanometers. This is the manner in which it has been demonstrated that the SNS junction also exhibits Josephson effects [316]. The extremely low internal resistance can be an advantage for some applications in measurement technology.

The film bridge (Fig. 175c) merely consists of a narrow constriction, which limits the exchange of Cooper pairs as a result of its very small cross section. Here too it is not very easy to prepare the desired width of only ca. 1 μm reproducibly. The double layer contact is a modification of the film bridge contact. A bridge is constructed here by a cross sputtered normal conductor. As a result of the proximity effect (see Section 8.4) this bridge contains appreciably fewer Cooper pairs. It is experimentally simpler to sputter a sufficiently narrow strip of normal conductor (width ca. 1 μm) than to prepare a well-defined film bridge. This is done most simply by preparing the normal conductor by a photoetching method and then sputtering the superconductor over this.

Point junctions are particularly simple. Here a sharp point is simply brought into contact with a flat surface. The area of the bridge is determined by the pressure of contact. This allows the desired contact to be set up very easily and it can be adjusted later, if necessary. The device illustrated in Fig. 175 f. is a modification of the point contact. Here a thin niobium wire ($d \approx 10^{-2}$ cm) is simply surrounded by a drop of solder made of superconducting material (Pb + Sn). The oxide film of the Nb wire stops the drop wetting it. Only a few point-like contacts are formed with the Nb and, thus, form a multiple contact. If there are more than two contact points the Josephson direct current exhibits several periods (see Section 3.4.1), which correspond to the various areas between the point contacts.

The current-voltage characteristics of the various coupling types differ in their stability behavior. The stability of a "descending" characteristic (Fig. 37) depends on the resistance, the capacity and the inductivity of the complete circuit. The devices in Fig. 175 differ very greatly in their capacitive behavior.

The device illustrated in Fig. 175 f. was developed by Clarke as a very sensitive galvanometer [316]. It is often referred to in the literature as a SLUG (superconducting low-inductance undulatory galvanometer). The measurement principle is evident from Fig. 176. The current I to be measured flows through the Nb wire and creates a circular magnetic field round the wire. Between the two point contacts A and B the magnetic field penetrates an area which is basically described by the distance l_{AB} between the point contacts and the penetration depths λ_{Nb} and λ_{PbSn}. The insulating oxide layer on the Nb wire can be regarded as thin compared with the penetration depth. With this assumption it is easy to estimate the current by which exactly one flux quantum Φ_0 is produced through the area between the point contacts. It is:

$$B = \mu_0 \frac{I}{2\pi r_{Nb}}; \quad \Phi = BA = \mu_0 \frac{I}{2\pi r_{Nb}} l_{AB} \left\{\lambda_{Nb} + \lambda_{PbSn}\right\} \tag{9-22}$$

$$I_{\Phi_0} = \Phi_0 \frac{2\pi r_{Nb}}{\mu_0 l_{AB} \{\lambda_{Nb} + \lambda_{PbSn}\}} . \tag{9-23}$$

With $\Phi_0 = 2 \times 10^{-7}$ G cm^2 $= 2 \times 10^{-15}$ Vs; $2r_{Nb} = 10^{-4}$ m; $l_{AB} = 5 \times 10^{-3}$ m
$\lambda_{Nb} = 5 \times 10^{-8}$ m; $\lambda_{PbSn} = 10^{-7}$ m and $\mu_0 = 4\pi \times 10^{-7}$ Vs/Am:
$I_{\Phi_0} = 7 \times 10^{-4}$ A.

If a current i_s is passed through the contacts 1 and 2 and the critical value of i_s is measured as a function of I then a periodic dependence is found as illustrated in Fig. 176b. This periodic change in the critical current corresponds completely in its physical causes to those illustrated in Fig. 43. The device employed here evidently only had two point contacts. The constancy of the amplitude reveals that these are very small point contacts. The area of an individual contact determines the envelope of the amplitude structure (see Section 3.4). The estimate given above was made from the geometric size of this device. The agreement is satisfactory at the quantitative level too.

The SLUGs often have several point contacts. Figure 176c illustrates the $i_c(I)$ dependence when at least three point contacts were present and the effective areas were evidently very different.

A sine-wave alternating current with a frequency of 20 kHz was applied to contacts 1 and 2 to determine the critical current. The amplitude of this alternating current was chosen to be somewhat greater than the maximum critical current. A voltage pulse then occurs at every half-cycle, whose duration in time depends on how long the critical cur-

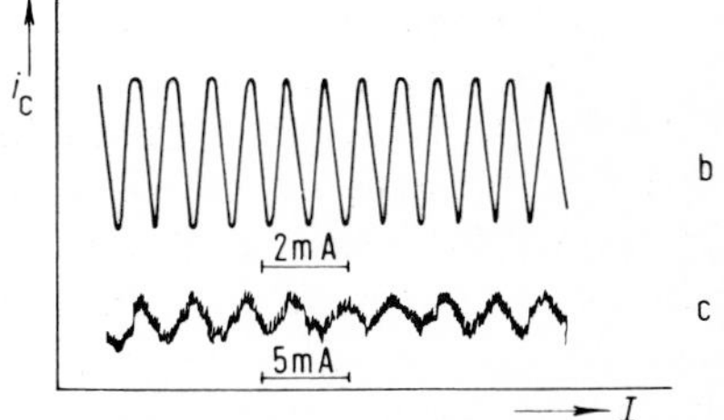

Fig. 176. a) SLUG with contacts for the measurement of the critical current i_c.
b) Critical current i_c as a function of the current through the Nb wire for a device with two effective point contacts.
c) $i_c = f(I)$ for a SLUG with at least 3 point contacts (after [315]).

rent is exceeded. As the critical current decreases the voltage pulses become ever longer if the amplitude of the alternating current is maintained equal. Now these changes in the voltage signal at junctions 3 and 4 can be measured in various ways. For example, the voltage pulses can be integrated after amplification. This would make it relatively simple to determine changes of 10^{-7} A in current I[1]. So here we already have a galvanometer with a sensitivity of 10^{-7} A with an internal resistance which can virtually be made arbitrarily small. Such an instrument can naturally measure very small voltages if the potential source itself also has a low internal resistance. For example the current voltage characteristic of a SNS junction were determined with a SLUG [316].

A SLUG becomes a very highly sensitive magnetometer when the two ends of the Nb wire are connected to produce a completely superconducting ring. Changes in the magnetic field of ca. 10^{-12} T can be detected with a ring area of ca. 2 cm^2. It is best to employ a compensation method. The signal from the magnetometer is amplifier and used to control a supplementary coil which compensates every change in the magnetic field. The compensation current is then a measure of the change in the magnetic field. The compensation method has the advantage that the measurements are then independent of the details of the $i_c(I)$ curve. SLUGs are characterized by their simplicity. But they are one to two orders of magnitude less sensitive than another group of magnetometers that have been given the acronym SQUID.

SQUIDs (Superconducting Quantum Interferometer Devices) are superconducting rings or cylinders with a region of weak linkage. Three possibilities are illustrated in Fig. 177 a-c. Devices a and b are thin superconducting cylinders sputtered or condensed onto support tubes. The region of weak linkage is created here by a film bridge or a double layer contact. Device c is constructed of compact material, e.g. niobium [317]. The cylinder is made up of two halves mechanically held together either by insulating plastic screws or by being wound round by nylon threads. The two halves of the cylinder are insulated from each other electrically by mylar film. The weak link is achieved by a point contact with screw A. Screw B, on the other hand, is tightened down and constitutes an ideal contact between the two halves of the cylinder.

In order to understand the SQUID it must be made clear how such a superconducting cylinder incorporating a "weak link" will react to application of an external magnetic field. Let us for this purpose consider the variation of the magnetic flux Φ_i through the cylinder as a function of the magnetic field B_e. A cylinder without the weak link would shield its interior from the external field until the critical current is reached in the cylinder. A further increase of B_e causes magnetic flux to enter the cylinder. This behavior is illustrated for a superconducting ring in Fig. 95.

The region of the weak link has a much lower critical current than the rest of the superconductor. This makes it possible – and that is its crucial purpose – for magnetic flux to enter the cylinder at very much lower fields B_e. The small variations of the flux Φ_i in the cylinder that are, thus, possible make the quantum conditions visible (see Section 3.2), in that a definite structure becomes evident in the $\Phi_i(B_e)$ dependence. This

1 Another method determines the phase at the point of occurrence of the voltage relative to the current signal and, thus, obtains the critical current.

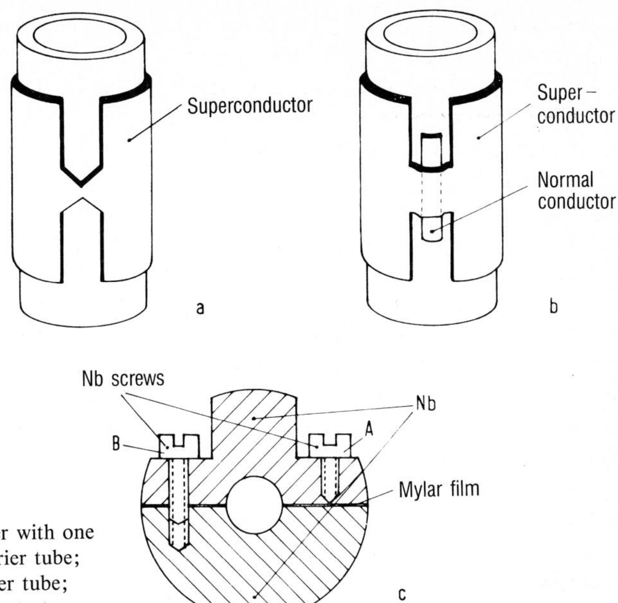

Fig. 177. Superconducting cylinder with one weak link. a) Film bridge on carrier tube; b) double layer junction on carrier tube; c) niobium cylinder with point contact.

structure can be used for the generation of electrical signals and, hence, for the determination of very small changes in the field B_e.

It is not just the critical current and the flux quantization that are responsible for the details of this variation of Φ_i. The specific current-potential behavior of the weak link is also important. This means that the relationships can be very complex. We will limit ourselves to a qualitative description of several typical examples.

Figure 178a-c illustrates the variation of Φ_i as a function of the external field B_e for three different values of the critical current i_c. The flux quantum Φ_0 is the decisive quantity in all treatments. It is, therefore, appropriate to relate both Φ_i and the external field B_e to Φ_0. Therefore, Φ_i/Φ_0 is plotted in Fig. 178 as a function of Φ_e/Φ_0. Φ_e is the magnetic flux generated in the cylinder by B_e when no current is flowing. It is given by $\Phi_e = B_e A_c$ where A_c = the area of cross section of the cylinder. In the case of diagrams a-c in Fig. 180 the following conditions have been chosen for the critical current i_c:

a) $L i_c = 3\,\Phi_0/4$
b) $L i_c = \Phi_0/2$
c) $L i_c = \Phi_0/4$.

L is the coefficient of self-induction of the cylinder, Li_c is then the maximum flux which can be generated in the cylinder by the supercurrent i_s[1].

Let us in Fig. 178a allow the flux Φ_e to increase from zero, then a screening current i_s will be set up in the cylinder which will screen the interior of the cylinder from the

1 A small cylinder 1 cm long and 1 mm diameter has a coefficient of self-induction L of ca. 10^{-10} H. A critical current of 2×10^{-5} A (= 20 μA) generates just one flux quantum Φ_0.

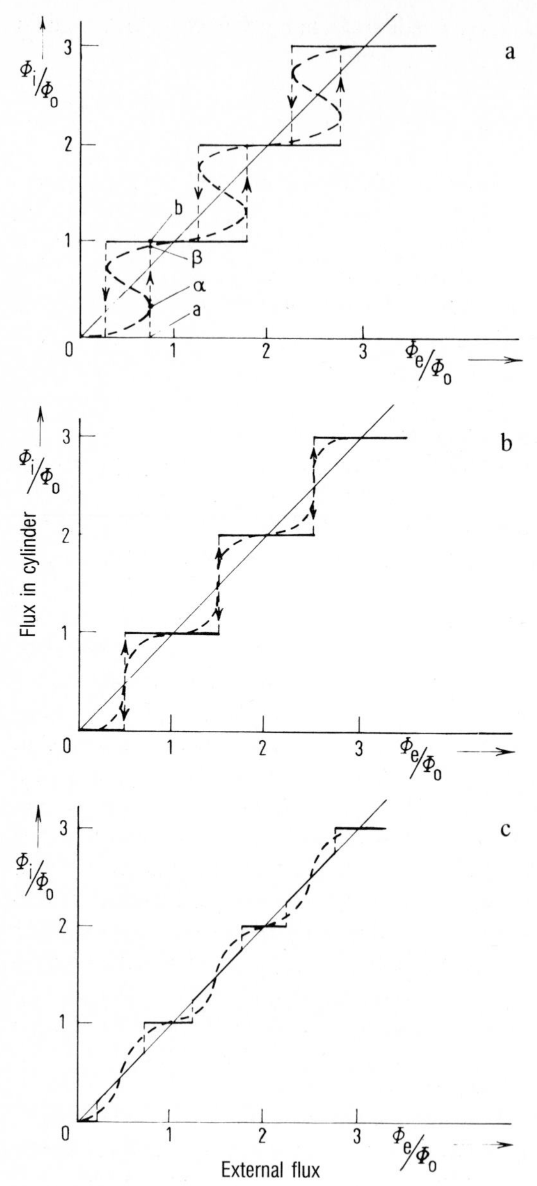

Fig. 178. Magnetic flux Φ_i in a cylinder with a weak link as a function of $\Phi_e = B_e A_c$. A_c = cylinder cross section, – – variation for a Josephson junction.

penetration of the flux. Let us first assume that the shielding is complete until the current reaches its critical value i_c. So that the value of Φ_i remains equal to zero until $\Phi_e = Li_c = 3\Phi_0/4$ (case a). When this value is exceeded the flux must enter the cylinder. This happens in such a manner that a complete flux quantum penetrates the cylinder. The cylinder enters the next quantum state with 1 flux quantum in its interior. Now a supercurrent i_s is flowing in this state, so that the external flux Φ_e is *increased* by exactly one flux quantum. We could also say that the supercurrent had changed from i_c to the value $-\Phi_0/L$. The new supercurrent at point b is given by:

$$i_s = i_c - \Phi_0/L = 3\Phi_0/4L - \Phi_0/L = -\Phi_0/4L\,. \tag{9-24}$$

As Φ_e is increased further i_s decreases with the cylinder remaining in the same quantum state. When $\Phi_e = \Phi_0$ i_s has fallen to zero. The cyclic repetition is then started. The transition to a state with two flux quanta occurs at $\Phi_e = 7\Phi_0/4$. We must mention here that a hysteresis occurs when Φ_e is reduced. The transition from the state with two flux quanta to that with only one takes place at $\Phi_e = 2\Phi_0 - 3\Phi_0/4 = 5\Phi_0/4$.

We assumed a behavior of the weak link here where the state of the cylinder at any time with $\Phi_i = n\Phi_0$ is stabilized up to a critical current of $\pm\, i_c$. Film bridges have this property by and large.

In the case of an ideal Josephson junction (see Section 3.4) with critical current $i_c = 3\Phi_0/4L$ it is possible to calculate the dependence illustrated by the broken line for Φ_i [317]. This dependence becomes unstable at point α, the cylinder goes to point β. The value of Φ_i can deviate at this point from $n\Phi_0$ on account of the phase relationships at the Josephson junction. The flux Φ_i determines the phase difference across the junction and, hence, the Josephson current.

Part b of Fig. 178 will now be understandable after this discussion. Here the superconducting ring current can hold the individual quantum states with $\Phi_i = n\Phi_0$ stable up to just $\Phi_e - \Phi_i = \pm\, \Phi_0/2$. No hysteresis occurs on increasing or decreasing Φ_e.

At even lower levels of the critical current – part c of Fig. 178 illustrates the case when $Li_c = \Phi_0/4$ – the Φ_i dependence of the cylinder with a Josephson junction approaches the straight line $\Phi_i = \Phi_e$ even more closely. The full line is obtained in the case of a junction with complete shielding up to i_c [317]. The weak link must be normalconducting between quantum states containing a whole number of flux quanta. This result is surprising at first sight. However, the following consideration will make it understandable. If the weak link were to be superconducting for $Li_c < \Phi_e < \Phi_0 - Li_c$ then, since complete screening has been assumed, it would be possible to find a path that went right round the cylinder that lay completely in the current-free region. But then according to Section 3.2 the condition $\Phi_i = n\Phi_0$ would have to apply, i.e. i_s must be greater than i_c. Since this is impossible the bridge (which is completely screening in the superconducting state) must be normalconducting. The energy of condensation of the Cooper pairs in the weak link is not sufficient to set up the required supercurrents.

It can naturally also be assumed that the concentration of Cooper pairs is very dependent on the current. The junction could then lose its complete shielding if this concentration becomes small. This would mean that the weak link would take on more of the character of a Josephson junction.

We have considered the behavior of a superconducting cylinder with a weak link in some detail because the importance of flux quantization is very apparent here once again. The details of the dependence on Φ_i are not of moment for the functioning of a SQUID. The important point is merely that the flux changes take place rather clearly one flux quantum Φ_0 at a time. So i_c should be larger than $\Phi_0/2L$ but not too large. If i_c is too large flux jumps of more than one quantum can occur. The sudden changes in Φ_i are transformed into voltage signals. This is done by introducing the cylinder into the coil of an oscillating circuit (Fig. 179). As strong a coupling as possible is particularly favorable for operation. The oscillating circuit is operated at its resonance frequency ω_R (in the range of a few 10^6 Hz). The coil generates an alternating field $\tilde{B}$ which we express as the normalized flux $\tilde{\Phi}_0$, as in Fig. 178. For

$$\tilde{\Phi} = \Phi^* \sin \omega_R t \, . \tag{9-25}$$

The total external flux Φ_e is then the sum of the flux Φ_d of the constant field to be determined and the flux of the alternating field $\tilde{\Phi}$

$$\Phi_e = \Phi_d + \Phi^* \sin \omega_R t \, . \tag{9-26}$$

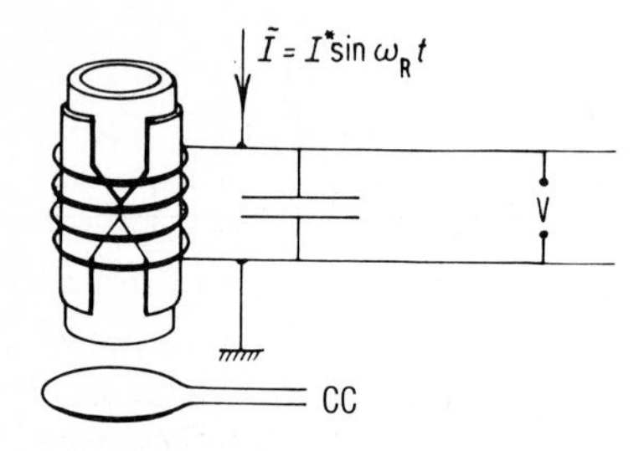

Fig. 179. Schematic representation of the resonance circuit of a SQUID. The voltage V is amplified and rectified. CC is a compensation coil. This is employed with feedback to compensate small field changes $\Delta\Phi_d$.

Whenever a flux quantum penetrates the cylinder there will be a voltage pulse. The sequence of these voltage pulses at a given Φ_d and Φ^* forms the signal. Only the component of this signal at ω_R is employed for the determination[1]. The amplitude of this component at ω_R depends on Φ^* and Φ_d in a very characteristic manner [318].

Figure 180 illustrates this periodic dependence schematically. Each cycle has the length Φ_0 in both cases. The periodic variation with Φ^* is easy to understand. If the amplitude of the alternating field is increased ever more flux quanta will penetrate the cylinder per period length of $T = 2\pi/\omega_R$, i.e. ever more voltage pulses will occur. This naturally means that the mean signal voltage V increases. This increase does not take place proportionately to Φ^*, since the penetration of a new additional flux quantum per period can only occur with an increase in Φ^* if Φ_e exceeds the next step in the Φ_i dependence according to Fig. 178.

It is somewhat more difficult to provide a graphic explanation of the periodicity of V in a variable direct field (measured in Φ_d/Φ_0). This periodicity can only be understood in terms of the Fourier analysis of the sequence of voltage peaks with time, since at constant alternating field amplitude Φ^* only a change in this sequence in time can alter the amplitude of the fundamental frequency.

The analytical presentation of these relationships is represented to a good approximation by the following expression for the voltage signal at ω_R (Fourier component with frequency ω_R) [319].

---

1 The voltage pulses can be very sharp. The Fourier analysis of the sequence of pulses naturally contains many other frequencies. Only the change in the amplitude of the fundamental frequency ω_R will be considered here.

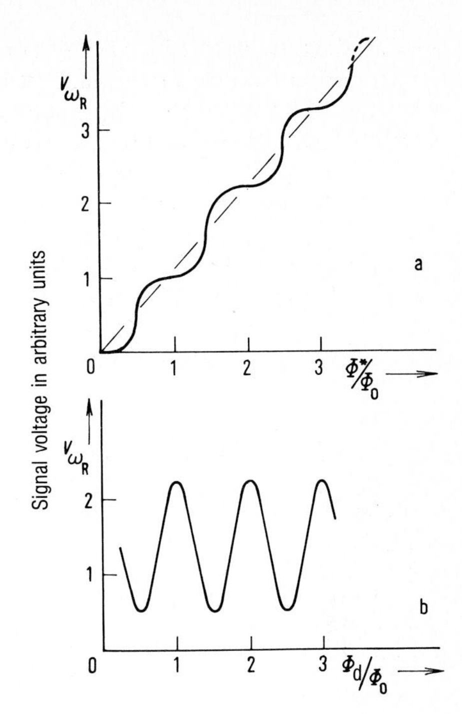

Fig. 180. Signal voltage of a SQUID as a function of a) the amplitude of the alternating field measured in Φ^*/Φ_0; b) of the steady field, measured in Φ_d/Φ_0.

$$\tilde{V}_{\omega R} \approx \Phi^* \omega_R I_1 \left(2\pi \frac{\Phi^*}{\Phi_0}\right) \cos \left(2\pi \frac{\Phi_d}{\Phi_0}\right) \sin \omega_R t , \qquad (9\text{-}27)$$

I_1 is the 1st order Bessel function.

In this presentation the periodicity is distinctly apparent with both Φ^* and Φ_d.

The periodicity in Φ_d is employed for the highly sensitive determination of the magnetic field [318]. If, for example, it is desired to measure very small changes in the magnetic field then a low frequency change is superimposed on the flux $\Phi_e = \Phi_d + \Phi^* \sin\omega_R t$. This small change in Φ_d is experienced by $dV_{\omega R}/d\Phi_d$. The resulting low frequency signal is amplified and used to drive a subsidiary coil (see Fig. 179) in such a manner that the effective flux Φ_d remains constant. The current through the compensation coil is a measure of the deviation of the external field from the value to which it was set. Such feedback allows the detection of flux changes of ca. $10^{-4}\,\Phi_0$. If the effective area of the SQUID is 0.1 cm^2 this implies sensitivity in field measurement of 2×10^{-14} T.

In order to measure greater changes in field it is possible to measure the number of cycles through which $V_{\omega R}$ passes with the change in Φ_d [318]. Devices are already in use which allow changes of Φ_d of ca. $10^4\,\Phi_0$ per second. It would seem entirely possible to increase this counting rate.

The decisive unit for all flux changes is Φ_0. This means, however, that for all mesurements of this type it is necessary to shield the SQUID itself very well from magnetic field variations. Superconducting shields are employed alongside the usual ferromagnetic materials such as μ-metal.

Since spaces with very low magnetic fields are of crucial importance for the operation of SQUIDs it is worth saying a few words on the subject of shielding from magnetic fields by means of superconductors. A space bounded by superconducting walls is shielded against changes in the magnetic field outside it, so long as the critical field strength is not exceeded. The walls must naturally be of sufficient thickness to be many times the penetration depth (see Section 5.1.2).

The shielding from field *changes* is solely the result of the fact that the ohmic resistance disappears in the superconducting state. The size of the residual field remaining within the cavity when superconductivity starts depends on the quality of the field exclusion, i.e. on how ideal the Meissner-Ochsenfeld effect is. It has been found that residual fields remain frozen in even with very slow cooling and excellent material homogeneity. A reduction of the external field by a factor of 100 has been demonstrated (even for very small fields of ca. 10^{-8} T) [320]. This means that the shielded room only contains fields of less than 10^{-10} T. The residual fields are the result of frozen-in flux. A proposal by W.M. Fairbank revealed the possibility of greatly reducing the frozen-in flux by using suitable cooling conditions [321].

On the basis of the quantum rules for magnetic flux in a superconducting cylinder it ought to be possible, in principle, to generate a completely field-free volume. For this purpose it is necessary to fold a cylinder (of say lead foil) very closely together. The lead must then be cooled down in a space with as small a residual field as possible. A certain arbitrary number of flux quanta will then remain frozen into the lead. If the lead cylinder is now unfolded carefully in the superconducting state, the number of flux quanta in its interior remains constant, i.e. the residual field is reduced by the ratio between the areas before and after unfolding. If this process is repeated with cylinders mounted one inside the other then a state can be reached, in principle, where the last cylinder can no longer freeze a flux quantum into it. After it is unfolded it will enclose an ideal field-free volume. It has been possible to reduce the field strength to 2×10^{-11} T in this manner [321]. Let us now return to our treatment of the SQUID (see also [XX]).

The changes in field that are to be measured are fed into the SQUID in the screened space by means of what is known as a flux transformer. The flux transformer consists of a completely superconducting wire loop, such as a niobium wire loop. Every flux change in one part of the loop leads to a shielding current which will generate flux in other parts of the loop. The only requirement is that the total flux through the superconducting loop remains constant. Suitable arrangement of the sensitive part of a flux transformer, e.g. in a figure of eight, can lead to only the gradient of the magnetic field being measured. The constant background is compensated in the two equally large loops of the eight. Figure 181 will make the application of a flux transformer clear.

Figure 181 illustrates a SQUID configuration with particularly favorable properties [322]. This SQUID is made up of two holes in a solid superconductor, they are connected by a fine slit. This slit is bridged by a weak link contact, here a point contact. Hole 1 contains coil CF of the flux transformer which inputs the measurement field. Hole 2 accommodates coil CD of the detection circuit in such a manner as to give as good a coupling of the coil with the superconductor [1] as possible.

---

1 The quality of the coupling can be achieved by a tight fit of the coil in the hole. The voltage $V_{\omega R}$ will then be basically determined by the quantized changes in the flux.

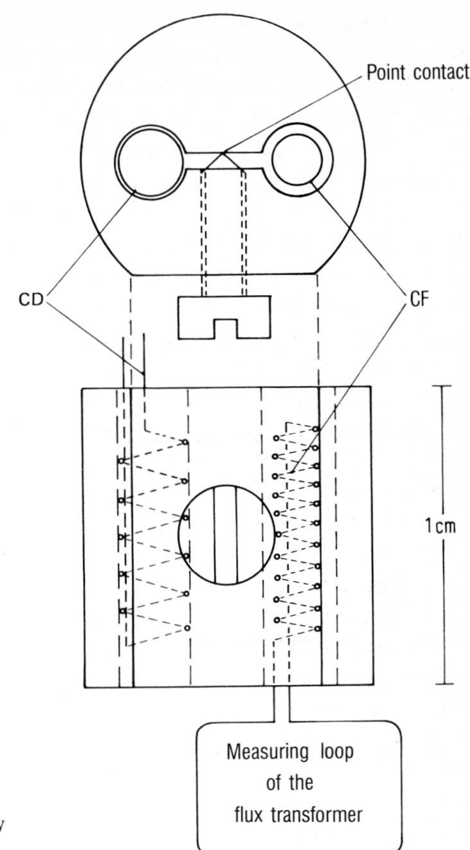

Fig. 181. Schematic representation of a SQUID with flux transformer and detection coil. The measuring loop can be increased appreciably in size to enhance the sensitivity (after [322]).

If an alternating field Φ^* is generated in the detection coil CD, then a screening current will be set up in the walls of hole 2 via the weak link. Whenever the critical current i_c of the point contact is exceeded then the flux in the hole will alter by one flux quantum. In this respect this device corresponds completely to that illustrated in Fig. 179.

The direct field, which, in Fig. 179, we simply assumed filled the whole space surrounding the SQUID, is now taken up outside by the flux transformer and partially transferred to hole 1. Screening currents are set up in the wall of this hole *and in the point contact*. Whenever this results in i_c being exceeded the flux penetrates into hole 2 from hole 1. This behavior also corresponds completely to that we discussed on the basis of Fig. 179. The exceeding of the critical current of the weak link is always the decisive process.

The device chosen here has the advantage that the thick superconducting mantle already provides a very good screening for the sensitive circuit (internal walls of the holes and the point contact). Here the magnetic shielding in the Nb mantle can only serve to hold the magnetic flux through the holes constant. So if the field gradient outside alters so will the field distribution in the holes. So this SQUID reacts to changes in the external field gradient in spite of its screening.

The sensitivity of such a device[1], which is at the moment approximately equaled only by optical magnetometers, is so great that new fields have been opened up for magnetic investigations. Just two examples should serve to illustrate this. SQUIDS can now be employed to record mangetocardiograms [322]. Here the sensing surface of a few cm^2 of the flux transformer is contained in a small cryostat situated a few cm from the left chest wall of the test subject. Very sharp signals resulting from the action of the heart can be recorded. It is not necessary here to attach electrodes to the body of the test subject as is the case for the electrocardiogram. The SQUID technique has acquired great importance for investigation of the human brain. Here the SQUID achieves far greater spacial resolution than do electrical sensors. A review of the state of the art in these biomagnetic investigations is given in [323]. Very recently a device has been constructed containing 37 SQUIDs which synchronously record the effect of a stimulation of the human brain. This made it possible to locate the site of the reaction in the brain stereoscopically (Dr. H. Hoenig, Siemens, Erlangen, private communication).

SQUIDs are ideal measurement instruments for the search for magnetic monopoles. Electrodynamics as we know it today and which we describe by means of the Maxwell equations, has an assymetry. We know single electrical charges (electrical monopoles) but

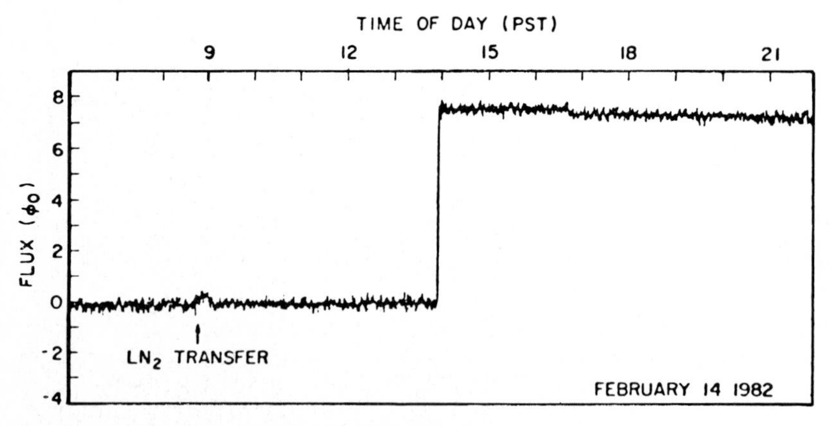

Fig. 182. Record of the magnetic flux through the sensing coil of a SQUID. The extraordinary constancy of the magnetic flux, which is only slightly disturbed when filling up with cooling liquid, could be observed for days. The flux jump on 14th February 1982 corresponds quantitatively to the change expected on the passage of a Dirac monopole through the device. (From Blas Cabrera: Phys. Rev. Lett. *48*, 1378 (1982) – I thank Mr. Cabrera for allowing me to reproduce the record.)

1 Another very sensitive superconducting magnetometer has been developed by Opfer [326]. The complete superconducting circuit consists, as in a field transformer, of a field-sensitive loop 1 and a loop 2 separated from it in space. This loop 2 is situated as close as possible to the surface of a superconductor. If the surface of the superconductor is now vibrated backwards and forwards with respect to loop 2 by means of a quartz oscillator then the inductivity of the loop will be altered with time. If a persistent current is now set up in the superconducting circuit, e.g. by means of an external magnetic field, then, because of the change in L with time, an alternating voltage will be set up across the ends of loop 2 which can be used to detect the supercurrent and, hence, the external field. One advantage of this magnetometer is that the measurement circuit remains completely superconducting and, in particular, there are no flux jumps. This is reported to reduce fluctuation phenomena.

not magnetic monopoles. Dirac[1] expressed the presumption that magnetic monopoles should also exist and should have the elementary pole strength $\Phi_D = h/e$ [324].

Every research center for high energy physics and particle physics in the world has sought these magnetic monopoles; but without success until now. We think we now know why this is so. Magnetic monopoles are probably very heavy particles (rest mass ca. 10^{14} (!) proton masses) and cannot, therefore, be produced at present-day accelerator energies.

The event where a magnetic monopole was perhaps detected took place on 14th February 1982. Blas Cabrera [325] had positioned a superconducting niobium coil with four turns in a space that was very well screened magnetically by superconducting shields (see p. 286) and was recording any changes in flux by means of a SQUID attached to the niobium coil. On account of the four turns the passage of a magnetic monopole of strength $\Phi_D = h/e$ would be expected to yield a flux change of $\Delta\Phi = 4h/e = 8\ \Phi_0$.

The recording of the output is illustrated in Fig. 182. The flux jump corresponds precisely to the expected change. Simple explanations, such as disturbing electrical signals or cosmic ray cascades etc., can be discounted. Nevertheless, one must today have doubts as to whether this observation really was the result of the passage of a magnetic monopole through the detection coil. Similar instruments have been operating in other laboratories for years. No second such event has yet been observed.

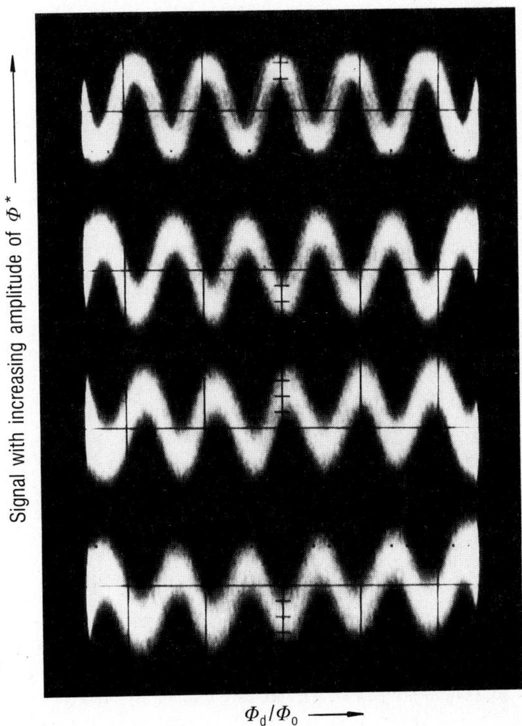

Fig. 183. The signal from a high-frequency SQUID made of $YBa_2Cu_3O_7$ at a temperature of 77 K as a function of the external magnetic field for four different amplitudes of the high-frequency field [327].

1 Paul Adrien Maurice Dirac, English physicist, 8.08.1902–20.10.1984, Nobel prize for physics 1933.

It may be said, in conclusion, that it is not yet possible to see the limits to the application of the superconducting magnetometer. These are instruments for measuring magnetic fields, whose sensitivity can, in principle, be increased right to the limits set by statistical noise. The use of such devices will increase enormously when the miniature refrigerators, that are being worked on intensively, become available.

The new oxide materials have already been used with success in the construction of SQUIDs [327]. Figure 183 illustrates the signal from a high frequency SQUID at a working temperature of 77 K. The various signals correspond to the voltage signal of a high frequency SQUID as shown in Fig. 180b. As the amplitude of the high frequency supply is increased the phase of the signal alters [327]. Even though the most extreme sensitivity cannot be reached on account of their higher working temperature, there are, nevertheless, many applications open to such magnetic field meters. The cooling problem is much simpler to solve at the temperature of liquid nitrogen and, if necessary, such meters can be cooled to He temperatures.

Outlook to 1st German Edition (1972)

There has been a particularly rapid development in superconductivity over the last 15 years. A physical phenomenon, still limited to a narrow temperature range over the absolute zero point, has found large-scale technical application in the construction of magnets. How will this development continue?

It is, of course, very difficult to forecast scientific and technological developments accurately. Surprises are always to be expected. What follows cannot, therefore, be anything more than the subjective opinion of the author.

The major focus of interest in superconductivity in the future will be in the field of technical applications. There are enticing possibilities here. Even in fields where it is perhaps not to be expected at first sight, superconductivity can acquire great importance in the course of general development. An example may serve to illustrate this.

There is no doubt that long years of development have brought the conventional technology for the construction of electric motors and generators to an almost optimal state. How could superconductivity offer advantages? There are certainly no advantages to be expected in the case of small motors. However, the situation is different in the case of large and very large units. The trend of development in both motors and generators is towards ever greater performance per unit. Conventional technology requires ever larger machines, that is greater both with respect to weight and to volume. This, however, makes the machines impossible to move because they are simply too large for normal means of transport. Superconducting machines may employ a rather unusual technology but they can be built to be much smaller and more readily transportable for a given performance. So that superconductivity makes it possible to increase the performance of single units considerably. Similar considerations will apply in the future to power transmission. This development has already taken place in magnet construction. Magnetic fields, such as are made available for high energy physics today with superconducting magnets, could simply not be generated by conventional means.

Materials science, a branch of basic research, will have many problems to solve in connection with these technological applications. We are a long way from an adequate understanding of the quantitative relations between metal parameters and superconductivity. It may well not be impossible to find materials that are superconducting at 30 K or even somewhat higher. Such an increase in transition temperature would be accompanied by enormous advantages for technological applications.

Such materials will certainly not be simple substances. There is a great deal of evidence that materials particularly favorable for superconductivity have a tendency towards instability. The attempt to improve further the parameters of importance for superconductivity could lead to these substances transforming into new structures with lower transition temperatures. It is possible that extremely high values of T_c will only be attainable with artificially stabilized materials.

These considerations apply to the electron-phonon interaction. It must also be remembered that it is possible that other interaction mechanisms will be discovered which lead to a condensation of the system of conduction electrons and can, thus, lead to superconductivity. Unfortunately the numerous theoretical hypotheses have not provided success that could be exploited. But this does not mean to say that great surprises may not be forthcoming. Superconductivity at room temperature is likely to remain wishful thinking.

Superconductivity will doubtless be employed in the future to an ever increasing extent in measurement technology and will yield a wealth of new results. Furthermore, the development of such superconducting measuring instruments is not completed.

The theoretical descriptions of many-particle systems developed in order to understand superconductivity will also be used with success in the future for the treatment of other systems. Further insights into superconductivity will be had in this manner. What other possibilities nature has to offer here will only be revealed by future developments.

Outlook to 2nd German Edition (1977)

An outlook is always a sort of preview and is, thus, accompanied by all the uncertainties necessarily attending a prediction concerning present scientific and technical developments. It, therefore, seems justifiable not simply to replace the "Outlook" of the first Edition in 1972 but to extend it.

There is no doubt that materials science has been a central field of superconductivity in past years and will attract a great deal of interest and require great efforts in the future. Even if no new interactions are discovered, which make higher transition temperatures possible, we can still hope to reach T_c levels of around 30 K by means of the electron-phonon interaction. Cautious extrapolations have yielded this for Nb_3Si in the β-tungsten structure. That, however, would yield superconductors that could be employed in technical instruments at the temperature of liquid hydrogen. Since hydrogen technology is anyway likely to acquire more and more importance in connection with energy problems – think only that automobiles running on hydrogen are already under development – a superconductor that was technically applicable in liquid hydrogen would be particularly attractive.

Technical applications are scarcely likely to lose their interest in the near future. The main focus at the moment is superconducting motors and generators. Research machines with considerable outputs are being built in several places (Section 9.1.4.4).

The advances made in the manufacture of miniaturized Josephson junctions have shown that these very small, low-power circuit elements can be competitive with the best transistors and are far superior to these for certain purposes. There are new possibilities for the future here.

All these developments reveal that superconductivity has lost none of its topicality.

Outlook to 3rd German Edition (1984)

The application of superconductivity to magnet construction and measurement technology is certain to continue to be developed in the future. We are only just at the start of application to medicine. It is evident, however, that the superconducting magnetic field meter will acquire crucial importance in the investigation of the brain and in the diagnosis of brain diseases.

Superconducting measuring instruments, such as bolometers and magnetic field meters, will increase greatly in importance when the developments underway in the construction of miniature refrigerators are ripe for mass production. The possibilities cannot even be guessed at, as yet.

Outlook to 4th German Edition (1989)

The work of Bednorz and Müller has opened up entirely new perspectives in superconductivity. The systematic search for new substances with higher transition temperatures will occupy us for at least a decade. The realization that this or that mechanism can yield transition temperatures above 100 K is a fantastic challenge to our theoretical colleagues. Experimental physicists will make efforts to determine the important parameters of the new superconductors unequivocally, so as to provide a solid basis for theoretical consideration. Efforts will be made in the development laboratories to construct suitable conductors from the new materials. This too is a tremendous challenge and will certainly require many new ideas.

We are entitled to await eagerly developments over the next few years in this old and now once again very young field.

Appendix A. The Josephson equations

The Josephson equations (3-21) and (3-53) follow from the basic equations for two weakly linked quantum mechanical systems [1]. The systems are described by wave functions ψ_1 and ψ_2. If the systems are completely separate (coupling zero), the variation in the wave functions with time is given by:

$$\frac{\partial \psi_1}{\partial t} = -\frac{\mathrm{i}}{\hbar} E_1 \psi_1 \, ; \quad \frac{\partial \psi_2}{\partial t} = -\frac{\mathrm{i}}{\hbar} E_2 \psi_2 \, . \tag{A-1}$$

1 This form of derivation is given in Volume 3 of the "Feynman Lectures on Physics", 21-14ff, Addison-Wesley Publ. Comp. New York 1965. See also [I], p. 398.

When the systems are weakly linked the change of ψ_1 with time affects ψ_2, just as that of ψ_2 affects ψ_1. This situation is described in the following equations:

$$\frac{\partial \psi_1}{\partial t} = -\frac{\mathrm{i}}{\hbar} \{E_1 \psi_1 + K \psi_2\} \tag{A-2}$$

$$\frac{\partial \psi_2}{\partial t} = -\frac{\mathrm{i}}{\hbar} \{E_2 \psi_2 + K \psi_1\} \, . \tag{A-3}$$

In our case the linkage means that Cooper pairs can be exchanged between superconductors 1 and 2. The level of exchange, which is symmetrical, is laid down by the coupling constant K.

A peculiarity of the two weakly linked superconductors in comparison with other two-state systems (e.g. H_2^+ molecule or NH_3 molecule) lies in the fact that ψ_1 and ψ_2 describe macroscopically occupied states. We can interpret the square of the amplitude as the particle density n_C (density of Cooper pairs). That is, we can write:

$$\psi_1 = \sqrt{n_{C1}} \, \mathrm{e}^{\mathrm{i}\varphi_1} \, ; \quad \psi_2 = \sqrt{n_{C2}} \, \mathrm{e}^{\mathrm{i}\varphi_2} \tag{A-4}$$

where φ_1 and φ_2 are the phases of the wave functions ψ_1 and ψ_2.

If we introduce these wave functions into (A-2) and (A-3) we get:

$$\frac{\dot{n}_{C1}}{2\sqrt{n_{C1}}} \mathrm{e}^{\mathrm{i}\varphi_1} + \mathrm{i}\sqrt{n_{C1}} \, \mathrm{e}^{\mathrm{i}\varphi_1} \dot{\varphi}_1 = -\frac{\mathrm{i}}{\hbar} \{E_1 \sqrt{n_{C1}} \, \mathrm{e}^{\mathrm{i}\varphi_1} + K \sqrt{n_{C2}} \, \mathrm{e}^{\mathrm{i}\varphi_2}\} \tag{A-5}$$

$$\frac{\dot{n}_{C2}}{2\sqrt{n_{C2}}} \mathrm{e}^{\mathrm{i}\varphi_2} + \mathrm{i}\sqrt{n_{C2}} \, \mathrm{e}^{\mathrm{i}\varphi_2} \dot{\varphi}_2 = -\frac{\mathrm{i}}{\hbar} \{E_2 \sqrt{n_{C2}} \, \mathrm{e}^{\mathrm{i}\varphi_2} + K \sqrt{n_{C1}} \, \mathrm{e}^{\mathrm{i}\varphi_1}\} \, . \tag{A-6}$$

The separation of real and imaginary parts yields:

$$\frac{1}{2}\frac{\dot{n}_{C1}}{\sqrt{n_{C1}}} = \frac{K}{\hbar}\sqrt{n_{C2}}\sin(\varphi_2 - \varphi_1)$$

$$\frac{1}{2}\frac{\dot{n}_{C2}}{\sqrt{n_{C2}}} = \frac{K}{\hbar}\sqrt{n_{C1}}\sin(\varphi_1 - \varphi_2) \tag{A-7}$$

$$\mathrm{i}\sqrt{n_{C1}}\,\dot{\varphi}_1 = -\frac{\mathrm{i}}{\hbar}\{E_1\sqrt{n_{C1}} + K\sqrt{n_{C2}}\cos(\varphi_2 - \varphi_1)\}$$

$$\mathrm{i}\sqrt{n_{C2}}\,\dot{\varphi}_2 = -\frac{\mathrm{i}}{\hbar}\{E_2\sqrt{n_{C2}} + K\sqrt{n_{C1}}\cos(\varphi_1 - \varphi_2)\}\,. \tag{A-8}$$

If we now consider that, during the exchange of Cooper pairs between 1 and 2, $\dot{n}_{C1} = -\dot{n}_{C2}$ must always apply and assume for simplicity that the two superconductors are the same (i.e. $n_{C1} = n_{C2}$), then from (A – 7) we get the differential equation:

$$\dot{n}_{C1} = \frac{2K}{\hbar}n_{C1}\sin(\varphi_2 - \varphi_1) = -\dot{n}_{C2}\,. \tag{A-9}$$

The change in time of the particle density in 1 multiplied by V, the volume of 1, yields the change in numbers of particles and, hence, the particle current through the junction. The electric current I_s is obtained by multiplication of the particle current by the charge $2e$ of each individual particle. This yields the Josephson equation (3–21):

$$\underset{2\rightarrow 1}{I_s} = I_{smax}\sin(\varphi_2 - \varphi_1) \tag{3-21}$$

where

$$I_{smax} = \frac{2K\,2e}{\hbar}Vn_C \tag{A-10}$$

where e = elementary charge; V = volume of superconductor.

In making this transition from $\dot{n}_C$ to current through the junction we have to remember that both superconductors are connected to a current source, which ensures that n_C remains constant by supplying or accepting charges as necessary.

Equation (A – 8) yields a differential equation for the alteration of the phase difference with time. When $n_{C1} = n_{C2}$ this becomes:

$$\frac{\mathrm{d}}{\mathrm{d}t}(\varphi_2 - \varphi_1) = \frac{1}{\hbar}(E_1 - E_2)\,. \tag{A-11}$$

When $E_1 = E_2$ the phase difference remains constant with time. But if an electrical voltage U is maintained between 1 and 2 then $E_1 \neq E_2$ and

$$E_1 - E_2 = 2eU\,. \tag{A-12}$$

The charge $2e$ appears here too because it is the exchange of Cooper pairs that is involved. So Equation (A – 11) becomes:

$$\frac{\mathrm{d}}{\mathrm{d}t}(\varphi_2 - \varphi_1) = \frac{2eU}{\hbar}\,. \tag{A-13}$$

The phase difference increases linearly with time.

$$\varphi_2 - \varphi_1 = \frac{2eU}{\hbar}\,t + \varphi_0\,. \tag{A-14}$$

However, this means that according to Equation (3–21) an alternating current occurs in the junction, the frequency of this is given by:

$$\nu = \frac{2eU}{h}\,. \tag{3-54}$$

Appendix B. On the Ginsburg-Landau theory

The London theory (*cf.* Section 5.1.2) is restricted to states where the density of superconducting charges – say Cooper pairs – can be assumed to be constant in the whole of the sample. It does not, therefore, cover the intermediate state. In order to describe this (see Section 5.1.4) it is necessary to assume, in addition, the existence of a surface energy for the interface between superconducting and normalconducting regions.

The theory published in 1950 by Ginsburg and Landau overcame this restriction to states where n_s remains constant spacially. With this they produced an important extension to the London theory. Astonishingly for a long time this theory did not receive the attention it deserved. It was only after the development of a microscopic theory that the importance of this extended phenomenological theory became known and it was realized how brilliantly the fundamental physical relationships had already been comprehended by the formulations of Ginsburg and Landau in 1950.

The theory assumes that the transition N → S in the absence of an external field is a 2nd order phase transition (see Section 4.1). Landau had developed a theory for such phase transitions. Here a parameter, the order parameter, is defined, which rises in the new phase – here the superconducting phase – continuously from zero at T_c to the value 1 at $T = 0$. Ginsburg and Landau introduced a function $\psi(\vec{r})$ as order parameter. The quantity $|\psi(\vec{r})|^2$ can be understood as the density of superconducting charges.

Since $\psi(\vec{r})$ in the superconducting phase must continually fall towards zero as T_c is approached, the Gibbs function g_s [1] of the superconducting phase in the neighborhood of T_c can be developed from $\psi(\vec{r})$, giving:

$$g_s = g_n + \alpha\,|\psi|^2 + \frac{1}{2}\,\beta\,|\psi|^4 + \ldots \tag{B-1}$$

The Gibbs function for the normalconducting phase g_n occurs here because when $\psi = 0$ the Gibbs function g_s equals g_n. Since when $T < T_c$ it also follows that $g_s < g_n$ (criterion of stability), then of necessity $\alpha < 0$.

An adequate approximation is obtained in a sufficiently small region near T_c if only the first two terms in the series are taken into account, that is if the series is ended with $|\psi^4|$. The development coefficients α and β are then simply related to the thermodynamic critical field B_{cth} and the equilibrium density $n_s = |\psi_\infty|^2$ of the Cooper pairs at field zero [2]. We get:

$$g_n - g_s = -\alpha\,|\psi_\infty|^2 - \frac{1}{2}\,\beta\,|\psi_\infty|^4 = \frac{1}{2\mu_0}\,B_{cth}^2\,. \tag{B-2}$$

1 We use the letter g for the Gibbs function here to indicate that an energy *density* is involved.

2 $|\psi_\infty|^2$ is intended to indicate that we are dealing with values of $|\psi|^2$ that are present at equilibrium sufficiently distant from any phase boundary.

The fact that at equilibrium $g_s(|\psi|^2)$ reaches a minimum provides another equation relating α and β. So it follows that $dg_s/d|\psi|^2 = 0$ at the equilibrium value $|\psi_\infty|^2$. This gives:

$$\alpha + \beta\,|\psi_\infty|^2 = 0\,. \tag{B-3}$$

The combination of Eqs. (B – 2) and (B – 3) yields:

$$\alpha = -\frac{1}{\mu_0}\,\frac{B^2_{\text{cth}}}{n_s}$$
$$\beta = \frac{1}{\mu_0}\,\frac{B^2_{\text{cth}}}{n_s^2}\,. \tag{B-4}$$

The decisive extension of the phenomenological description now follows by the expression of the Gibbs function of the superconductor in the magnetic field assuming a feasible spacial variation of ψ. This would yield:

$$g_s(B) = g_n + \alpha|\psi|^2 + \frac{1}{2}\beta|\psi|^4 + \frac{1}{2\mu_0}|B_e - B_i|^2 + \frac{1}{2m'}|(-i\hbar\nabla - e'\vec{A})\psi|^2\,. \tag{B-5}$$

m' and e' are the mass and charge of the particles described by ψ, that is of the Cooper pairs [1]; ∇ is a differential operation to be applied to the function ψ [2].

Thus, in the magnetic field two new terms have to be added. The first term includes the energy necessary to change the magnetic field from B_e, the external field without superconductor, to B_i. When $B_i = 0$ (Meissner phase) this term yields the full exclusion energy. The second term takes account of any local variation of B_i and ψ in the superconductor. It encompasses the supercurrents which are necessary for a variation of the magnetic field (see Section 5.2.2). It also includes the energy that is necessary for the local variation of the Cooper pair density. This particular contribution which introduces a "stiffness" of the wave function (see also [I], p. 324), is of especial importance, as we saw in Section 5.1.5. For instance, it is the source of the positive surface energy for the phase boundary in type I superconductors.

The Gibbs function for the whole superconducting sample is obtained by integration of Eq. (B – 5) over the volume of the sample V:

---

1 Since we now know that Cooper pairs with mass 2 m and charge 2 e are the particles concerned we can substitute this mass and charge into the equation. Here we keep to the neutral descriptions m' and e' because the G.-L. theory was developed without any knowledge of the existence of Cooper pairs.

2 It is:

$$\nabla\psi = \operatorname{grad}\psi = \frac{\partial\psi}{\partial x}\vec{e}_x + \frac{\partial\psi}{\partial y}\vec{e}_y + \frac{\partial\psi}{\partial z}\vec{e}_z\,.$$

$$G_s(B) = \int_V \left\{ g_n + \alpha|\psi|^2 + \frac{1}{2}\beta|\psi|^4 + \frac{1}{2\mu_0}|B_e - B_i|^2 + \frac{1}{2m'} |(-i\hbar\nabla - e'\vec{A})\psi|^2 \right\} dV . \qquad (B-6)$$

This function G_s must be minimized by variation of ψ and $\vec{A}$. The variation process yields the two equations of the Ginsburg-Landau theory:

$$\frac{1}{2m'} |-i\hbar\nabla - e'\vec{A}|^2 \psi + \alpha\psi + \beta |\psi|^2 \psi = 0 \qquad (B-7)$$

$$j_s = \frac{e' i\hbar}{2m'} (\psi^* \nabla\psi - \psi\nabla\psi^*) - \frac{e'^2}{m'} |\psi|^2 \vec{A}\,[1] . \qquad (B-8)$$

This system of differential equations has to be complemented by a suitable boundary condition. This boundary condition is provided by the requirement that the current perpendicular to the surface of the superconductor shall disappear. So that:

$$\vec{n}(-i\hbar\nabla - e'\vec{A})\psi = 0 \qquad (B-9)$$

where $\vec{n}$ = unit vector perpendicular to the surface.

The solutions of these equations for particular cases now become a mathematical problem. The Shubnikov phase (see Section 5.2.2) is, as Abrikosov was able to demonstrate, one such solution.

1 ψ^* is the conjugated complex function to ψ.

Monographs and Textbooks

[I] R. D. Parks: "Superconductivity", 2 volumes, Marcel Dekker, Inc., New York 1969. Comprehensive monographic treatment of the current knowledge of superconductivity.

[II] D. Shoenberg: "Superconductivity", Cambridge, The University Press 1952. Excellent compilation of the knowledge until 1952.

[III] A. C. Rose-Innes, and E. H. Rhoderick: "Introduction to Superconductivity", Pergamon Press, 1969. Elementary introduction to the field.

[IV] E. A. Lynton: "Superconductivity", Methuen's Monographs on Physical Subjects, London 1964 (2nd edition).

[V] J. E. C. Williams: "Superconductivity and Its Application" Pion Limited, 207 Brondesburg Park, London NW 2 1970.

[VI] F. London: "Superfluids" Volume 1, "Macroscopic Theory of Superconductivity", John Wiley & Sons, Inc., New York 1950. Introduction to the phenomenological theory of superconductivity.

[VII] P. G. De Gennes: "Superconductivity of Metals and Alloys", W. A. Benjamin, Inc., New York 1966. Excellent introduction to the modern theory of superconductivity.

[VIII] G. Rickayzen: "Theory of Superconductivity", Interscience Publishers, John Wiley & Sons, New York 1965.

[IX] D. Saint-James, G. Sarma, and E. J. Thomas: "Type II Superconductivity", Pergamon Press, 1969.

[X] V. L. Newhouse: "Applied Superconductivity", Wiley & Sons, New York 1964. Introduction to the applications of superconductivity.

[XI] J. W. Bremer: "Superconducting Devices", McGraw-Hill Book Company, New York 1962. Introduction to the applications.

[XII] D. Fishlock: "A Guide to Superconductivity", McDonald, London 1969. Articles on special applications of superconductivity.

[XIII] "Vorträge über Supraleitung" ed. by Physikalische Gesellschaft, Zürich, Birkhäuser Verlag, Basel and Stuttgart 1968.

[XIV] H. Ullmaier: "Irreversible Properties of Type II Superconductors", Springer Tracts in Modern Physics 76, Springer, Berlin, Heidelberg, New York 1975. Short monograph on hard superconductors.

[XV] L. Solymar: "Superconductive Tunnelling and Applications", Chapman and Hall Ltd., London 1972.

[XVI] M. Tinkham: "Introduction to Superconductivity", McGraw-Hill Book Company, New York 1975.

[XVII] V. L. Ginsburg, and D. A. Kirzhuits: "High Temperature Superconductivity", Consultants Bureau, New York and London 1982.

[XVIII] Otto Henkel, and E. M. Sawitzkij: "Supraleitende Werkstoffe", VEB Deutscher Verlag f. Grundstoffindustrie, Leipzig 1982.

[XIX] J. Hinken: "Supraleiter-Elektronik", Grundlagen – Anwendungen in der Elektronik, Springer, Heidelberg 1988.

[XX] O. V. Lounasmaa: "Experimental Principles and Methods below 1K", Academic Press, London and New York 1974, p. 140ff.

References

[1] H. K. Onnes: Comm. Leiden Nr. 108, Proc. Roy. Acad. Amsterdam **11,** 168 (1908).
[2] H. K. Onnes: Comm. Leiden, Suppl. Nr. 34 (1913).
[3] H. K. Onnes: Comm. Leiden 120b (1911).
[4] H. K. Onnes: Comm. Leiden 140b, c, and 141b (1914).
[5] H. K. Onnes: Reports and Comm. 4th Intern. Congress on Cryogenics, London 1924, Seite 175. W. Tuyn: Comm. Leiden 198 (1929).
[6] D. J. Quinn, and W. B. Ittner: J. Appl. Phys. **33,** 748 (1962).
[7] B. W. Roberts: "Superconducting Materials and Some of Their Properties". IV. Progress in Cryogenics, page 160–231 (1964), and National Bureau of Standards, Technical Note 482.
[8] R. F. Hoyt, and A. C. Mota: Solid State Commun. **18,** 139 (1975); Solid State Commun, **20,** 1025 (1976).
[9] K. Ulmer: Solid State Commun. **2,** 327 (1964).
[10] W. Buckel, and B. Stritzker: Phys. Lett. **43A,** 403 (1973).
[11] R. L. Greene, G. B. Street, and L. J. Suter: Phys. Rev. Lett. **34,** 577 (1975).
[12] W. A. Little: Phys. Rev. **134A,** 1416 (1964).
[12a] L. R. Testardi, J. H. Wernick, and W. A. Royer: Sol. State Comm. **15,** 1 (1974) – J. R. Gavaler, M. H. Janoko, and C. K. Jones: J. Appl. Phys. **45,** 7 (1974).
[13] B. T. Matthias: Phys. Rev. **97,** 74 (1955), and Prog. Low Temp. Phys. **2,** 138 (1957), ed. by C. J. Gorter, North Holland, Amsterdam 1957.
[14] J. K. Hulm, and R. D. Blaugher: Phys. Rev. **123,** 1569 (1961) – V. B. Compton, E. Corenzwit, E. Maita, B. T. Matthias, and F. J. Morin: Phys. Rev. **123,** 1567 (1961) – W. Gey: Z. Phys. **229,** 85 (1969).
[15] J. G. Bednorz, and K. A. Müller: Z. Physik **B64,** 189 (1986).
[16] J. G. Bednorz, M. Takashige, and K. A. Müller: Europhysics Lett. **3,** 379 (1987).
[17] Ch. J. Raub: J. Less-Common Met. **137,** 287 (1988).
[18] R. J. Cava, R. B. van Dover, B. Batlogg, and E. A. Rietmann: Phys. Rev. Lett. **58,** 408 (1987). C. W. Chu, P. H. Hor, R. L. Meng, L. Gao, Z. J. Huang, and Y. O. Wang: Phys. Rev. Lett. **58,** 405 (1987).
[19] M. K. Wu, J. R. Asburn, C. J. Torng, P. H. Hor, R. L. Meng, L. Gao, Z. J. Huang, and C. W. Chu: Phys. Rev. Lett. **58,** 908 (1987).
[20] Z. X. Zhao: Int. J. Mod. Phys. **B1** (187), Lecture held at the Spring Meeting of the American Phys. Soc. in New York, March 1987.
[21] C. Politis, M. Geerk, M. Dietrich, B. Obst, and H. L. Luo: Z. Physik **B66,** 279 (1987).
[22] H. Maeda, Y. Tanaka, M. Fukutomi, and T. Asano: Jap. J. Phys. Lett. **27,** L 209 (1988).
[23] S. S. P. Parkin, V. Y. Lee, E. M. Engler, A. I. Nazzal, T. C. Huang, G. Gorman, R. Savoy, and R. Bayers: Phys. Rev. Lett. **60,** 2539 (1988); Z. Z. Sheng, A. M. Herman, A. El Ali, L. Almasan, J. Estrada, T. Datta, and R. J. Matsoun: Phys. Rev. Lett. **60,** 937 (1988).
[24] W. Heisenberg: Z. Naturforsch. **2a,** 185 (1947).
[25] H. Welker: Z. Phys. **114,** 525 (1939).
[26] M. Born, and K. C. Cheng: Nature, (London) **161,** 968 and 1017 (1948).
[27] H. Fröhlich: Phys. Rev. **79,** 845 (1950).
[28] J. Bardeen: Phys. Rev. **80,** 567 (1950).
[29] J. Bardenn, L. N. Cooper, and J. R. Schrieffer: Phys. Rev. **108,** 1175 (1957).
[30] L. N. Cooper: Phys. Rev. **104,** 1189 (1956).
[31] "Feynman-Lectures on Physics", Vol. 3, 10-1, and ff. Addison-Wesley Publ. Comp. 1965.
[32] Ch. Kittel: "Introduction to Solid State Physics", 4th Edition, John Wiley & Sons Inc. New York 1971.
[33] F. London: "Superfluids", Vol. I, page 152, Wiley 1950.

[34] H. K. Onnes, and W. Tuyn: Comm. Leiden **160b** (1922).
[35] E. Justi: Phys. Z. **42**, 325 (1941).
[36] E. Maxwell: Phys. Rev. **78**, 477 (1950).
[37] C. A. Reynolds, B. Serin, W. H. Wright, and L. B. Nesbitt: Phys. Rev. **78**, 487 (1950).
[38] E. Maxwell: Phys. Rev. **86**, 235 (1952) – B. Serin, C. A. Reynolds, and C. Lohman: Phys. Rev. **86**, 162 (1952) – J. M. Lock, A. B. Pippard, and D. Shoenberg: Proc. Cambridge Phil. Soc. **47**, 811 (1951).
[39] R. D. Fowler, J. D. G. Lindsay, R. W. White, H. H. Hill, and B. T. Matthias: Phys. Rev. Lett. **19**, 892 (1967). – W. E. Gardner, and T. F. Smith: Phys. Rev. **154**, 309 (1967).
[40] W. L. McMillan: Phys. Rev. **167**, 331 (1968).
[41] J. J. Engelhardt, G. W. Webb, and B. T. Matthias: Science **155**, 191 (1967).
[42] B. Batlogg, R. J. Cava, A. Jayaraman, R. B. van Dover, G. A. Kourouklis, S. Sunshine, D. W. Murphy, L. W. Rupp, H. S. Chen, A. White, A. M. Mujsce, and E. A. Rietman: Phys. Rev. Lett. **58**, 2333 (1987).
[43] B. Batlogg, G. Kourouklis, W. Weber, J. R. Cava, A. Jayaraman, A. E. White, K. T. Short, L. W. Rupp, and E. A. Rietman: Phys. Rev. Lett. **59**, 912 (1987).
[44] R. Doll, and M. Näbauer: Phys. Rev. Lett. **7**, 51 (1961).
[45] B. S. Deaver Jr., and W. M. Fairbank: Phys. Rev. Lett. **7**, 43 (1961).
[46] D. Estere, J. M. Martinis, C. Urbina, M. H. Deveret, G. Collin, P. Monod, and M. Ribault: Europhysics Lett. **3**, 1237 (1987).
[47] C. E. Gough, M. S. Colclough, E. M. Forgan, R. G. Jordan, M. Keene, C. M. Muirhead, A. I. M. Rae, N. Thomas, J. S. Abell, and S. Sutton: Nature 326, 855 (1987), see also C. E. Gough: Physica **C153–155**, 1569 (1988).
[48] R. E. Glover III, and M. Tinkham: Phys. Rev. **108**, 243 (1957).
[49] P. L. Richards, and M. Tinkham: Phys. Rev. **119**, 575 (1960).
[50] J. G. Daunt, and K. Mendelssohn: Proc. R. Soc. London, Ser. A **185**, 225 (1946).
[51] R. W. Morse, and H. V. Bohm: Phys. Rev. **108**, 1094 (1957).
[52] J. C. Fisher, and I. Giaever: J. Appl. Phys. **32**, 172 (1961).
[53] I. Giaever: Phys. Rev. Lett. **5**, 464 (1960).
[54] I. Giaever, and K. Megerle: Phys. Rev. **122**, 1101 (1961).
[55] J. Sutten, and P. Townsed: Proc. LT 8, 182 (1963).
[56] I. Giaever: Proc. LT 8, 171 (1963).
[57] A. Barone: Physica **C153–155**, 1712 (1988).
[58] B. D. Josephson: Phys. Rev. Lett. **1**, 251 (1962).
[59] D. N. Langenberg, D. J. Scalapino, and B. N. Taylor: Proc. IEEE **54**, 560 (1966).
[60] G. Möllenstedt, and W. Bayh; Phys. Bl. **18**, 299 (1962).
[61] R. C. Jaklevic, J. Lambe, J. E. Mercereau, and A. H. Silver: Phys. Rev. **140**A, 1628 (1965).
[62] A. Th. A. M. de Waele, W. H. Kraan, and R. de Bruyn Ouboter: Physica **40**, 302 (1968) – see also Prog. Low Temp. Phys. **6**, (1969), ed. by C. J. Gorter, North Holland, Amsterdam.
[63] S. Shapiro: Phys. Rev. Lett. **11**, 80 (1963).
[64] M. D. Fiske: Rev. Mod. Phys. **36**, 221 (1964).
[65] I. Giaever: Phys. Rev. Lett. **14**, 904 (1965).
[66] A. H. Dayem, and R. J. Martin: Phys. Rev. Lett. **8**, 246 (1962).
[67] D. N. Langenberg, D. J. Scalapino, B. N. Taylor, and R. E. Eck: Phys. Rev. Lett. **15**, 294 (1965).
[68] D. N. Langenberg, D. J. Scalapino, and B. N. Taylor: Sci. Am. **214**, May 1966.
[69] L. K. Yanson, V. M. Svistunov, and J. M. Dmitrenko: Zh. Eksperim. Teor. Fiz. **48**, 976 (1965) – Sov. Phys. JETP **21**, 650 (1966).
[70] D. N. Langenberg, W. H. Parker, and B. N. Taylor: Phys. Rev. **150**, 186 (1966), and Phys. Rev. Lett. **18**, 287 (1967).
[71] Y. Zhao, S. Sun, Z. Su, W. Kuan, H. Zhang, Z. Chen, and Qu. Zhang: Physica **C153–155**, 304 (1988).

[72] N. V. Zavaritsky, and V. N. Zavaritsky: Physica **C153–155,** 1405 (1988).
[73] F. London, and H. London: Z. Phys. **96,** 359 (1935) – F. London: "Une conception nouvelle de la supraconductivité", Hermann and Cie, Paris 1937.
[74] W. H. Keesom: IV. Congr. Phys. Solvay (1924), Rapp. et Disc. page 288.
[75] C. J. Gorter, and H. B. G. Casimir: Physica **1,** 306 (1934).
[76] W. Meissner, and R. Ochsenfeld: Naturwissenschaften **21,** 787 (1933).
[77] L. D. Landau: Phys. Z. Sowjet. **11,** 545 (1937), L. D. Landau, and E. M. Lifshitz: "Course of Theoretical Physics" Vol. **5,** 430, Pergamon Press 1959.
[78] K. Mendelssohn, and J. R. Moore: Nature **133,** 413 (1934). K. Mendelssohn: "Cryophysics", Interscience Publisher Ltd. London 1960 – K. Mendelssohn: Nature (London) **169,** 336 (1952).
[79] A. J. Rutgers: Physica **3,** 999 (1936).
[80] W. H. Keesom, and P. H. van Laer: Physica **5,** 193 (1938).
[81] M. A. Biondi, A. T. Forester, M. P. Garfunkel, and C. B. Satterthwaite: Rev. Mod. Phys. **30,** 1109 (1958).
[82] A. Junod: Physica **C153–155,** 1078 (1988).
[83] D. M. Ginsberg, S. E. Inderhees, M. B. Salamon, N. Goldenfeld, J. Rice, S. Tajima, S. Tanaka, and B. G. Pazol: Physica **C153–155,** 1082 (1988).
[84] K. Kitazawa, H. Takagi, K. Kishio, T. Hasegawa, S. Uchida, S. Tajima, S. Tanaka, and K. Fueki: Physica **C153–155,** 9 (1988).
[85] G. J. Sizoo, and H. K. Onnes: Comm. Leiden 180 b.
[86] L. D. Jennings, and C. A. Swenson: Phys. Rev. **112,** 31 (1958).
[87] N. B. Brandt, and N. I. Ginsburg: Sov. Phys. USPEKHI **12,** 344 (1969).
[88] R. I. Boughton, J. L. Olsen, and C. Palmy, Prog. Low Temp. Phys. **6,** 163 (1970), ed. by C. J. Gorter, North Holland, Amsterdam.
[89] N. E. Alekseevskii, Y. and P. Gaidukov: Sov. Phys. JETP **2,** 762 (1956).
[90] B. G. Lasarev, and L. S. Kan: Zh. Eksp. Teor. Fiz. **18,** 825 (1944).
[91] J. L. Olsen, and H. Rohrer: Helv. Phys. Acta **30,** 49 (1957).
[92] F. P. Bundy: J. Chem. Phys. **41,** 3809 (1964).
[93] W. Buckel, and J. Wittig: Phys. Lett. **17,** 187 (1965). J. Wittig: Z. Phys. **195,** 215 (1966).
[94] J. Wittig: Z. Phys. **195,** 288 (1966).
[95] A. Eichler, and J. Wittig: Z. Angew. Physik **25,** 319 (1968).
[96] C. W. Chu, P. H. Hor, R. L. Meng, L. Gao, and Z. J. Huang: Science **235,** 567 (1987).
[97] N. Lotter, J. Wittig, W. Assmus, and J. Kowalevsky: Physica **C153–155,** 1355 (1988).
[98] M. B. Maple, Y. Dalichaouch, J. M. Ferreira, R. R. Hake, S. E. Lambert, B. W. Lee, J. J. Neumeier, M. S. Torikachvili, K. N. Yang, and H. Zhou, Z. Fisk, M. W. McElfresh, and J. L. Smith: "Novel Superconductivity", ed. by A. Wolf, and V. Z. Kresin, Plenum Press, New York 1987, p. 839.
[99] A. Driessen, R. Griessen, N. Koeman, E. Salomons, R. Brouwer, D. G. deGroot, K. Heck, H. Hemmes, and J. Rector: Phys. Rev. **B36,** 5602 (1987).
[100] J. K. Hulm: Proc. R. Soc. London, Ser. A**204,** 98 (1950).
[101] K. Mendelssohn, and J. L. Olsen: Proc. Phys. Soc. London, Sect. A**63,** 1182 (1950).
[102] H. K. Onnes: Comm. Leiden Suppl. 50 a (1924).
[103] J. C. Gorter: Arch. Mus. Teyler **7,** 378 (1933).
[104] A. B. Pippard: Proc. R. Soc. London, Ser. A**203,** 210 (1950).
[105] D. Shoenberg: Nature (London) **143,** 434 (1939), and Proc. R. Soc. London, Ser. A**175,** 49 (1940).
[106] P. C. L. Tai, M. R. Beasley, and M. Tinkham: Phys. Rev. **B11,** 411 (1975).
[107] J. M. Lock: Proc. R. Soc. London, Ser. A**208,** 391 (1951).
[108] A. G. Meshkovsky, and A. I. Shalnikov: Zh. Eksp. Teor. Fiz. **17,** 851 (1947). (Detailed discussion in [II].)

[109] A. G. Meshkovsky: Zh. Eksp. Teor. Fiz. **19**, 1 (1949).
[110] A. L. Schawlow, B. T. Matthias, H. W. Lewis, and G. E. Delvin: Phys. Rev. **95**, 1345 (1954).
[111] H. Kirchner: Phys. Lett. **26A**, 651 (1968). P. B. Alers: Phys. Rev. **116**, 1483 (1959), and phys. stat. sol. (a) **4**, 531 (1971). W. DeSorbo: Phil. Mag. **11**, 853 (1965).
[112] A. B. Pippard: Proc. R. Soc. London, Ser. **A216**, 547 (1953), and Proc. Camb. Phil. Soc. **47**, 617 (1951).
[113] V. L. Ginsburg, and L. D. Landau: Zh. Eksp. Teor. Fiz. **20**, 1044 (1950).
[114] W. J. DeHaas, and J. Voogd: Comm. Leiden 208b (1930), and 214b (1931).
[115] K. Mendelssohn: Proc. R. Soc. London, Ser. **A152**, 34 (1935).
[116] A. A. Abrikosov: Zh. Eksp. Teor. Fiz. **32**, 1442 (1957) – Sov. Phys. JETP **5**, 1174 (1957).
[117] L. P. Gorkov: Zh. Eksp. Teor. Fiz. **36**, 1918 (1959); Sov. Phys. JETP **9**, 1364 (1960).
[118] J. D. Livingston: Phys. Rev. **129**, 1943 (1963).
[119] L. P. Gorkov: Zh. Eksp. Teor. Fiz. **37**, 1407 (1959) – Sov. Phys. JETP **10**, 998 (1960).
[120] T. Kinsel, E. A. Lynton, and B. Serin: Rev. Mod. Phys. **36**, 105 (1964).
[121] G. Otto, E. Saur, and H. Witzgall: J. Low Temp. Physics **1**, 19 (1969).
[122] W. DeSorbo: Phys. Rev. **A104**, 914 (1965).
[123] R. Chevrel, M. Sergent, and J. Prigent: J. Sol. State Chem. **3**, 515 (1971).
[124] R. Odermatt, Ø. Fischer, H. Jones, and G. Bongi: J. Phys. **C7**, L13 (1974).
[125] Ø. Fischer, H. Jones, G. Bongi, M. Sergent, and R. Chevrel: J. Phys. **C7**, L450 (1974).
[126] S. Foner, E. J. McNiff, and E. J. Alexander: Phys. Lett. **49A**, 269 (1974).
[127] D. Cribier, B. Jcrot, L. Madhav Rao, and B. Farnoux: Phys. Lett. **9**, 106 (1964). See also Progress Low Temp., Phys. Vol. 5, ed. by C. J. Gorter, North Holland Publishing Comp. Amsterdam (1967).
[128] J. Schelten, H. Ullmaier, and W. Schmatz: Phys. Status Solidi **48**, 619 (1971).
[129] U. Eßmann, and H. Träuble: Phys. Lett. **24A**, 526 (1967), and J. Sci. Instrum. **43**, 344 (1966).
[130] L. Neumann, and L. Tewordt: Z. Phys. **191**, 73 (1966).
[131] F. B. Silsbee: J. Wash. Acad. Sci. **6**, 597 (1916).
[132] B. K. Mukherjee, and J. F. Allen: Proc. LT 11, StAndrews 1968, p. 827. B. K. Mukherjee, and D. C. Baird: Phys. Rev. Lett. **21**, 996 (1968).
[133] F. Haenssler, and L. Rinderer: Helv. Phys. Acta **40**, 659 (1967).
[134] W. Buckel, and R. Hilsch: Z. Phys. **149**, 1 (1957).
[135] C. J. Gorter: Physica **23**, 45 (1957). C. J. Gorter, and M. L. Potters: Physica **24**, 169 (1958). B. S. Chandrasekhar, I. J. Dinewitz, and D. E. Farrell: Phys. Lett. **20**, 321 (1966).
[136] H. Meißner: Phys. Rev. **97**, 1627 (1955). K. Steiner, and H. Schoeneck: Phys. Z. **38**, 887 (1937).
[137] W. Klose: Phys. Lett. **8**, 12 (1964).
[138] C. J. Gorter: Phys. Lett. **1**, 69 (1962).
[139] I. Giaever: Phys. Rev. Lett. **15**, 825 (1966).
[140] G. J. van Gurp: Phys. Lett. **24A**, 528 (1967). P. R. Solomon: Phys. Rev. **179**, 475 (1969).
[141] J. W. Heaton, and A. C. Rose-Innes: Cryogenics (G.B.) **4**, 85 (1965).
[142] A. Campell, J. E. Evetts, and D. DewHughes: Phil. Mag. **10**, 333 (1964), and Phil. Mag. **10**, 339 (1964).
[143] A. R. Strnad, C. F. Hempstead, and Y. B. Kim: Phys. Rev. Lett. **13**, 794 (1964).
[144] G. Bogner: Elektrotech. Z **89**, 321 (1968).
[145] J. Petermann: Z. Metallk. **61**, 724 (1970).
[146] C. P. Bean: Phys. Rev. Lett. **8**, 250 (1962), and Rev. Mod. Phys. **36**, 31 (1964).
[147] Y. B. Kim, C. F. Hempstead, and A. R. Strnad: Phys. Rev. **129**, 528 (1963).
[148] E. V. Thuneberg, J. Kurkijärvi, and D. Rainer: Phys. Rev. **B29**, 3913 (1984).
[149] K. Yamafuji, and F. Irie: Phys. Lett. **25A**, 387 (1967).
[150] R. D. Dunlap, C. F. Hempstead, and Y. B. Kim: J. Appl. Phys. **34**, 3147 (1963).

[151] A. Umezawa, G. W. Crabtree, and J. Z. Liu: Physica **C153–155**, 1461 (1988).
[152] H. Küpfer, I. Apfelstedt, W. Schauer, R. Flückiger, R. Meier-Hirmer, and H. Wühl: Z. Physik **B69**, 159 (1987).
[153] T. K. Worthington, W. J. Gallagher, D. L. Kaiser, F.H. Holtzberg, and T. R. Dinger: Physica **C153–155**, 32 (1988).
[154] M. Mitchell Waldorf: Science **238**, 1656 (1987).
[155] P. Chaudhari, R. H. Koch, R. B. Laibowitz, T. R. McGuire, and R. J. Gambino: Phys. Rev. Lett. **58**, 2684 (1987).
[156] B. Roas, G. Endres, and L. Schultze: Appl. Phys. Lett. **53**, 1557 (1988).
[157] J. Fröhlingsdorf, W. Zander, and B. Stritzker: Proc. EMRS-Fall Meeting 10.11.1988, Straßbourg, J. of the Less Common Metals **151**, 407 (1989).
[158] Rolfe E. Glover III: Prog. Low Temp. Phys. **6**, p. 291, ed. by C. J. Gorter, North Holland Publishing Comp., Amsterdam – Phys. Lett. **25**A, 542 (1967).
[159] L. G. Aslamazov, and A. I. Larkin: Phys. Lett. **26**A, 238 (1968). H. Schmidt: Z. Phys. **216**, 336 (1968). A. Schmid: Z. Phys. **215**, 210 (1968).
[160] J. P. Gollub, M. R. Beasley, R. S. Newbower, and M. Tinkham: Phys. Rev. Lett. **22**, 1288 (1969).
[161] G. D. Zally, and J. M. Mochel: Phys. Rev. Lett. **27**, 1710 (1971).
[162] A. Kapitulnik: Physica **C153–155**, 520 (1988).
P. P. Freitas, C. C. Tsuei, and T. S. Plaskett: Phys. Rev. **B36**, 833 (1987).
M. A. Dubson, J. J. Calabrese, S. T. Herbert, D. C. Harris, B. R. Patton, and J. C. Garland: "Novel Superconductivity", ed. by S. A. Wolf, and V. Z. Kresin, Plenum Press, New York 1987, p. 981.
[163] D. N. Langenberg: Proceedings of LT 14, Otaniemi, August 1975, North Holland Publishing Comp. Amsterdam 1975, pp. 223 and ff.
W. Eisenmenger, K. Laßmann, H. J. Trumpp, and R. Krauß: Appl. Phys. **11**, 307 (1976).
[164] D. M. Ginsberg: Phys. Rev. Lett. **8**, 204 (1962).
[165] K. F. Gray, A. R. Long, and C. J. Adkins: Phil. Mag. **20**, 273 (1969); K. F. Gray: J. Phys. **F.1**, 290 (1971).
[166] G. A. Sai-Halasz, C. C. Chi, A. Denenstein, and D. N. Langenberg: Phys. Rev. Lett. **33**, 215 (1974).
[167] Yu. I. Latysher, and F. Ya. Nad': JETP Lett. **19**, 380 (1974) – A. F. Wyatt, V. M. Dmitriev, W. S. Hoote, and F. W. Sheard: Phys. Rev. Lett. **16**, 1166 (1966).
A. H. Dayem, and J. Wiegand: Phys. Rev. **155**, 419 (1967).
[168] T. M. Klapwijk, and J. E. Moij: Physica **81B**, 132 (1976).
[169] R. H. Parmenter: Phys. Rev. Lett. **7**, 274 (1961).
[170] G. M. Eliashberg: JETP Lett. **11**, 114 (1970) – B. I. Ivler, S. G. Lisistsyn, and G. M. Eliashberg: J. Low Temp. Phys. **10**, 449 (1973).
[171] J. Meyer, and G. v. Minnigerode: Phys. Lett. **38A**, 529 (1972); J. D. Meyer: Appl. Phys. **2**, 303 (1973).
[172] W. J. Skopol, M. R. Beasley, and M. Tinkham: J. Low Temp. Phys. **16**, 145 (1974).
[173] G. J. Dolan, and L. D. Jackel: Phys. Rev. Lett. **39**, 1628 (1977).
[174] M. Tinkham: Festkörperprobleme **XIX**, 363 (1979).
[175] H. London: Proc. R. Soc. London, Ser. **A176**, 522 (1940).
[176] M. A. Biondi, and M. P. Garfunkel: Phys. Rev. **116**, 853 and 862 (1959), and Phys. Rev. Lett. **2**, 143 (1959).
[177] S. L. Lehoczky, and C. V. Briscoe: Phys. Rev. **B4**, 3938 (1971).
[178] Ch. Fuchs, and J. Hasse: Z. Phys. **B28**, 183 (1977).
[179] A. D. Misener, and J. O. Wilhelm: Trans. Roy. Soc. Canada **29**, (III) 5 (1935).
[180] P. Hilsch, R. Hilsch, and G. v. Minnigerode: Proc. 8th Int. Conf. Low Temp. Phys. (LT8), p. 381, London 1963, Butterworth. – P. Hilsch: Z. Phys. **167**, 511 (1962).

[181] G. v. Minnigerode: Z. Phys. **192,** 379 (1966).
[182] J. J. Hauser, H. C. Theurer, and N. R. Werthammer: Phys. Lett. **18,** 222 (1965). – See also P. Hilsch, and R. Hilsch: Z. Phys. **180,** 10 (1964).
[183] G. Bergmann: Z. Phys. **187,** 395 (1965).
[184] W. Rühl: Z. Phys. **186,** 190 (1965).
[185] D. G. Naugle, and R. E. Glover III: Phys. Lett. **28**A, 611 (1969). – W. Kessel, and W. Rühl: PTB-Mitteilungen **79,** 258 (1969).
[186] B. Abeles, R. W. Cohen, and G. W. Cullen: Phys. Rev. Lett. **17,** 632 (1966).
[187] E. A. Lynton, B. Serin, and M. Zucker: J. Phys. Chem. Solids **3,** 165 (1957).
[188] J. Hasse, and K. Lüders: Z. Phys. **173,** 413 (1963).
[189] W. Gey: Phys. Rev. **153,** 422 (1967).
[190] W. Buckel, and R. Hilsch: Z. Phys. **132,** 420 (1952).
[191] W. Buckel: Z. Phys. **138,** 136 (1954).
[192] A. Schertel: Phys. Verh. **2,** 102 (1951).
[193] J. Fortmann, and W. Buckel: Z. Phys. **162,** 93 (1961).
[194] W. Buckel, and R. Hilsch: Z. Phys. **138,** 109 (1954).
[195] R. Hilsch, and A. Schertel: Phys. Verh. **1,** 104 (1950); H. Leitz, H.-J. Nowak, S. El-Desouki, V. K. Srivastava, and W. Buckel: Z. Phys. **B29,** 199 (1978).
[196] F. Meunier, J. J. Hauser, J. P. Burger, E. Guyon, and M. Hesse: Phys. Lett. **28A,** 37 (1968); A. M. Lamoise, J. Chaumont, and F. Meunier: J. Phys. Lett. (Orsay, Fr.) **36,** L271 (1975); A. M. Lamoise, J. Chaumont, F. Meunier, H. Barnas, and F. Lalu: J. Phys. Lett (Orsay, Fr.) **37,** L287 (1976).
[197] R. E. Glover III, F. Baumann, and S. Moser: Proc. 12th Int. Conf. Low Temp. Phys. (LT 12), Kyoto 1970, p. 337, Academic Press of Japan.
[198] G. Bergmann: Phys. Rev. B**3**, 3797 (1971), and Phys. Reports **27C,** 159 (1976); B. Keck, and A. Schmid: J. Low Temp. Phys. **24,** 611 (1976).
[199] W. Buckel: Z. Phys. **154,** 474 (1959).
[200] P. F. Chester, and G. O. Jones: Phil. Mag. **44,** 1281 (1953). N. B. Brandt, and N. I. Ginsburg: Zh. Eksp. Teor. Fiz. **44,** 478 (1963) – Sov. Phys. JETP **17,** 326 (1963).
[201] W. Buckel, and W. Gey: Z. Phys. **176,** 336 (1963).
[202] H.-J. Güntherodt, and H. Beck: Top. Appl. Phys. **46** (1981).
[203] W. L. Johnson: Top. Appl. Phys. **46,** 191 (1981).
[204] R. C. Zeller, and R. O. Pohl: Phys. Rev. **B4,** 2029 (1971).
[205] G. Kämpf, H. Selisky, and W. Buckel: Physica **108B,** 1263 (1981); H. v. Löhneysen: Phys. Reports **79,** 161 (1981).
[206] B. T. Matthias, H. Suhl, and E. Corenzwit: Phys. Rev. Lett. **1,** 92 (1958).
[207] A. A. Abrikosov, and L. P. Gorkov: Zh. Eksp. Teor. Fiz. **39,** 1781 (1960) – Sov. Phys. JETP **12,** 1243 (1961).
[208] E. Wassermann: Z. Phys. **187,** 369 (1965).
[209] N. Barth: Z. Phys. **148,** 646 (1957).
[210] W. Buckel, M. Dietrich, G. Heim, and J. Keßler: Z. Phys. **245,** 283 (1971); W. Bauriedl, P. Ziemann, and W. Buckel: Phys. Rev. Lett. **47,** 1163 (1981).
[211] Kl. Schwidtal: Z. Phys. **158,** 563 (1960).
[212] T. H. Geballe, B. T. Matthias, E. Corenzwit, and G. W. Hull Jr.: Phys. Rev. Lett. **8,** 313 (1962).
[213] G. Boato, G. Gallinaro, and C. Rizutto: Phys. Rev. **148,** 353 (1966).
[214] Kl. Schwidtal: Z. Phys. **169,** 564 (1962).
[215] N. Falke, N. P. Jablonski, J. Kästner, and E. Wassermann: Z. Phys. **220,** 6 (1969).
[216] P. Ziemann: Festkörperprobleme **XXIII,** 93 (1983).
[217] W. Opitz: Z. Phys. **141,** 263 (1955); A. W. Bjerkaas, D. M. Ginsberg, and B. J. Mostik: Phys. Rev. **B5,** 854 (1972).
[218] A. Hofmann, W. Bauriedl, and P. Ziemann: Z. Phys. **B46,** 1117 (1982).

[219] E. Müller-Hartmann, and J. Zittartz: Phys. Rev. Lett. **26**, 428 (1970).
[220] K. Winzer: Z. Physik **265**, 139 (1973) – G. Riblet, and K. Winzer: Solid State Commun. **9**, 1663 (1971) – M. Brian Maple: Appl. Phys. **9**, 179 (1976).
[221] K. Winzer: Solid State Comm. **24**, 551 (1977). R. Dreyer, T. Krug, and K. Winzer: J. Low Temp. Phys. **48**, 111 (1982).
[222] V. Ambegaokar, and A. Griffin: Phys. Rev. **137**A, 1151 (1965).
[223] M. A. Woolf, and F. Reif: Phys. Rev. **137**A, 557 (1965).
[224] J. Zittartz, A. Bringer, and E. Müller-Hartmann: Solid State Commun. **10**, 513 (1972).
[225] A. M. Clogston, and V. Jaccarino: Phys. Rev. **121**, 1357 (1961) – M. Weger: Rev. Mod. Phys. **36**, 175 (1964) – J. Labbé, and J. Friedel: J. Phys. (Paris) **27**, 153 and 303 (1966).
[226] L. R. Testardi, and T. B. Bateman: Phys. Rev. **154**, 402 (1967).
[227] B. W. Batterman, and C. S. Barrett: Phys. Rev. Lett. **13**, 390 (1964).
[228] P. Schweiss: Dissertation Karlsruhe 1976.
[229] R. Flückiger, J. L. Staudemann, A. Treyvaud, and P. Fischer: Proceedings of LT 14, Otaniemi 1975, Vol. 2, p. 1, North Holland Publishing Comp. 1975.
[230] M. Weger, and I. B. Goldberg: Solid State Phys. **28**, 1 (1973).
[231] W. Weber: Proc. IV. Conf. Superconductivity in d- and f-Band Metals, 1982, p. 15, Karlsruhe FRG: Ed. W. Buckel, and W. Weber, Kernforschungszentrum Karlsruhe 1982.
[232] D. Dew-Hughes, and V. G. Rivlin: Nature (London) **250**, 723 (1974) – S. Geller: Appl. Phys. **7**, 322 (1975).
[233] J. Noolandi, and L. R. Testardi: Bell Laboratories, Murray Hill, N.J. 07974, private communication 1976.
[234] J. Muller: Rep. Prog. Phys. **43**, 641 (1980) (G.B.).
[235] T. Skoskiewicz: Phys. Status Solidi **A11K**, 123 (1972).
[236] B. Stritzker, and W. Buckel: Z. Phys. **257**, 1 (1972).
[237] G. Heim, and B. Stritzker: Appl. Phys. **7**, 239 (1975).
[238] R. J. Miller, and C. B. Satterthwaite: Phys. Rev. Lett. **34**, 144 (1975).
[239] B. Stritzker: Z. Phys. **268**, 261 (1974).
[240] B. N. Ganguly: Z. Phys. **265**, 433 (1973); Phys. Lett. **46A**, 23 (1973).
[241] A. Eichler, H. Wühl, and B. Stritzker: Solid State Commun. **17**, 213 (1975).
[242] J. M. Rowe, J. J. Rush, H. G. Smith, M. Mostoller, and H. E. Flotow: Phys. Rev. Lett. **33**, 1297 (1974).
[243] F. Ochmann, and B. Stritzker: Nucl. Instrum. & Methods **209**, 831 (1983); B. Stritzker: "Electronic Structure and Properties of Hydrogen in Metals", ed. by P. Jena, and C. B. Satterthwaite, Plenum Publishing Corp. (1983).
[244] D. Jerome, A. Mazaud, M. Ribault, and K. Bechgaard: C. R. Hebd. Seances Acad. Sci. Sér. **B290**, (1980).
[245] K. Bechgaard, K. Carneiro, M. Olsen, F. B. Rasmussen, and C. B. Jacobsen: Phys. Rev. Lett. **46**, 852 (1981).
[246] K. Oshima, H. Urayama, H. Yamochi, and G. Saito: Physica **C153–155**, 1148 (1988).
[247] J. D. Jörgensen, H. Shaked, D. G. Hinks, B. Dabrowski, B. W. Veal, A. P. Panlikas, L. J. Nowicki, G. W. Grabtree, W. K. Kwok, L. H. Numez, and H. Claus: Physica **C153–155**, 578 (1988); R. Beyers, G. Lim, E. M. Engler, V. Y. Lee, M. L. Ramirez, R. Y. Savoy, R. D. Jacowitz, T. M. Shaw, S. LaPlaca, R. Boehme, C. D. Tsuei, Sung I. Park, M. W. Shafer, and W. J. Gallagher: Appl. Phys. Lett. **51**, 614 (1987).
[248] G. Deutscher: Physica **C153–155**, 15 (1988).
[249] L. Forro, M. Raki, C. Ayache, P. C. E. Stamp, J. Y. Henry, and J. Rossat-Mignod: Physica **C153–155**, 1357 (1988).
[250] H. Rietschel, J. Fink, E. Gering, F. Gompf, N. Nücker, L. Pinschovius, B. Renker, W. Reichardt, H. Schmidt, and W. Weber: Physica **C153–155**, 1067 (1988).

[251] B. Batlogg, R. J. Cava, G. H. Chen, G. Kourouklis, W. Weber, A. Jayarama, E. A. White, K. T. Short, E. A. Rietman, L. W. Rupp, D. Werder, and S. M. Zahurak: "Novel Superconductivity", ed. by S. A. Wolf, and V. Z. Kresin, Plenum Press, New York 1987, p. 653.
[252] J. M. Tarascon, L. H. Greene, B. G. Bagley, W. R. McKinnon, P. Barboux, and G. W. Hull: "Novel Superconductivity", ed. by S. A. Wolf, and V. Z. Kresin, Plenum Press, New York 1987, p. 705.
[253] C. Politis, and H. Luo: Mod. Phys. Lett. **B2,** 793 (1988); Z. X. Zhao, L. Q. Chen, Z. H. Mai, Y. Z. Huang, Z. L. Xiáo, X. Chu, D. N. Zheng, S. L. Jla, J. H. Wang, G. H. Chen, Y. M. Ni, J. Q. Bi, Q. S. Yang, D. H. Shen, and L. Z. Wang: Mod. Phys. Lett. **2,** 477 (1988).
[254] R. J. Cava, B. Batlogg, J. J. Krajewski, R. Farrow, L. W. Rupp Jr., A. E. White, K. Short, W. F. Peck, and T. Kometani: Nature **332,** 814 (1988).
[255] P. Fulde: Physica **C153–155,** 1769 (1988).
J. R. Schrieffer, X.-G. Wen, and S.-C. Zhang: Physica **C153–155,** 21 (1988);
P. W. Anderson, G. Baskaran, Z. Zou, J. Wheatley, T. Hsu, B. S. Sastry, B. Doucot, and S. Liang: Physica **C153–155,** 527 (1988).
[256] F. Steglich, J. Aarts, C. D. Bredl, W. Lieke, D. Meschede, W. Franz, and H. Schäfer: Phys. Rev. Lett. **43,** 1892 (1979).
[257] H. Ott, H. Rüdiger, Z. Fisk, and J. L. Smith: Phys. Rev. Lett. **50,** 1595 (1983).
[258] Z. Fisk, H. Borges, M. McElfresh, J. L. Smith, J. D. Thompson, H. R. Ott, G. Aeppli, E. Bucher, S. E. Lambert, M. B. Maple, C. Borholm, and J. K. Kjems: Physica **C153–155,** 1728 (1988).
[259] "Physics today" December 83, p. 20.
[260] P. F. Smith: "The Technology of Large Magnets" in "A Guide to Superconductivity", ed. by D. Fishlock, McDonnald & Co. Ltd. 1969.
[261] P. Komarek, and H. Krauth: Nachrichten d. Kernforschungszentrums Karlsruhe **14,** 75 (1982).
[262] Z. J. Stekly: J.B.N.L. 50155 (c-55) p. 750; A. R. Kontorowitz, and Z. J. Stekly: Appl. Phys. Lett. **6,** 56 (1965). – D. B. Montgomery, and L. Rinderer: Cryophysics **8,** August 1968.
[263] K. Brennfleck, M. Dietrich, E. Fitzer, and D. Kehr: Proc. 7th Int. Conf. Vapor Deposition 1979, Proc. Electrochem. Soc. **79-3,** 300 (1979).
[264] A. Lösche: "Kerninduktion", VEB Deutscher Verlag d. Wissenschaften, Berlin 1957.
[265] Cryophysics newsletter **8,** Januar 1970.
[266] K. P. Jüngst, and P. Komarek: KfK Nachrichten **20,** 36 (1988).
[267] K. van Hulst, C. J. M. Aarts, A. R. de Vroomen, and P. Wyder: J. Magn. Magn. Mater. **11,** 317 (1979). M. J. Leupold, Y. Iwasa, and R. J. Weggel: to be published in Proceedings of MT8, Grenoble, Sept. 1983. J. Physique, Colloque C1, Suppl. No 1 Band **45,** C-41 (1984).
[268] E. R. Lady, and D. L. Call: "Fusion Technology", Intersociety Energy Conversion Engineering Conference, Energy 70, Las Vegas, Nevada, N.M. USA, p. 86. – C. D. Henning, ibid., p. 93.
[269] T. A. Buchold: Cryogenics **1,** 203 (1961).
[270] J. T. Harding, and R. H. Tuffias: Cryogenics Engineering **6,** 95 (1961).
[271] A. Lichtenberg: Verkehrstechnik/Der Verkehrsingenieur **26,** 73 (1942) 2; C. Albrecht, W. Elsel, H. Franksen, C. P. Parsch, and K. Wilhelm: ICEC 5 International Cryogenic Engineering Conference, May 1974, Kyoto, Vortrag B2.
[272] S. W. Kash, and R. F. Tooper: II T Frontier 1964, p. 20.
[273] A. D. Appleton: "Superconductors in Motion" in "A Guide to Superconductivity", ed. by D. Fislock, McDonald & Co. Ltd. 1969, p. 78.
[274] D. Lambrecht: ETG-Fachberichte **20,** 15 (1986). Lecture hold on the meeting of Energietechnische Gesellschaft (VDE/GTE) "Mit Energie in die Zukunft", October, 86. M. Liese, M. Duffert, H. W. Emshoff, M. Jungmaier, W. Schier, and E. Weghaupt: "Interna-

tional Conference on Large High Voltage Electric Systems", 112 boulevard Hausmann, F-75008 Paris, 28.8.-3.9.88.
[275] P. R. Wiederhold, and D. L. Ameen: Electronics **37,** 75 (1964).
[276] P. S. Swartz, and C. H. Rosner: J. Appl. Phys. **33,** 2292 (1962).
[277] A. F. Hildebrandt, D. D. Eleeman, F. C. Whitmore, and R. Simpkins: J. Appl. Phys. **33,** 2375 (1962); see also [X].
[278] H. L. Laquer: Cryogenics **3,** 27 (1963).
[279] J. Volger, and P. A. Admiraal: Phys. Lett. **2,** 257 (1962), see also J. van Suchtelen, J. Volger, and D. van Houwelingen: Cryogenics **5,** 256 (1965).
[280] H. Voigt: Z. Naturforsch. **A21A,** 510 (1966).
[281] R. Weber: Z. Angew. Phys. **6,** 449 (1967).
[282] W. F. Gauster, D. C. Freeman, and H. M. Long: paper 56, Proc. World Power Conf. 1964, p. 1954.
[283] R. L. Garvin, and J. Martisoo: Proc. IEEE **55,** (4) (1967) p. 538.
[284] P. A. Klaudy: Eletrotechniý Časopis **XXI,** 370 (1970) (with hints for further literature); G. Bogner, and P. Penczynski: Cryogenics 1976, p. 355.
[285] R. McFee: Elect. Eng. (N.Y.), Feb. 1962, p. 122.
[286] P. A. Klaudy: Patentschrift Nr. 256956, 11. Sept. 1967, Austria. Patentamt.
[287] P. Klaudy, J. Gerold, A. Beck, P. Rohner, E. Scheffler, and G. Ziemeck: IEEE Trans. Magn. **MAG-17,** 153 (1981).
[288] D. A. Buck: Proc. IRE **44,** 482 (1956).
[289] V. L. Newhouse, and J. W. Brenner: J. Appl. Phys. **30,** 1458 (1959).
[290] M. Mück, H. Rogalla, and B. David: Phys. Stat. Solid. **a87,** K105 (1985).
[291] IBMJ. Res. Dev. **24,** 107 (1980). Overview of the state of Josephson elements for large computers.
[292] R. T. Miller: IBM Research Reports Vol. 9, Nr. 1 (1973).
[293] J. M. Lock: Cryogenics, **2,** 65 (1961).
[294] Phys. Bl. **39,** 297 (1983); S. Faris, S. Raider, W. Gallagher, and R. Drake: IEEE Trans. Magn. **MAG-19,** 1293 (1983).
[295] V. L. Newhouse, and H. H. Edwards: Proc. IEEE **52,** 1191 (1964).
[296] I. M. Templeton: J. Sci. Instrum. **32,** 314 (1955).
[297] I. M. Templeton: J. Sci. Instrum. **32,** 172 (1955).
[298] A. R. de Vrooman, and C. van Baarle: Physica **23,** 785 (1957).
[299] R. Poerschke, and H. Wollenberger: Cryogenics **10,** 333 (1970).
[300] R. Fasel, and J. L. Olsen: Z. Klimatechnik – Klimatisierung **19,** 274 (1967).
[301] G. Aschermann, E. Friedrich, E. Justi, and J. Kramer: Phys. Z. **42,** 349 (1941).
[302] D. H. Martin, and D. Bloor: Cryogenics **1,** 159 (1961).
[303] C. V. Heer, C. B. Barnes, and J. G. Daunt: Phys. Rev. **91,** 412 (1953).
[304] R. M. Mueller, C. Buchal, T. Oversluizen, and F. Pobell: Rev. Sci. Instrum. **49,** 516 (1978).
[305] C. Passow; Eletrotechnikÿ Časopis **XXI,** 419 (1970).
[306] J. Halbritter: Z. Phys. **238,** 466 (1970).
[307] W. Bauer, A. Brandelik, A. Citron, F. Graf, L. Szecsi, and D. Proch: Nucl. Instrum. & Methods **214,** 189 (1983).
[308] A. Critron, G. Dammertz, M. Gründner, L. Husson, R. Lehm, and H. Lengeler: Nucl. Instrum. & Methods **164,** 31 (1979).
[309] W. Eisenmenger, and A. H. Dayem: Phys. Rev. Lett. **18,** 125 (1967).
[310] H. Kinder, K. Laszmann, and W. Eisenmenger: Phys. Lett. **31**A, 475 (1970).
[311] H. Kinder: Phys. Rev. Lett. **28,** 1564 (1972).
[312] P. Berberich, R. Buemann, and H. Kinder: Phys. Rev. Lett. **49,** 1500 (1982).
[313] C. C. Grimes, P. L. Richards, and S. Shapiro: Phys. Lett. **17,** 431 (1966).

[314] P. Fulde: Adv. Phys. **22**, 667 (1973).
[314] R. Meservey, P. M. Tedrow, and P. Fulde: Phys. Rev. Lett. **25**, 1270 (1970).
[316] J. Clarke: Rev. Phys. Appl. **5**, 32 (1970).
[317] A. H. Silver, and J. E. Zimmerman: Phys. Rev. **157**, 317 (1967).
[318] M. Nisenoff: Rev. Phys. Appl. **5**, 21 (1970).
[319] J. E. Mercereau: Rev. Phys. Appl. **5**, 13 (1970).
[320] A. F. Hildebrandt: Rev. Phys. Appl. **5**, 49 (1970).
[321] W. O. Hamilton: Rev. Phys. Appl. **5**, 41 (1970).
[322] J. E. Zimmerman: J. Appl. Phys. **42**, 30 (1971).
[323] "Biomagnetism" Proc. Third Int. Workshop on Biomagnetism, Berlin 1980, ed. S.N. Erné, H.-D. Hahlbohm, and H. Lübbig, Walter de Gruyter, Berlin 1981.
[324] P. A. M. Dirac: Proc. R. Soc. London, Ser. A**133**, 60 (1931), and Phys. Rev. **74**, 817 (1948).
[325] Blas Cabrera: Phys. Rev. Lett. **48**, 1378 (1982).
[326] J. E. Opfer: Rev. Phys. Appl. **5**, 37 (1970).
[327] Y. Zhang, M. Diegel, and C. Heiden: IEEE Trans. Magn. **MAG 25**, 1989.
[328] P. Turowski, and Th. Schneider: IEEE Trans. Magn. **MAG 24**, 1063 (1988).

Author Index

Subject Index

LIBRARY OF THE UNIVERSITY OF CALIFORNIA
Berkeley
1-MONTH